Adenovirus Methods and Protocols, Second Edition
VOLUME 2

METHODS IN MOLECULAR MEDICINE™

John M. Walker, SERIES EDITOR

METHODS IN MOLECULAR MEDICINE™

Adenovirus Methods and Protocols

SECOND EDITION

Volume 2: Ad Proteins, RNA, Lifecycle,
Host Interactions, and Phylogenetics

Edited by

William S. M. Wold
Ann E. Tollefson

Department of Molecular Microbiology and Immunology
Saint Louis University School of Medicine
St. Louis, Missouri

HUMANA PRESS ✳ TOTOWA, NEW JERSEY

© 2007 Humana Press Inc.
999 Riverview Drive, Suite 208
Totowa, New Jersey 07512

humanapress.com

This publication is printed on acid-free paper. ∞
ANSI Z39.48-1984 (American Standards Institute)

Permanence of Paper for Printed Library Materials.

Cover illustration: Chapter 6, vol. 1, "Production and Release Testing of Ovine Atadenovirus Vectors," by Gerald W. Both, Fiona Cameron, Anne Collins, Linda J. Lockett, and Jan Shaw.

Production Editor: Amy Thau

Cover design by Donna Niethe

For additional copies, pricing for bulk purchases, and/or information about other Humana titles, contact Humana at the above address or at any of the following numbers: Tel.: 973-256-1699; Fax: 973-256-8341; E-mail: orders@humanapr.com; or visit our Website: www.humanapress.com

Printed in the United States of America. 10 9 8 7 6 5 4 3 2 1

eISBN: 1-59745-277-7

ISSN: 1543-1894

ISBN13: 978-1-58829-901-7

Library of Congress Cataloging-in-Publication Data

Adenovirus methods and protocols / edited by William S.M. Wold, Ann E. Tollefson. -- 2nd ed.
    v. ; cm. -- (Methods in molecular medicine, ISSN 1543-1894 ; 130-131)
  Includes bibliographical references and indexes.
  Contents: v. 1. Adenoviruses, ad vectors, quantitation, and animal models -- v. 2. Ad proteins, RNA, lifecycle, host interactions, and phylogenetics.
  ISBN 1-58829-598-2 (v. 1 : alk. paper) -- ISBN 1-58829-901-5 (v. 2 : alk. paper)
  1. Adenoviruses--Laboratory manuals. 2. Molecular biology--Laboratory manuals. I. Wold, William S. M. II. Tollefson, Ann E. III. Series.
  [DNLM: 1. Adenoviridae--Laboratory Manuals. W1 ME9616JM v.130-131 2007 / QW 25 A232 2007]
  QR396.A336 2007
  579.2'443--dc22
                              2006012284

Preface

Since their discovery in 1954, adenoviruses (Ads) have become a model for studying virology, as well as molecular and cellular biology. Ads are easily grown and manipulated, stable, and versatile. Ads replicate reproducibly, can transform rodent cells to an oncogenic state, can induce tumors in certain animals, and have been instrumental in defining key cellular proteins and mechanisms such as splicing, transcriptional regulation through transcription adaptor proteins, and regulation of cell division and apoptosis. About half of the approximately 35 Ad proteins physically interact with cellular proteins and subvert them for use by the virus. In recent years, Ads have become premier tools in vector technology and in experimental gene therapy research.

The *Adenovirus Methods and Protocols* volumes are designed to help new researchers to conduct studies involving Ads and to help established researchers to branch into new areas. The chapters, which are written by prominent investigators, provide a brief general introduction to a topic, followed by tried and true step-by-step methods pertinent to the subject. We thank returning contributors for their updated and new chapters, and thank new contributors who have expanded the content of this book.

Adenovirus Methods and Protocols, Second Edition, Volume 2: Ad Proteins, RNA, Lifecycle, Host Interactions, and Phylogenetics, focuses on methods that elucidate and quantitate the interactions of Ad with the host. This volume provides methods for analysis of transcription, splicing, RNA interference, subcellular localization of proteins during infection, and cell cycle effects. Four chapters are devoted to definition of interactions of viral and cellular proteins (by co-immunoprecipitation or tandem mass spectrometry) or protein interactions with viral DNA (by chromatin immunoprecipitation and electrophoretic mobility shift assays). Other chapters provide thorough descriptions of the use of microinjection procedures, transformation assays, and NK cell-mediated cytolysis. Several chapters describe characterization of specific Ad proteins (hexon, fiber, or protease/proteinase). Two chapters are devoted to defining the phylogenetic relationships of Ads.

We thank contributors for sharing their secrets, John Walker for his patience, and especially Dawn Schwartz, without whose expert assistance this work would not have been possible.

William S. M. Wold
Ann E. Tollefson

Contents

CONTENTS OF THE COMPANION VOLUME
Volume 1: Adenoviruses, Ad Vectors, Quantitation, and Animal Models

Contributors

GÖRAN AKUSJÄRVI, PhD • *Department of Medical Biochemistry and Microbiology, Uppsala Biomedical Center (BMC), Uppsala, Sweden*

GUNNAR ANDERSSON, PhD • *Department of Medical Biochemistry and Microbiology, Uppsala Biomedical Center (BMC), Uppsala, Sweden*

G. ERIC BLAIR, PhD • *School of Biochemistry and Molecular Biology, University of Leeds, Leeds, United Kingdom*

MÁRIA BENKŐ, DVM, PhD • *Veterinary Medical Research Institute, Hungarian Academy of Sciences, Budapest, Hungary*

GRAHAM BOTTLEY, MSc, PhD • *School of Biochemistry and Molecular Biology, University of Leeds, Leeds, United Kingdom*

EILEEN BRIDGE, PhD • *Department of Microbiology, Miami University, Oxford, OH*

ROBIN N. BROUGHTON-SHEPARD, PhD • *Department of Cell and Developmental Biology, University of North Carolina at Chapel Hill, Chapel Hill, NC*

ROGER M. BURNETT, PhD • *The Wistar Institute, Philadelphia, PA*

GRAHAM P. COOK, BSc, PhD • *School of Biochemistry and Molecular Biology, University of Leeds, Leeds, United Kingdom*

THOMAS DOBNER, PhD • *Institut fuer Medizinische Mikrobiologie und Hygiene, Universitaet Regensburg, Regensburg, Germany*

WALTER DOERFLER, MD • *Institute for Clinical and Molecular Virology, Erlangen University, Erlangen, Germany*

JEFFREY A. ENGLER, PhD • *University of Alabama at Birmingham, Department of Biochemistry and Molecular Genetics, Birmingham, AL*

DEAN D. ERDMAN, PhD • *Respiratory Virus Diagnostic Program, Respiratory and Gastroenteritis Virus Branch, Division of Viral Diseases, Centers for Disease Control and Prevention, Atlanta, GA*

S. J. FLINT, PhD • *Department of Molecular Biology, Princeton University, Princeton, NJ*

ANUJ GAGGAR, PhD • *Division of Medical Genetics, University of Washington, Seattle, WA*

FELICIA D. GOODRUM, PhD • *Department of Microbiology and Immunology BIO5 Institute, University of Arizona, Tuscon, AZ*

MAURICE GREEN, PhD • *Institute for Molecular Virology, Saint Louis University School of Medicine, St. Louis, MO*

BALÁZS HARRACH, DVM, PhD, DSc • *Veterinary Medical Research Institute, Hungarian Academy of Sciences, Budapest, Hungary*

PATRICK HEARING, PhD • *Department of Molecular Genetics and Microbiology, School of Medicine, Stony Brook University, Stony Brook, NY*

JEONG SHIN HONG, PhD • *Department of Cell Biology, University of Alabama at Birmingham, Birmingham, AL*

CRISTINA IFTODE, PhD • *Department of Biological Sciences, Rowan University, Glassboro, NJ*

MICHAEL J. IMPERIALE, PhD • *Department of Microbiology and Immunology, University of Michigan Medical School, Ann Arbor, MI*

ADRIANA E. KAJON, PhD • *Infectious Disease Program, Lovelace Respiratory Research Institute, Albuquerque, NM*

ROBERT KERN, BA • *Eppendorf, North America, Westbury, NY*

ANDRÉ LIEBER, MD, PhD • *Division of Medical Genetics, University of Washington, Seattle, WA*

PAUL M. LOEWENSTEIN, BS • *Institute for Molecular Virology, Saint Louis University School of Medicine, St. Louis, MO*

WALTER F. MANGEL, PhD • *Biology Department, Brookhaven National Laboratory, Upton, NY*

DAVID A. MATTHEWS, PhD • *Department of Cellular and Molecular Medicine, School of Medical Sciences, University of Bristol, Bristol, United Kingdom*

WILLIAM J. MCGRATH, PhD • *Biology Department, Brookhaven National Laboratory, Upton, NY*

OLIVER MÜHLEMANN, PhD • *Institute of Cell Biology, University of Bern, Bern, Switzerland*

MICHAEL NEVELS, PhD • *Institut fuer Medizinische Mikrobiologie und Hygiene, Universitaet Regensburg, Regensburg, Germany*

DAVID A. ORNELLES, PhD • *Department of Microbiology and Immunology, Wake Forest University School of Medicine, Winston-Salem, NC*

PILAR PEREZ-ROMERO, PhD • *Department of Microbiology and Immunology, University of Michigan Medical School, Ann Arbor, MI*

JOHN M. ROUTES, MD • *Department of Asthma, Allergy, and Immunology, Children's Hospital of Wisconsin, Medical College of Wisconsin, Milwaukee, WI*

JOHN J. RUX, PhD • *The Wistar Institute, Philadelphia, PA*

DMITRY SHAYAKHMETOV, PhD • *Division of Medical Genetics, University of Washington, Seattle, WA*

CHAO-ZHONG SONG, PhD • *Institute for Molecular Virology, Saint Louis University School of Medicine, St. Louis, MO*

ANDREW THORBURN, PhD • *University of Colorado Comprehensive Cancer Center, University of Colorado Health Science Center, Aurora, Colorado*

JOSEPH M. WEBER, PhD • *Département de Microbiologie et d'Infectiologie, Université de Sherbrooke, Sherbrooke, Québec, Canada*

NING XU, BMS • *Department of Medical Biochemistry and Microbiology, Uppsala Biomedical Center (BMC), Uppsala, Sweden*

PETER YACIUK, PhD • *Department of Molecular Microbiology and Immunology, Saint Louis University School of Medicine, St. Louis, MO*

JIHONG YANG, MD • *Department of Cell Biology and Molecular Medicine, New Jersey Medical School, Newark, NJ*

1

Analysis of the Efficiency of Adenovirus Transcription

Cristina Iftode and S. J. Flint

Summary

This method is designed to measure rates of transcription from adenoviral promoters as a function of the concentrations within infected cells of the promoter(s) of interest. The latter parameter is assessed by quantification of viral DNA by hybridization of membrane-bound DNA following purification of DNA from nuclear fractions of adenovirus-infected cells. Two alternative protocols, primer extension and quantitative reverse transcription polymerase chain reaction, are described for determination of the concentrations of viral mRNAs purified from the cytoplasmic fractions of the same infected cell samples. An alternative procedure to measure rates of transcription directly using run-on transcription in isolated nuclei is also presented.

Key Words: Adenovirus transcription; run-on transcription; quantification of adenoviral DNA; primer extension; quantitative RT-PCR; mRNA isolation; isolation of nuclei; DNA purification; promoter utilization efficiency.

1. Introduction

This protocol was designed to assess the effects of IVa_2 promoter mutations that impair binding of the cellular repressor of IVa_2 transcription (1) on activation of IVa_2 transcription during entry into the late phase of infection (2). The goal was to assess the efficiency with which IVa_2 templates were utilized after increasing periods of infection. Consequently, it was crucial to examine IVa_2 transcription as a function of viral DNA template concentration. To this end we measured viral DNA concentrations by slot-blotting and estimated rates of transcription either directly using run-on transcription in isolated nuclei or indirectly via measurement of mRNA concentration. As illustrated in **Fig. 1**, the efficiency with which templates are utilized can then be calculated as the ratios of IVa_2 RNA:DNA under each condition.

From: *Methods in Molecular Medicine, Vol. 131:*
Adenovirus Methods and Protocols, Second Edition, vol. 2:
Ad Proteins, RNA, Lifecycle, Host Interactions, and Phylogenetics
Edited by: W. S. M. Wold and A. E. Tollefson © Humana Press Inc., Totowa, NJ

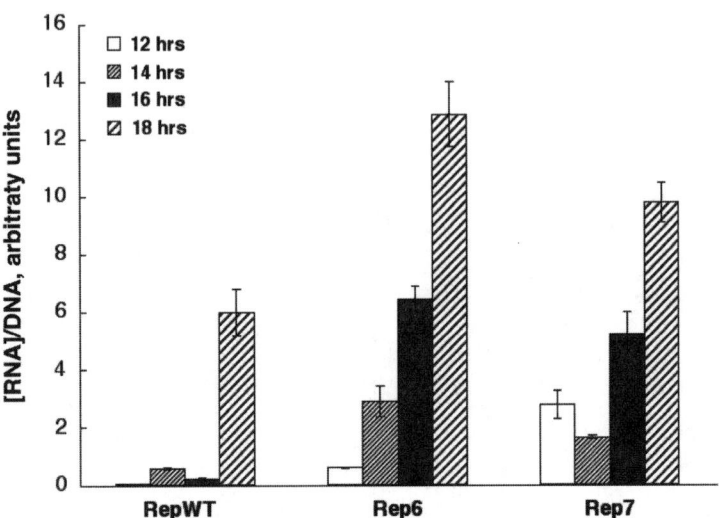

Fig. 1. Mutations that increase the efficiency of utilization of the adenoviral IVa_2 promoter. HeLa cells were infected in parallel with 10 PFU/cell of Ad2/Ad5 recombinant viruses carrying a wild-type Ad2 IVa_2 promoter or promoters with precise substitutions that impair binding of the cellular repressor of IVa_2 transcription *(18)*. The Rep7 mutation introduced no change into the overlapping coding sequence for the viral DNA polymerases, whereas the Rep 6 mutation resulted in substitution of a Trp residue by Phe *(18)*. At the times indicated, infected cells were harvested and the concentrations of nuclear viral DNA and cytoplasmic IVa_2 mRNA determined as described in **Heading 3.** These values were then used to calculate ratios of IVa_2 transcript concentration to the viral DNA template concentration, a measure of the efficiency with which IVa_2 transcription templates, and promoters, are utilized. The results shown are the mean of two independent experiments in which each parameter was measured at least twice. They indicated that the mutant IVa_2 promoters are utilized much more efficiently than the wild type during the initial portion of the late phase of infection, for example, up to 60-fold more efficiently at 12 h postinfection (p.i.) However, this difference was no more than 2-fold by 18 h p.i. This temporal pattern is consistent with the hypothesis that activation of IVa_2 transcription is the direct result of viral DNA synthesis-dependent titration of the cellular repressor of IVa_2 transcription.

This approach is necessary whenever it is likely (or possible) that viral gene expression is affected by differences in concentrations of viral DNA molecules in infected cell nuclei. One example is cells infected by viruses carrying mutations that affect both early and late events in the infectious cycle, such as mutations that prevent production of the E4 Orf 6 protein *(3,4)*. However, the methods described below can be used independently when this constraint does not apply.

2. Materials

2.1. Analysis of Viral DNA

2.1.1. Isolation of Nuclear and Cellular Fractions From Infected Cells

1. NP-40 lysis buffer: 50 mM Tris-HCl, pH 8.0, 150 mM NaCl, 5 mM MgCl$_2$, 0.6% (v/v) NP-40.

2.1.2. Preparation of Total DNA

1. Digestion buffer: 10 mM Tris-HCl, pH 8.0, 100 mM NaCl, 25 mM ethylene diamine tetraacetic acid (EDTA), pH 8.0, 0.5% (w/v) sodium dodecyl sulfate (SDS), 0.1 mg/mL Proteinase K.
2. Phenol:chloroform mix: 1:1 (v/v), with phenol buffered with 10 mM Tris-HCl, pH 7.6.
3. TE: 10 mM Tris-HCl, pH 7.4, 1 mM EDTA.
4. RNase solution: 0.1% (w/v) SDS, 4 µg/mL DNase-free RNase.

2.1.3. Quantification of Nuclear Viral DNA

1. 20X Sodium saline citrate (SSC): 0.3 M sodium citrate-2H$_2$O adjusted to pH 7.0 with 1 M HCl, 3 M NaCl.
2. 100X Denhardt solution: 2% (w/v) Ficoll 400, 2% (w/v) polyvinylpyrolidine, 2% (w/v) bovine serum. Store at –20°C.
3. Aqueous prehybridization/hybridization (APH) solution: 5X SSC, 5X Denhardt solution, 1% (w/v) SDS. Add 100 µg/mL denatured salmon sperm DNA just before use.
4. Membrane: positively charged nylon membrane that binds DNA covalently at high pH.

2.2. Analysis of Newly Transcribed RNA

1. Glycerol storage buffer: 50 mM Tris-HCl, pH 8.3, 5 mM MgCl$_2$, 0.1 mM EDTA, 40% (v/v) glycerol.
2. Transcription buffer: 40 mM Tris-HCl, pH 7.9, 300 mM KCl, 10 mM MgCl$_2$, 40% (v/v) glycerol, 2 mM dithiotheitol (DTT; add immediately prior to use).
3. ACG mix: 10 mM each of ATP, CTP, GTP.
4. HSB: 10 mM Tris-HCl, pH 7.5, 0.5 M NaCl, 10 mM MgCl$_2$.
5. Proteinase K mix (made fresh): 0.4% SDS, 20 µg/mL Proteinase K.
6. DNase buffer: 10 mM Tris-HCl, pH 7.5, 10 mM MgCl$_2$.
7. Church buffer: 0.5 M Na$_2$HPO$_4$, pH 7.5 (adjust pH with 85% H$_3$PO$_4$), 1% (w/v) bovine serum albumin, 7% (w/v) SDS, 1 mM EDTA.

2.3. Analysis of mRNA

2.3.1. Preparation of Cytoplasmic RNA

1. Diethyl pyrocarbonate (DEPC)-treated water: stir 0.1% (v/v) DEPC in water, incubate for at least 12 h at 37°C, and autoclave for 20 min.

2. Resuspension buffer: 10 mM Tris-HCl, pH 7.4, 5 mM NaCl, 2 mM EDTA, made in DEPC water.

2.3.2. Quantification of Cytoplasmic RNA by Primer Extension

1. Hybridization buffer: 40 mM piperazine-N,N'-bis(2-ethanesulfonate) (PIPES), pH 6.8, 400 mM NaCl, 1 mM EDTA, pH 8.0, 40% (v/v) formamide, ultrapure.
2. Reverse transcription cocktail: first strand buffer (1X), 1 mM dATP, 1 mM dCTP, 1 mM dGTP, 1 mM TTP, 10 mM DTT, 1 U RNasin/reaction, 10 U reverse transcriptase/reaction.
3. RNA loading buffer: 10 mM Tris-HCl, pH 7.4, 80% (v/v) formamide, 0.02% (w/v) each of bromophenol blue and xylene cyanol.
4. 10X TBE: 0.9 M Tris-borate, 0.02 M EDTA, pH 8.0.

2.3.3. Quantitative Reverse Transcription Polymerase Chain Reaction

1. DNA suspension buffer: 0.01 M Tris-HCl, pH 7.4, 1 mM EDTA, 80% (v/v) formamide, 0.02% (w/v) xylene cyanol, 0.02% (w/v) bromphenol blue.
2. C18 Sep-Pak cartridges (Waters).
3. Polymerase chain reaction (PCR) mix: 1X Universal PCR Master Mix (ABI or equivalent), 0.2 μM major late (ML) and β-actin probes, 0.04 μM ML forward and reverse primers, 0.02 μM β-actin forward and reverse primers.

3. Methods

3.1. Measurement of Template Viral DNA Concentration

This procedure requires separation of nuclear and cytoplasmic fractions *(5)* from infected cells, purification of DNA from the former *(6)*, and measurement of relative viral DNA concentrations by hybridization of labeled viral DNA probes to test DNAs bound to membranes (slot-blotting) *(6,7)*.

3.1.1. Isolation of Nuclear and Cytoplasmic Fractions From Infected HeLa Cells (see **Note 1**)

1. Centrifuge cells out of their medium for 5 min at 500g at 4°C and discard the supernatant.
2. Resuspend the cell pellet in 10 mL of ice-cold phosphate-buffered saline and centrifuge at 500g for 5 min at 4°C.
3. Suspend cells in 0.5 mL (*see* **Note 2**) of ice-cold NP-40 lysis buffer and incubate on ice for 3 min.
4. Centrifuge suspended cells at 500g for 3 min at 4°C.
5. Repeat **steps 3** and **4**, and pool supernatants (*see* **Note 3**).
6. Immediately use the pooled supernatants to isolate cytoplasmic mRNA (**Subheading 3.3.1.**), and the pellet to extract nuclear DNA (**Subheading 3.1.2.**), or store fractions at –80°C.

3.1.2. Extraction of Total Nuclear DNA

1. Suspend nuclei in 500 µL of digestion buffer and incubate at 37°C for 30 min. The digestion can be also performed overnight (*see* **Notes 4** and **5**).
2. Extract nucleic acids with an equal volume of phenol:chloroform.
3. Centrifuge the mix for 10 min at 1700*g* in a swinging bucket rotor and save the top aqueous layer (*see* **Note 6**).
4. Precipitate the DNA by addition of half volume (of the aqueous layer) of 7.5 *M* ammonium acetate and 2 vol of absolute ethanol. The DNA should be observed immediately as a string. Centrifuge the mix at 1700*g* for 20 min, and remove the organic solvent.
5. Rinse the DNA pellet with 70% ethanol. Decant ethanol and air-dry the pellet.
6. Suspend the DNA in 200 µL TE for the pellet obtained from 4×10^7 cells.
7. Remove residual RNA by treatment with RNase solution for 1 h at 37°C, and repeat Proteinase K digestion, organic extraction, and ethanol precipitation (**Subheading 3.1.2.**, **steps 1–6**).

3.1.3. Quantification of Viral DNA by Hybridization of DNA Blots

To prepare the blotting manifold as indicated below, use a positively charged membrane.

1. Prewet the membrane in distilled water for 10 min.
2. Adjust each DNA sample (*see* **Note 7**) to final concentrations of 0.4 *M* NaOH and 10 m*M* EDTA, pH 8.2. Boil for 10 min, then place on ice.
3. Apply samples to the membrane under vacuum and then rinse each slot with 500 µL of 0.4 *M* NaOH.
4. After dismantling the manifold, rinse the membrane in 2X SSC and air dry.
5. UV-crosslink the DNA at $1200 \times 100 \ \mu\text{J/cm}^2$ (*see* **Note 8**).
6. Wet the membrane with 6X SSC and prehybridize with 1 mL/cm² of APH solution for 1 h at 65°C with rotation.
7. Remove the prehybridization solution and add 1 mg/cm² of APH solution containing 10 ng/mL denatured probe (*see* **Note 9**). Hybridize the membrane at 65°C overnight.
8. Remove the hybridization solution and wash the membrane twice with each of the following: 2X SSC, 0.1% SDS for 5 min at room temperature; 0.2X SSC, 0.1% SDS for 5 min at room temperature; 0.2X SSC, 0.1% SDS for 15 min, at 42°C; 0.1X SSC, 0.1% SDS for 15 min at 65°C.
9. Rinse the membrane with 2X SSC, air-dry, and expose to X-ray film, or a PhosphorImager screen.

3.2. Measurement of Transcription Rates by Run-On Transcription in Isolated Nuclei

The run-on transcription procedure *(5,8)* is described below in the context of transcription of adenoviral genes by human RNA polymerase II, but this procedure can also be applied to analysis of transcription of cellular genes by this enzyme. Furthermore, both RNA polymerase I (used to provide an internal control) and RNA polymerase III *(8,9)* are active under the conditions described in **Subheading 3.2.1.** If necessary, transcription by these three cellular RNA polymerases can be distinguished by the addition to transcription reactions of specific inhibitors of RNA polymerase II (1–2 µg/mL α-amanitin *[10]*) or RNA polymerase III (40 µ*M* tagetitoxin *[11]*).

The protocol comprises isolation of nuclei, run-on transcription, purification of RNA labeled in isolated nuclei, and its hybridization to membrane-bound viral DNA (or RNA) probes.

3.2.1. Isolation of Nuclei

1. Isolate infected cell nuclei from 1×10^7 cells as described in **Subheading 3.1.1.**
2. Suspend nuclei in 200 µL of glycerol storage buffer, and use immediately or store in liquid nitrogen.

3.2.2. Run-On Transcription

3.2.2.1. REACTION

1. Thaw 50 µL nuclei and immediately add 50 µL of transcription buffer containing 2 m*M* DTT.
2. Add 3 µL of 10 m*M* ACG mix.
3. Add 5 µL of [α-^{32}P]UTP (800 Ci/mmol; 20 mCi/mL) and incubate for 40 min at 30°C.
4. Spin in a microcentrifuge at maximum speed for 30 s and remove the supernatant.

3.2.2.2. RNA ISOLATION

1. Suspend the pellet in 200 µL of cold HSB (*see* **Note 10**).
2. Add 30 U of RNase-free DNase and incubate for 30 min at 30°C.
3. Add 200 µL of Proteinase K mix and incubate for 30 min at 37°C (*see* **Note 11**).
4. Extract the RNA with one volume of phenol/chloroform and transfer the top layer to a new tube.
5. Add LiCl to 0.4 *M* and precipitate with two volumes of 100% ethanol.
6. Spin in a microcentrifuge at maximum speed for 30 min at 4°C and remove supernatant.
7. Suspend the pellet in 200 µL of DNase buffer containing 10 U DNase and incubate for 20 min at 37°C (*see* **Note 12**).
8. Extract RNA with 1 vol of phenol/chloroform, add LiCl to 0.4 *M*, and precipitate with three volumes of prechilled 100% ethanol.
9. Spin in a microcentrifuge at maximum speed for 30 min at 4°C and remove the supernatant.
10. Wash the pellet with 70% ethanol and spin again for 10 min at 4°C. Remove the ethanol.
11. Dry the pellet in a speed-vac concentrator and suspend in 60 µL of DEPC-treated water (*see* **Note 13**).

12. The RNA can be stored at –20°C or used immediately in hybridization reactions (*see* **Note 14**).

3.2.3. Hybridization

1. Prepare the slot blots (2 pmol probe/slot) (*see* **Note 15**) as described in **Subheading 3.1.3.**, **steps 1–5**.
2. Place each membrane strip in hybridization vials containing 2 mL of prewarmed (65°C) Church buffer containing 5 µg/mL tRNA.
3. Prehybridize overnight at 65°C with rotation.
4. Replace the prehybridization solution with 2 mL prewarmed (65°C) Church buffer containing 5 µg/mL tRNA and 2.5 m*M* ribonucleotide mix.
5. Add 15 µL 1 *M* NaOH to 60 µL [^{32}P-labeled] RNA and place on ice for 5 min.
6. Neutralize with 30 µL 0.5 *M* Tris containing 0.5 *M* HCl.
7. Transfer the RNA mix to the vial containing the hybridization solution.
8. Allow the sample to hybridize to the probe for 40 h at 65°C.
9. Remove the hybridization solution and wash the strip sequentially as follows (*see* **Note 16**):
   a. 2X SSC, 1 h at 65°C
   b. 2X SSC, 0.1% SDS, 1 h at 65°C
   c. 2X SSC, 10 µg/mL RNase A, 30 min at 37°C, no rotation
   d. 2X SSC, 5 µg/mL tRNA, 1 h at 37°C
10. Blot the membrane dry on Whatman 3MM filter paper.
11. Place the membrane strips between sheets of plastic wrap and expose to X-ray film or a PhosphorImager screen.

3.3. Measurement of mRNA Concentration in Cytoplasmic Fractions

In addition to the run-on transcription assay, we have used this alternative approach to examine the efficiency of mRNA production in infected cells. The advantage of this method is that the concentration of both the DNA template and the mRNA product of transcription can be determined in the same batch of cells. Primer extension and reverse transcription-PCR (RT-PCR) are the two specific assays that we utilize to measure the concentrations of viral mRNAs in cytoplasmic fractions.

3.3.1. Isolation of Cytoplasmic mRNA

1. Digest the cytoplasmic fraction isolated as described in **Subheading 3.1.1.** with 40 U/mL RNase-free DNase and 5 m*M* MgCl$_2$ for 30 min at 37°C.
2. Add Proteinase K to 20 µg/mL and 0.5% (w/v) SDS to 0.5% and incubate for 30 min at 37°C.
3. Extract cytoplasmic RNA with an equal volume of phenol/chloroform and centrifuge for 5 min at 1300*g*.
4. Precipitate RNA by addition to the aqueous phase of 2.5 vol ethanol and dissolve in 50–100 µL resuspension buffer (*see* **Note 17**).

3.3.2. Primer Extension Analysis

This assay permits direct measurement of mRNA products and, unlike the run-on transcription, allows the accuracy of initiation of transcription to be assessed. This protocol was optimized for the determination of E2E, IVa$_2$, and ML mRNA concentrations, but it can be applied to other viral transcripts if appropriate primers are used.

3.3.2.1. PRIMERS

1. Design oligonucleotides 28–30 nucleotides (nt) long that are complementary to sequences located 60–100 nt downstream of the transcription initiation site. Avoid regions that are part of introns as they will be absent from cytoplasmic mRNA species.
2. Design or purchase a primer for an internal control mRNA (*see* **Note 18**).
3. Obtain primer(s) from your source of choice and, if necessary, purify by electrophoresis under denaturing conditions.

3.3.2.2. PURIFICATION OF REVERSE TRANSCRIPTION PRIMERS BY ELECTROPHORESIS UNDER DENATURING CONDITIONS

1. Dissolve oligonucleotide primers in 100–500 µL of DNA suspension buffer.
2. Heat samples at 95°C for 5 min and immediately cool on ice.
3. Load 10- to 50-µL portions onto individual wells of a 12% polyacrylamide gel cast in 1X TBE and containing 8 M urea. Run the gel in 0.5X TBE at 40 mA until the bromophenol blue dye has migrated three-quarters of the length of the gel.
4. Place the gel on a sheet of plastic wrap on a TLC plate and view under UV light.
5. Excise the gel slice containing the full-length oligonucleotide (*see* **Note 19**) and crush by passing through a 10-mL syringe (without a needle) into 3.0 mL of 0.3 M sodium acetate (*see* **Note 20**).
6. Extract the DNA by rotating the tube at 37°C for 12–16 h.
7. Remove the solution and repeat extraction for 3–4 h with an additional 3.0 mL of sodium acetate.
8. Place the pooled extractions into the barrel of a 10-mL syringe connected to a C18 cartridge that has been prewetted by passage of 10 mL of 100% methanol followed by 10 mL of H$_2$O.
9. Load the DNA onto the C18 column using gentle pressure.
10. Wash the column with 10 mL of 0.3 M sodium acetate.
11. Elute the DNA with three successive washes of 1 mL of 100% methanol, collecting each eluate in a 1.5-mL screw-cap microcentrifuge tube.
12. Dry the DNA using a speed-vac concentrator and store dry at –20°C until use.

3.3.2.3. HYBRIDIZATION

1. Label 5 pmol primer with 150 µCi [γ-^{32}P]ATP (3000 Ci/mmol) (*see* **Note 21**).
2. Mix 10 µg RNA sample with 20 fmol of labeled primer in DEPC water.

3. Add 30 µL of 3 *M* sodium acetate and 800 µL of 100% ethanol to precipitate RNA and DNA. Dry-pellet in a speed-vac concentrator.
4. Resuspend the pellet in 30 µL of hybridization buffer and incubate for 5 min at 80°C.
5. Allow primer to hybridize to template overnight at 30°C.

3.3.2.4. PRIMER EXTENSION

1. Add 130 µL of H_2O and 40 µL of 5 *M* ammonium acetate to each hybridization reaction.
2. Add 800 µL of 100% ethanol, precipitate, wash once with 70% ethanol, and allow to dry.
3. Suspend the pellet in 20 µL of reverse transcription cocktail and incubate for 1 h at 37°C.
4. Add 130 µL of TE, pH 7.8, and 15 µL of 3 *M* sodium acetate.
5. Add 500 µL of 100% ethanol, precipitate, wash once with 70% ethanol, and allow to dry.
6. Resuspend the pellet in 10 µL of RNA loading buffer and incubate for 5 min at 95°C.
7. Load onto an 8% denaturing (urea) polyacrylamide sequencing gel cast in 0.5X TBE buffer and run in the same buffer at 90 W, 55°C for 1.5 h.
8. Visualize products by autoradiography and quantify by using a PhosphorImager.

3.3.3. Quantitative Two-Step RT-PCR Using TaqMan® Chemistry

This method permits the accurate, and simultaneous, measurement of the relative concentrations of viral late (or early) mRNAs in many samples. It is also much more sensitive than methods such as Northern blotting or primer extension and far less time-consuming and labor-intensive. However, before the assay can be applied routinely, it is necessary to identify empirically and validate PCR conditions appropriate for quantification of the mRNA(s) of interest. This protocol was developed for assessment of ML L2 penton and L5 fiber mRNA concentrations, but with appropriate reverse transcription primers it can be applied to any ML mRNA.

3.3.3.1. REVERSE TRANSCRIPTION
3.3.3.1.1. Primers

1. Design mRNA-specific primers complementary to sequences on either side of the splice junction between the third exon of the ML tripartite leader sequence and the coding exon of the ML mRNA(s) of interest (*see* **Note 22**). Include complementarity to 10–12 bases upstream and downstream of the splice junction(s).
2. Design or purchase a primer for internal control mRNA (*see* **Note 18**).
3. Obtain primer(s) from your source of choice and, if necessary, purify them by electrophoresis under denaturing conditions (**Subheading 3.3.2.2.**).

3.3.3.1.2. Synthesis of cDNA

1. Purify cytoplasmic RNA (**Subheading 3.3.1.**) from the infected cells of interest and determine concentrations. Divide RNA samples into 5- to 10-µg portions and store at –80°C (*see* **Note 14**).
2. In 0.6-mL microcentrifuge tubes and a total volume of 10 µL, mix 1–5 µg cytoplasmic RNA samples, 15 pmol L2 penton (or L5) fiber primer, and 10 pmol β-actin primer. Include additional reactions for some or all of the RNA samples as "no-template controls": these samples will be incubated *without* reverse transcriptase (*see* **Note 23**).
3. Spin down samples, incubate at 65°C for 5 min in a heat-block, turn off heat, and continue incubation until samples cool to 50°C.
4. To each reaction add 9 µL of reverse transcription cocktail without reverse transcriptase.
5. Add 200 U (1 µL) reverse transcriptase or, in the case of no-template controls, 1 µL H_2O.
6. Mix reactants, spin down samples and incubate at 42°C for 60 min.
7. Analyze cDNA samples immediately, or store at –20°C.

3.3.2.2. REAL-TIME PCR

3.3.2.2.1. Amplicons

Amplicons for real-time PCR using TaqMan chemistry (<150 bp) comprise forward and reverse primers and a probe containing an internal sequence of the amplicon that is labeled with a fluorescent reporter dye and a quencher at its 5'- and 3'-ends, respectively. In the case of ML mRNAs, a single amplicon corresponding to a sequence within the tripartite leader sequence allows detection and quantification of all ML mRNAs. **Figure 2** illustrates the ML tripartite leader sequence amplicon to which this protocol applies. It is used in conjunction with a commercially available β-actin amplicon (ABI; *see* **Notes 24** and **25**).

3.3.2.2.2. PCR

The PCR conditions described were optimized for simultaneous detection of viral ML and human β-actin cDNAs. Nonlimiting concentrations of the ML amplicon primers (**Fig. 2**) that yielded maximal sensitivity were determined empirically, as was a limiting concentration of the β-amplicon primers (ABI).

1. Choose the cDNA sample that will serve as the standard (*see* **Note 26**) and prepare five dilutions (in H_2O) that span two orders of magnitude in cDNA concentration.
2. Assign each dilution an arbitrary, but appropriate, number of copies of the cDNA template.
3. Dilute samples of all other cDNAs 1:5 in H_2O.
4. Run duplicate or triplicate PCR reactions for each cDNA sample using a 96-well reaction plate. In each well, place 2 µL of standard or test cDNA and 23 µL of PCR mix (*see* **Note 27**).

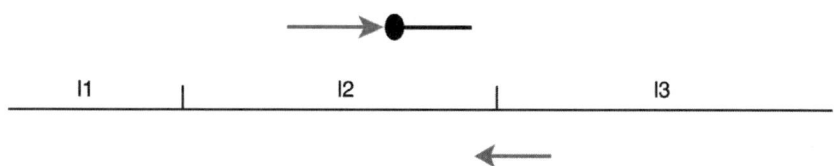

Fig. 2. The major late (ML) amplicon for quantitative, real time reverse transcriptase polymerase chain reaction (PCR). The horizontal line at the center represents the 202-nucleotide tripartite leader sequence of Ad5, drawn to scale in the 5'-to-3' direction. The vertical lines indicate the positions of the splice junctions between the l1, l2, and l3 exons. The positions of the forward and reverse PCR primers of the amplicon are indicated by the gray arrows and that of the probe by the black line. The black circle at the 5'-end of the probe represents the fluorescent label (VIC).

5. Spin the plates at 2000g for 2 min at 4°C.
6. Run the PCR reactions in a sequence-detection system (e.g., ABI Prism 7700) using the software provided by the manufacturer.
7. When the reactions are complete, use the software to select and analyze all reactions.
8. View the β-actin amplification plots of log ΔRn (the difference in normalized emission intensities of the reaction and an unreacted sample) vs cycle number. Check that the no template controls yielded no significant amplification.
9. Consult troubleshooting guides if no linear increase in ΔRn is evident in the plot.
10. Otherwise, set the background and cycle in which a statistically significant increase in the ΔRn value is first detected (the threshold cycle, C_t) according to the manufacturer's instructions, or use the default settings. The program will then use the standard curve to calculate relative, arbitrary concentrations of all unknown samples.
11. View the standard curve and check that the C_t values for the unknown samples lie within the range of the values of the standards. If they do not, repeat the analysis with appropriately adjusted standards.
12. Repeat **steps 7–10** for the ML amplification plots.
13. To ensure accuracy, repeat the RT-PCR analysis at least once.
14. To ensure reproducibility, analyze two or three independent preparations of the RNA samples.

4. Notes

1. This separation method yields cytoplasmic fractions that contain no nuclear components, such as intron-containing pre-mRNA *(12)*. However, nuclear lysis will occur if too large a volume of NP-40 lysis buffer for the number of cells is used, or if cells or nuclei are incubated in this buffer for too long a period.
2. For NP-40 lysis, the indicated volumes apply to 1–5 × 10^7 cells.
3. Because of differences in the lipid composition of their membranes, the initial cell pellet appears light brown or tan, whereas the nuclei are white. This change

in color, which is quite obvious, provides a simple way to monitor the effectiveness of the separation.

4. For digestion buffer, use 0.3 to 0.5 mL for 3×10^7 to 5×10^7 cells; for 10^8 cells, use 1 mL buffer.

5. The samples will be somewhat viscous, and care should be taken while processing them further to avoid loss of material.

6. If the phases do not resolve well, centrifugation can be repeated following addition of another volume of digestion buffer minus Proteinase K. If a white interphase is noticeable, the organic extraction may be repeated.

7. For slot blots, the amount of DNA should be determined relative to the abundance of the target sequence.

8. The specifications for charged membranes indicate that baking or UV-crosslinking is not required (DNA binds covalently to them). However, we have never tried using the membranes without crosslinking.

9. Any plasmid containing a segment of the adenoviral genome can be used as probe. Any convenient cellular DNA can be used as the internal control to correct for variations in cell number or loading. We routinely use a segment of the external transcribed spacer of the ribosomal RNA transcription unit, as the high copy number of rRNA genes *(13)* results in a strong hybridization signal. The specific activity of the probes should be at least 1×10^8 cpm/μg, labeled by random priming *(14)*.

10. The pellet is very difficult to suspend and frequently does not fully solubilize; one could skip this step if the nuclei are very fragile.

11. After this step, the RNA becomes completely soluble.

12. This step is useful to overcome background problems.

13. For RNA manipulations it is necessary to use RNase-free solutions (DEPC-treated) and RNase-free tips (alternatively, nonautoclaved, new tips, may be used).

14. It is important to avoid repeated freezing and thawing of RNAs.

15. Both DNA and antisense RNA probes can be used. For the application described here, the DNA probes were linearized plasmids carrying either a fragment of the gene for human ribosomal RNA (control) or the Ad2 IVa$_2$ coding sequence. The antisense RNA probes can be synthesized with either SP6 or T7 RNA polymerase, according to the manufacturer's instructions. We generated our probes to be complementary to either the E2E transcription unit or the ML L3 region.

16. This sequence of washes is essential to ensure a clean background.

17. The resuspension volumes apply to pellets obtained from $1–5 \times 10^7$ cells.

18. Any cellular RNA that does not increase or decrease in concentration in Ad5-infected cells can serve as the internal control. We generally use β-actin mRNA *(15–17)* and the reverse primer of a commercially available human β-actin amplicon (ABI).

19. If the oligonucleotide is not fully deprotected after synthesis, molecules that migrate more slowly than the full length oligonucleotide, as well as $n - 1$, $n - 2$, . . . $n - i$ products that migrate more quickly, will be observed.

20. The volumes listed apply to extraction of oligonucleotides from a gel slice of approx $5 \times 0.5 \times 0.15$ cm and should be adjusted according to the dimensions of the gel slice.
21. Incorporation should be at least 50% to obtain a reasonably good signal.
22. This protocol can be adapted readily for the specific detection of only pre-mRNAs by using primers complementary to exon–intron junctions. Primers complementary to only intronic sequences will also detect excised introns and therefore are not an appropriate choice. Alternatively, the total concentration of an RNA can be assessed using a primer complementary to a sequence within an exon.
23. The no-template control reactions will give no PCR signal if the sample contains no viral DNA. Hence, they confirm specificity for RNA detection. Such controls are particularly important when the goal is to assess the concentrations of nuclear RNAs. If a signal substantially above background is obtained with no-template controls, the RNAs should be repurified.
24. When the test and internal control RNAs are to be amplified in the same reaction, the emission maximum of the fluorescent reporter dye chosen to label the internal control probe must be distinct from that of the ML probe. One appropriate combination is a FAM-labeled β-actin probe and the VIC-labeled ML probe shown in **Fig. 2**.
25. A large number of fluorescent reporter dyes that can be covalently linked to oligonucleotides are available. Synthesis and purification of such labeled oligonucleotides are offered by various companies.
26. The cDNA chosen as the standard should be of high concentration, relative to other samples. For example, when comparing viral mRNAs synthesized in cells infected by wild-type Ad5 or mutant viruses, choose the wild-type sample as the standard. If the standard contains too low a concentration of the cDNA, the concentrations of many unknowns will lie outside the range defined by the standard curve, reducing the accuracy of quantification.
27. The PCR mix includes a passive reference dye, such as ROX. This dye provides a signal that is used for normalization of the emission intensities of each reporter for each reaction well and hence correction for variations in volume or concentration. Reporter emission intensities are calculated and reported for each PCR cycle as the ratio (Rn) of the reporter intensity to that of the passive reference.

References

1. Chen, H., Vinnakota, R., and Flint, S. J. (1994) Intragenic activating and repressing elements control transcription from the adenovirus IVa$_2$ initiator. *Mol. Cell. Biol.* **14**, 676–685.
2. Iftode, C. and Flint, S. J. (2004) Viral synthesis-dependent titration of a cellular repressor activates transcription of the human adenovirus type 2 IVa2 Gene. *Proc. Natl. Acad. Sci. USA* **101**, 17,831–17,836.
3. Leppard, K. N. (1997) E4 gene function in adenovirus, adenovirus vector and adeno-associated virus infections. *J. Gen. Virol.* **78**, 2131–2138.

4. Tauber, B. and Dobner, T. (2001) Molecular regulation and biological function of adenovirus early genes: the E4 ORFs. *Gene* **278,** 1–23.
5. Greenberg, M. and Bender, T. (1997) Identification of newly transcribed RNA, in *Current Protocols in Molecular Biology* (Ausubel, F. M., Brent, R., Kingston, R. E., et al., eds.), John Wiley and Sons, New York.
6. Strauss, M. (1998) Preparation of genomic DNA from mammalian tissue, in *Current Protocols in Molecular Biology* (Ausubel, F. M., Brent, R., Kingston, R. E., et al., eds.), John Wiley and Sons, New York.
7. Thomas, G. B. and Mathews, M. B. (1980) DNA replication and the early to late transition in adenovirus infection. *Cell* **22,** 523–533.
8. Huang, W., Pruzan, R., and Flint, S. J. (1994) *In vivo* transcription from the adenovirus E2E promoter by RNA polymerase III. *Proc. Natl. Acad. Sci. USA* **91,** 1265–1269.
9. Huang, W. and Flint, S. J. (2003) Unusual properties of adenoviral E2E transcription by RNA polymerase III. *J. Virol.* **77,** 4015–4024.
10. Roeder, R. G. (1976) Eukaryotic nuclear RNA polymerases, in *RNA Polymerase* (Losick, R. and Chamberlin, M., eds.), Cold Spring Harbor Laboratory, Cold Spring Harbor, New York, pp. 285–329.
11. Steinberg, T. H., Mathews, D. E., Durbin, R. D., and Burgess, R. R. (1990) Tagetitoxin: a new inhibitor of eukaryotic transcription by RNA polymerase III. *J. Biol. Chem.* **265,** 499–505.
12. Yang, U.-C., Huang, W., and Flint, S. J. (1996) mRNA export correlates with activation of transcription in human subgroup C adenovirus-infected cells. *J. Virol.* **70,** 4071–4080.
13. Young, B. D., Hell, A., and Birnie, G. D. (1976) A new estimate of human ribosomal gene number. *Biochim. Biophys. Acta* **454,** 539–548.
14. Feinberg, A. P. and Vogelstein, B. (1983) A technique for radiolabeling DNA restriction endonulease fragments to high specific activity. *Anal. Biochem.* **137,** 266–267.
15. Babich, A., Feldman, C. T., Nevins, J. R., Darnell, J. E., and Weinberger, C. (1983) Effect of adenovirus on metabolism of specific host mRNAs: transport and specific translational discrimination. *Mol. Cell. Biol.* **3,** 1212–1221.
16. Pilder, S., Moore, M., Logan, J., and Shenk, T. (1986) The adenovirus E1B-55kd transforming polypeptide modulates transport or cytoplasmic stablization of viral and host cell mRNAs. *Mol. Cell. Biol.* **6,** 470–476.
17. Williams, J., Karger, B. D., Ho, Y. S., Castiglia, C. L., Mann, T., and Flint, S. J. (1986) The adenovirus E1B 495R protein plays a role in regulating the transport and stability of the viral late messages. *Cancer Cells* **4,** 275–284.
18. Lin, H. J. and Flint, S. J. (2000) Identification of a cellular repressor of transcription of the adenoviral late IVa$_2$ gene that is unaltered in activity in infected cells. *Virology* **277,** 397–410.

2

The Use of In Vitro Transcription to Probe Regulatory Functions of Viral Protein Domains

Paul M. Loewenstein, Chao-Zhong Song, and Maurice Green

Summary

Adenoviruses (Ads), like other DNA tumor viruses, have evolved specific regulatory genes that facilitate virus replication by controlling the transcription of other viral genes as well as that of key cellular genes. In this regard, the E1A transcription unit contains multiple protein domains that can transcriptionally activate or repress cellular genes involved in the regulation of cell proliferation and cell differentiation. Studies using in vitro transcription have provided a basis for a molecular understanding of the interaction of viral regulatory proteins with the transcriptional machinery of the cell and continue to inform our understanding of transcription regulation. This chapter provides examples of the use of in vitro transcription to analyze transcriptional activation and transcriptional repression by purified, recombinant Ad E1A protein domains and single amino acid substitution mutants as well as the use of protein-affinity chromatography to identify host cell transcription factors involved in viral transcriptional regulation. A detailed description is provided of the methodology to prepare nuclear transcription extract, to prepare biologically active protein domains, to prepare affinity depleted transcription extracts, and to analyze transcription by primer extension and by run-off assay using naked DNA templates.

Key Words: In vitro transcription; nuclear transcription extract; protein affinity; chromatography; run-off transcription assay; primer extension assay; purification of E1A protein mutants; adenovirus protein domains; gene regulation; transcription activation; transcription repression; early gene 1A.

1. Introduction

Adenoviruses (Ads), like other DNA tumor viruses, have evolved specific regulatory genes that facilitate virus replication by controlling the transcription of other viral genes as well as that of key cellular genes that regulate cell cycle progression and cellular DNA synthesis. In this regard, the early gene 1

From: *Methods in Molecular Medicine, Vol. 131:*
Adenovirus Methods and Protocols, Second Edition, vol. 2:
Ad Proteins, RNA, Lifecycle, Host Interactions, and Phylogenetics
Edited by: W. S. M. Wold and A. E. Tollefson © Humana Press Inc., Totowa, NJ

(E1A) transcription unit, which encodes two major proteins of 243 and 289 amino acids, is of particular interest. The 289R protein is identical to the 243R protein except that it contains in addition conserved region 3 (CR3, residues 140-188), a powerful transcriptional activator of Ad early genes (*see* **ref. *1*** for review). The 243R protein encodes diverse biological functions, including the ability to induce cell cycle progression, immortalize cells, transform cells in cooperation with other oncogenes, and paradoxically to inhibit tumorigenicity and cell differentiation *(1)*. These E1A 243R functions are encoded within multiple protein domains that can transcriptionally activate or repress cellular genes involved in the regulation of cell proliferation and cell differentiation. Conserved region (CR)1 (amino acids 40–80) and CR2 (amino acids 120–139) are common to both 243R and 289R and together with the common nonconserved N terminus (amino acids 1 to 39) are required for the growth-regulatory functions of E1A. Exon 2 possesses an autonomous transformation suppression activity within CR4 localized within a 14-amino-acid region near the common C terminus of E1A *(2)*. An interesting function that maps to within the E1A N-terminal 80 amino acids is the ability to repress transcription from genes involved in cell proliferation and cell differentiation. To investigate the mechanisms of E1A transcriptional regulation, our laboratory developed an in vitro transcription assay that accurately measures transcriptional activation and transcriptional repression by E1A protein domains *(3)*.

Protein–protein binding studies have shown that viral regulatory proteins such as E1A can interact with sequence-specific transcriptional activators as well as with several general transcription factors (GTFs) of the cellular transcription machinery. The significance of these protein–protein interactions in transcriptional regulation can be addressed mechanistically by the application of in vitro transcription methodology using purified recombinant viral proteins.

Transcription in mammalian cells by RNA polymerase II is a complex process that involves the formation of preinitiation complexes composed of at least 44 distinct polypeptides that can be classified into several groups (*see* **ref. *4*** for review). A first group of polypeptides constitutes the general transcription machinery and includes RNA polymerase II and the GTFs TFIIA, TFIIB, TFIID, TFIIE, TFIIF, and TFIIH. A second group consists of nucleic acid sequence-specific transcriptional activators that stimulate transcription, at least in part, by increasing the number of functional transcription complexes. A third group consists of a continually growing number of proteins classified by function as coactivators, positive cofactors, negative cofactors, and general repressors of transcription. This third group activates or represses transcription through protein–protein interactions with transcription factors, often components of the general transcription machinery. The E1A proteins do not bind to specific DNA sequences, and their transcriptional activities best fit into the third group of transcription factors that function through protein–protein interaction.

The molecular mechanisms of transcriptional regulation by viral protein domains can be effectively explored using in vitro transcription systems. Of particular importance was the development of a procedure to prepare extracts from nuclei of HeLa cells that can direct accurate transcription initiation in vitro from RNA polymerase II promoters, as described by Dignam et al. *(5)*. Transcriptionally active nuclear extracts can be prepared from different cell types and from as few as 3×10^7 cells *(6)*. The preparation of transcription extracts and the assay conditions can be modified to optimize transcription of specific genes (*see* **Note 1**). Of additional value is the use of protein-affinity and antibody-depletion experiments to define components within nuclear extracts that interact with specific viral or cellular proteins and are involved in transcriptional regulation. Finally, given the cloning and expression of the transcription factors that comprise the general transcription machinery, it is possible to analyze the functions of viral proteins in a defined reconstituted transcription system. Such studies have provided a basis for a molecular understanding of the interaction of viral regulatory proteins with the transcriptional machinery of the cells and continue to inform our understanding of transcription regulation.

More recent developments relevant for the analysis of viral transcriptional regulators is the recognition that chromatin remodeling plays an important role in transcription regulation. Transcription in vivo is regulated at multiple levels from chromatin templates *(7)*. The packaging of DNA into chromatin affects its accessibility by the transcription machinery. Specific patterns of modifications at the histone tails serve as markers for the recruitment of different protein complexes that regulate chromatin structure and gene expression *(8)*. Promoters in which the chromatin–histone structure is unaltered are completely repressed transcriptionally. Interaction with chromatin-modifying activities, such as methylases, acetylases, and phosphorylases, allows for exposure of the promoter so that basal transcription can occur. This basal activity may be in turn modulated by the large number of gene-specific as well as global transcription factors that may activate or repress basal transcription, i.e., transcriptional activators and transcriptional repressors, discussed above. The usual analysis of in vitro transcription using "naked" DNA templates most likely reflects the transcription of promoters that are in a "basal state." The ability of E1A to interact with several histone acetylases and components of the Swi/Snf complex suggests that E1A may regulate transcription at the chromatin level as well as at the basal level. Chromatin assembly in vitro has been described *(7,9)*. Development of in vitro transcriptions systems that utilize chromatin templates in addition to naked DNA allows for another level for understanding transcriptional regulation in a setting that may more accurately reflect that physiological state of cellular gene promoters. In vitro transcription using chromatin templates has been developed to recapitulate transcriptional regulation, which occurs on chromatin in vivo. In vitro transcription studies using chromatin templates will facilitate the understanding of the regulation of gene expression by viral regulatory proteins.

This chapter provides brief examples of the use of in vitro transcription to analyze transcriptional activation and transcriptional repression by purified, recombinant E1A protein domains and single amino acid substitution mutants, as well as the use of protein-affinity chromatography to identify host cell transcription factors involved in viral transcriptional regulation. Following these is a detailed description of the methodology to prepare the nuclear transcription extract, to prepare biologically active protein domains, to prepare affinity-depleted transcription extracts, and to analyze transcription by primer extension and by run-off assay using naked DNA templates.

1.1. Using In Vitro Transcription to Demonstrate the Autonomous Transactivation Activity of Ad E1A Conserved Domain 3

Analysis of E1A mutants by transient expression has demonstrated that CR3 is essential for transactivation of early viral genes by E1A 289R (for review, *see* **ref.** *1*). To determine whether CR3 is sufficient for transactivation, a 49-amino-acid peptide encoding CR3 plus 3 amino acids in exon 2 (termed PD3) was chemically synthesized and tested for its ability to transactivate Ad promoters in vitro. In vitro transcription products were analyzed by primer extension analysis to ensure that transcription initiated at the correct start site. PD3 (100–500 ng), when added to a reaction mixture containing 500 ng of DNA template, either the Ad early region 3 (E3) promoter or the Ad major late promoter, stimulated transcription 5- to 20-fold (*see* **Fig. 1**). These results show directly that the sequences within CR3 are responsible for E1A transactivation *(10)*. Further, these results demonstrate that E1A protein domain CR3 is sufficient for transactivation activity.

1.2. Using In Vitro Transcription to Analyze Transcriptional Repression by the Ad E1A N-Terminal 80-Amino-Acid Sequence

Analysis of E1A mutants by transient expression had indicated that the transcriptional repression function of E1A 243R requires sequences within the nonconserved N-terminal 40 amino acids, within CR1, and also within CR2 in some cases (for review, *see* **ref.** *1*). In order to permit a biochemical definition of the viral and cellular components involved in the E1A repression function, it was important to establish a system that faithfully recapitulates E1A repression in vitro. A recombinant protein containing only the E1A 80 N-terminal

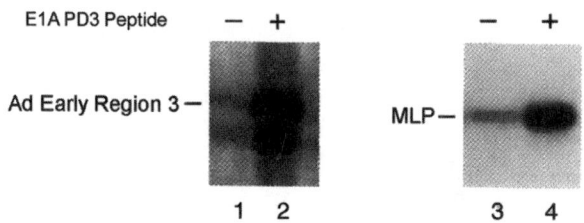

Fig. 1. In vitro transcription of the adenovirus (Ad)2 E3 promoter and the Ad2 major-late promoter (MLP) in the presence (+) and absence (–) of E1A PD3 peptide (CR3 plus three amino acids of Exon2). In vitro transcription was performed with reaction mixtures containing E3 or MLP plasmids as templates and subjected to primer extension analysis. Primers consisted of 5'-end-labeled synthetic oligonucleotides complementary to the E3 mRNA (positions +108 to +137) or MLP mRNA (positions +67 to +86) from the start site of transcription. Labeled cDNA products were analyzed by electrophoresis on a 6% polyacrylamide—*7M* urea gel followed by autoradiography.

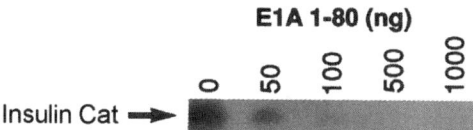

Fig. 2. Transcription repression in vitro of the insulin II promoter by purified recombinant E1A 1-80 protein. pInsulin-CAT was used as template for the in vitro transcription reaction. Transcripts were analyzed by primer extension using a CAT primer followed by autoradiography. Reaction mixtures contained from 0 to 1000 ng of E1A 1-80, as indicated.

amino acids (E1A 1-80), which includes the N terminus plus all of CR1, as well as a series of E1A 1-80 deletion mutant proteins were prepared (*see* **Subheading 3.2.**). These proteins were used to study in vitro transcriptional repression of promoters previously reported by transient expression to be E1A repressible, including those of insulin, interstitial collagenase, simian virus 40, and HIV-long terminal repeat (LTR). Assay conditions by primer extension were first established for in vitro transcription of each promoter fused to the CAT reporter gene *(3)*. The addition of E1A 1-80 protein to the transcription mixture strongly represses these promoters in a dose-dependent manner (*see* **Fig. 2**). Repression is promoter-specific because promoters not repressed by E1A in vivo are not repressed by E1A in vitro. Further, repression is E1A sequence specific, as shown by the analysis of E1A 1-80 deletion mutant proteins *(3)*. Thus, the in vitro transcription-repression assay faithfully reflects E1A repression and provides a valuable system to study molecular mechanism, as described below.

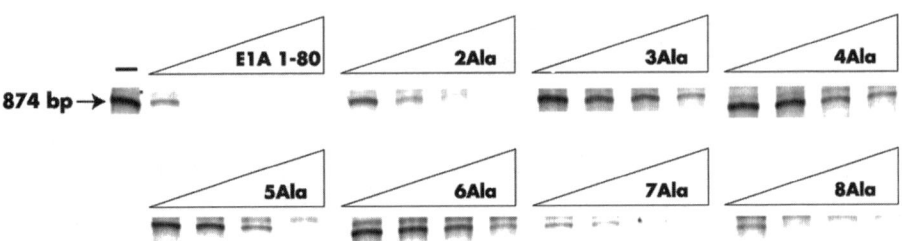

Fig. 3. In vitro transcription repression by single amino acid substitution mutants within the Ad E1A 1-80 polypeptide. Shown are run-off in vitro transcription reactions measuring the transcription repression abilities of E1A 1-80 polypeptide and the first seven E1A 1-80 mutants with single substitutions of alanine for the indicated residue. Reaction mixtures contained between 62.5 and 500 ng of polypeptide. Substitution of 3His, 4Ile, 5Ile, or 6Cys with Ala severely interfered with the ability of E1A 1-80 to repress transcription from the HIV long tandem repeat promoter.

1.3. Using In Vitro Transcription to Identify Amino Acids Critical for the Repression Function Within the E1A 1-80 Polypeptide

Single amino-acid substitution in each of the first 30 amino acids of E1A 1-80 were constructed and the mutant polypeptides were expressed in *Escherichia coli*, purified and renatured as described in **Subheading 3.2.** Each amino acid was substituted in turn with alanine except where alanine was the natural occurring amino acid; in this case the amino acid was substituted with glycine. Alanine scanning mutants have the advantage of removing all amino acid side-chain elements and have been successfully used to identify critical amino acids in several transcriptional regulatory proteins (*see* **ref.** *11* for example). E1A 1-80 and the alanine scanning mutant polypeptides were used in run-off transcription reactions with an HIV LTR promoter as template. **Figure 3** shows that E1A 1-80 efficiently represses transcription over the range of concentrations tested, whereas substitution of amino acids 3, 4, 5, and 6 (and 20 not shown in this figure) with alanine resulted in polypeptides that are deficient in transcription repression activity. Of significance, the amino acid substitution mutants defective for in vitro repression activity are also defective for in vivo repression activity as measured by cell microinjection *(12)*.

1.4. Using Protein-Affinity Depletion of Nuclear Extracts to Identify Transcription Factors That Interact With the E1A Repression Domain

The ability of E1A 1-80 to repress transcription in vitro strongly implies that E1A interacts with a cellular protein(s), presumably a transcription factor(s). In order to understand the mechanism of E1A repression, it is important to identify the cellular target(s) of E1A functional domains. E1A 1-80 protein-

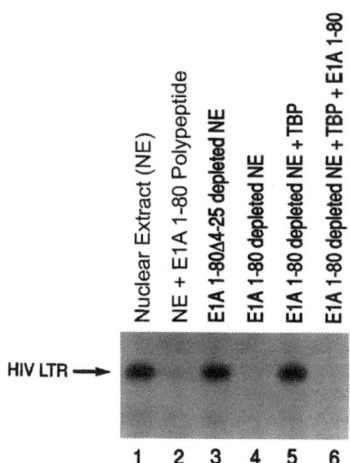

Fig. 4. TATA binding protein (TBP) can restore transcriptional activity to an E1A 1-80-depleted nuclear extract. In vitro transcription and primer extension analysis was performed with pBennCAT (HIV long tandem repeat) as template and a CAT primer. The transcriptional activity of the original nuclear extract (NE) (lane 1) is repressed by added E1A 1-80 protein (400 ng, lane 2). Transcriptional activity is lost by passage through an E1A 1-80 affinity column (lane 4) but not an E1A 1-80Δ4-25 column (lane 3). The addition of TBP (5 ng, lane 5) restores transcriptional activity to the E1A 1-80 affinity depleted NE. The activity restored by TBP can be repressed by addition of E1A 1-80 (400 ng, lane 6).

affinity chromatography was performed to sequester and identify cellular factor(s) that interact with the E1A N-terminal sequence *(13)*. A complete loss of transcription activity occurs when a nuclear extract is passed through a column containing immobilized E1A 1-80 protein, but no loss of activity occurs when the extract is passed through a column containing the repression-defective mutant protein, E1A 1-80Δ4-25 (**Fig. 4**). These results provide strong presumptive evidence that the E1A repression domain interacts specifically with and depletes an essential transcription factor(s) from the nuclear extract. Western analysis demonstrated that the GTF, TBP (TFIID), was depleted from the extract and was bound to E1A 1-80 *(13)*. Of significance, activity of the depleted extract is completely restored by the addition of TBP, thus providing strong evidence that TBP (TFIID) is a target of E1A repression (**Fig. 4**).

2. Materials

To minimize RNase contamination, all reagents should be made with water that is known to contain less that 20 ppm total organics. Alternately, DEPC-treated water may be used: add 1 mL of DEPC per L of water, shake well, and autoclave after 1 h. Wear gloves during all operations to avoid contamination with finger RNases. Pipetting devices should be wiped down with ethanol and never used with solutions containing RNase. All reagents can be stored at –20°C.

1. Phosphate-buffered saline (PBS): 1.54 mM KH$_2$PO$_4$, 155.17 mM NaCl, 2.71 mM Na$_2$HPO$_4$.

2. Buffer A: 10 mM N-2-hydroxyethylpiperazine-N'-2-ethanesulfonate (HEPES) (pH 7.9 at 4°C), 1.5 mM MgCl$_2$, 10 mM KCl, 0.5 mM dithiothreitrol (DTT).

3. Low-salt buffer: 20 mM HEPES (pH 7.9 at 4°C), 1.5 mM MgCl$_2$, 20% glycerol, 20 mM KCl, 0.2 mM ethylene diamine tetraacetic acid (EDTA), 0.2 mM phenylmethylsulfonyl fluoride (PMSF), 0.5 mM DTT.

4. High-salt buffer: 20 mM HEPES (pH 7.9 at 4°C), 1.5 mM MgCl$_2$, 20% glycerol, 1.2 M KCl, 0.2 mM EDTA, 0.2 mM PMSF, 0.5 mM DTT.

5. Buffer D: 20 mM HEPES (pH 7.9 at 4°C), 20% glycerol, 100 mM KCl, 0.2 mM EDTA, 0.2 mM PMSF, 0.5 mM DTT.

6. QIAGEN buffer A: 100 mM NaH$_2$PO$_4$, 10 mM Tris-HCl, pH 8.0, 6 M guanidine-HCl (adjusted to pH 8.0 with NaOH).

7. 0.5X buffer D: 10 mM HEPES (pH 7.9 at 4°C), 10% glycerol, 50 mM KCl, 0.1 mM EDTA, 0.1 mM PMSF, 0.25 mM DTT.

8. 20 mM HEPES, pH 7.2.

9. 1 M Ethanolamine, pH 8.0.

10. Forward exchange buffer (10X): 500 mM Tris-HCl (pH 7.5), 100 mM MgCl$_2$, 50 mM DTT, 1 mM spermidine.

11. Transcription buffer (10X): 40 mM HEPES (pH 7.9 at 4°C), 40 mM creatine phosphate, 100 mM MgCl$_2$, 200 mM KCl, 5 mM DTT, 0.2 mM EDTA.

12. Stop mix: 20 mM EDTA, pH 8.0, 200 mM NaCl, 1% sodium dodecyl sulfate, 0.2 mg/mL glycogen.

13. PE buffer (2X): 100 mM Tris-HCl (pH 8.3 at 42°C), 100 mM KCl, 20 mM MgCl$_2$, 20 mM DTT, 2 mM dNTPs, 1 mM spermidine.

14. TBE (10X): 1 M Tris base, 900 mM boric acid, 10 mM EDTA.

15. Formamide loading mix: 98% formamide, 10 mM EDTA (pH 8.0), 0.01% xylene cyanol, 0.01% bromophenol blue.

16. Acrylamide gel mix: 40% acrylamide/*bis*-acrylamide (29:1).

17. Corning polypropylene centrifuge tubes, 250-mL (cat. no. 25350-250).

18. Corning polypropylene disposable centrifuge tubes, 50-mL (cat. no. 430291).

19. Corning polypropylene disposable centrifuge tubes, 15-mL (cat. no. 430766).

20. Kontes B pestle (VWR).

21. Spectro-Por dialysis tubing (18-mm flat width, molecular-weight [MW] cutoff: 2000) (VWR).

22. Slide-A-Lyzer, 3000 MW cut-off (Pierce).

23. Centiprep YM-3, 3000 MW cut-off (Millipore).

24. Centricon YM-3, 3000 MW cut-off (Millipore).

25. Affi-Gel 10 (Biorad).

26. ^{32}P-ATP (approx 1000 Ci/mmol).

27. T4 DNA ligase.
28. ATP, GTP, CTP, UTP: 5 m*M* each.
29. 0.3 *M* Sodium acetate.
30. Phenol:chloroform:isoamyl alcohol (50:50:2), saturated with RNase-free water.
31. 80% Ethanol.
32. 40 m*M* Sodium pyrophosphate.
33. AMV reverse transcriptase (Promega).
34. Siliconizing reagent: Rain-X (Unelko Corp.).
35. Urea (Invitrogen).
36. 3 MM filter paper sheets.
37. ^{32}P-UTP (800–1000 Ci/mmol).

3. Methods

3.1. Preparation of the Nuclear Transcription Extract

Nuclear extracts can be made from virtually any volume of cells grown in monolayer or in suspension culture *(6)*. For reproducibility and convenience, the majority of nuclear extracts are from liter quantities of HeLa cell suspension culture. The protocol used in our laboratory is a modification of that of Dignam et al. *(5)*. We use suspension cultures of HeLa cells grown in Joklik's minimum essential medium (ICN) supplemented with 10% calf serum.

1. Six liters of HeLa cells in suspension are grown by feeding cells every day while maintaining a cell density of about 6×10^5 per mL *(14)*. Cells should be harvested for the preparation of nuclear extracts when they are growing well, i.e., doubling nearly every 24 h.
2. Harvest cells in six 250-mL Corning polypropylene centrifuge tubes by repeated centrifugation at 4°C for 10 min at 180*g* (Beckman J6-HC centrifuge), i.e., carefully pour off the supernatant and add fresh suspension culture on top of the existing cell pellet and repeat centrifugation until the entire 6 L of suspension culture are harvested.
3. All subsequent steps are done at 4°C using precooled buffers. Gently resuspend cell pellets into 40 mL of PBS by pipetting, combine the suspended pellets in a single 50-mL Corning polypropylene centrifuge tube, and centrifuge again. Gently resuspend the packed cell pellet (usually 5–8 mL; note exact volume) in five times the pellet volume of hypotonic buffer A and incubate on ice for 10 min. Centrifuge the swollen cells for 10 min at 180*g* and carefully withdraw the supernatant with a pipet so as not to disturb the soft, swollen cell pellet. Using buffer A, resuspend the cells in twice the *original* volume of packed cells. A microscopic examination of a small aliquot should reveal that cells are greatly swollen but largely intact.
4. Dounce-homogenize the cells using 10 strokes of a Kontes B pestle. A microscopic examination should show that more than 90% of the cells have been disrupted and that the vast majority of nuclei are intact.

5. Centrifuge the nuclear preparation in a 50-mL polypropylene disposable centrifuge tube at 1000g for 10 min (Beckman Avanti J-E centrifuge using a JA-12 rotor). Remove the supernatant and recentrifuge at 1000g for 5 min. Remove and discard the small amount of remaining supernatant above the nuclear pellet. Note the volume of the nuclear pellet.

6. To prepare the nuclear extract, resuspend the nuclear pellet in a volume of low-salt buffer representing *exactly* one-half the nuclear pellet volume. Add dropwise the same volume of high-salt buffer as low-salt buffer while gently vortexing the tube. Tightly cap the tube and continue to extract the pellet by rotation for 30 min.

7. Centrifuge the extract at 9000g for 30 min (Beckman Avanti J-E centrifuge using a JA-12 rotor). Transfer the supernatant into Spectra-Por dialysis tubing and dialyze for 3–5 h against 1 L of buffer D.

8. Remove the extract from the dialysis tubing and clarify it by centrifugation for 10 min at 10,000g. Aliquot the supernatant (100–500 µL) into precooled microfuge tubes (4°C) and freeze by immersion in liquid nitrogen. Store in a –70°C freezer or in a vapor-phase nitrogen freezer for long-term storage.

3.2. Preparation of Biologically Active E1A Functional Domains

1. Biologically functional His6-tagged E1A 243R and E1A 1-80 polypeptides are prepared by a protocol modified from that suggested by QIAGEN (QIA expressionist). Harvest a culture of isopropylthio-β-D-galactoside-induced bacterial cells (500 mL) expressing an appropriate pQE (QIAGEN) E1A construct *(12)* by centrifugation at 2000g for 10 min at 4°C (Beckman J6-HC centrifuge), and freeze the cell pellet at –20°C.

2. Thaw the pellet at room temperature and lyse in 40 mL of QIAGEN buffer A with gentle mixing for 1 h. Clarify the supernatant by centrifugation at 9000g for 30 min at 4°C (Beckman Avanti J-E centrifuge; JA-12 rotor).

3. Bind the His6-tagged polypeptide to 2 mL of Ni-NTA resin (QIAGEN) by rotating the clarified lysate with the resin overnight at 4°C. Batch wash the resin five times with 20 mL of buffer A and five times with buffer A adjusted to pH 6.3.

4. Load the resin into two 5-mL columns (Image Molding) and wash each column with 20 mL (10 column volumes) of the same buffer and then with 50 mL of buffer A adjusted to pH 5.9.

5. Elute the His6-tagged polypeptide from the resin with 20 mL of buffer A adjusted to pH 4.5. Collect 2-mL fractions. Generally, sufficient polypeptide is produced so that protein-containing fractions can be identified by adding 2 µL of each fraction to 100 µL of BIO-RAD DC Protein Assay Reagent B. Detectable color develops after 10 min at room temperature.

6. To prepare biologically active E1A 1-80 polypeptides, it is necessary to remove guanidine-HCl slowly from the preparation to facilitate proper folding. Pool fractions containing eluted polypeptide and adjust to 6 mL with elution buffer (buffer A at pH 4.5 containing 6 M guanidine-HCl). Dilute the sample 1:1 with 0.5X buffer D.

7. Dialyze the diluted sample (now at 3 M guanidine-HCl) in a 3000 MW cutoff Slide-A-Lyzer dialysis cassette against 0.5X buffer D containing 2 M guanidine-HCl. After 6–8 h of dialysis, remove one-half of the dialysis buffer and replace it with fresh 0.5X buffer D without guanidine-HCl, thereby reducing the guanidine-HCl concentration by half.

8. Continue dialysis with buffer replacement in the manner described above until the guanidine-HCl concentrations is reduced to 50–100 mM. Complete the dialysis against several changes of 0.5X buffer D for 8 h.

9. Concentrate the E1A polypeptides by size exclusion centrifugation using Centriprep YM-3 followed by Centricon YM-3 to a final concentration of about 1 mg per mL of polypeptide.

3.3. Preparation of Nuclear Transcription Extracts Affinity Depleted With E1A Functional Domain Polypeptides

1. Affi-Gel 10 immobilized E1A polypeptides are prepared as follows. Exchange by dialysis into 20 mM HEPES, pH 7.2, the E1A 1-80 polypeptides purified as described above (**Subheading 3.2.**).

2. Prepare 1.5 mL of packed Affi-Gel 10 beads immediately prior to use as follows. Take 3 mL of Affi-Gel suspension (comes as a 1:1 slurry in isopropanol) and centrifuge in a 15-mL centrifuge tube (Corning) at 2000 rpm (1000g) for 5 min. Wash beads three times with 12 mL of cold water by centrifugation at 2500 rpm (1700g) for 2 min. Incubate the packed beads with 1.5 mg of polypeptide in 1.5 mL of buffer for 4 h at 4°C with rotation. Retain 10 µL of polypeptide prior to addition to beads and after 4 h of incubation to determine the efficiency of polypeptide binding as follows.

3. Centrifuge the aliquots in a microfuge tube at 1700g for 5 min and add 5 µL of 1 N HCl to the supernatant. Add 40 µL of water and read the absorbance at 280 nm. Efficient binding is reflected by a reduction in absorbance at 280 nm of 80–90%. If efficient binding is not attained, continue rotation of the beads with the polypeptide.

4. When satisfactory binding of polypeptide to beads is obtained, centrifuge the beads at 2000 rpm for 5 min and resuspend the beads in 800 µL of PBS. Add 200 µL of 1 M ethanolamine, pH 8.0, and rotate for 1 h at 4°C to block reactive groups remaining on the beads. Centrifuge to remove the ethanolamine and wash three times with buffer D. Store the polypeptide-immobilized beads at 4°C.

5. Prepare affinity column (5-mL, Image Molding) by loading 250 µL of the packed polypeptide-immobilized beads. Recirculate 1 mL of HeLa cell nuclear extract through each column for 2 h at 4°C at a flow rate of 0.1 mL per min. Aliquots of the nuclear extract before and after affinity chromatography (as well as the proteins bound to the beads) can be analyzed by immunoblot analysis with specific antibodies to candidate E1A cellular partners by use of a sensitive Western blotting kits utilizing chemiluminescent substrates (Amersham Pharmacia, Pierce). The depleted nuclear extracts can now be used by in vitro transcription to probe the function of the E1A polypeptide interacting domain(s).

3.4. Analysis of In Vitro Transcription by Primer Extension and by Run-Off Assay

Inasmuch as RNA polymerase II does not efficiently terminate transcription in vitro, two assays have been developed that yield products of discrete length. The first assay is the run-off assay, which uses a DNA template cut with a restriction enzyme downstream of the transcription start site. This simple solution creates a site where the polymerase will fall off the template, thus terminating transcription. By use of a ^{32}P-labeled rNTP precursor, an RNA transcription product of specific length is synthesized and is resolved as a discrete band by denaturing polyacrylamide gel electrophoresis (PAGE) followed by autoradiography or phosphor image analysis. The second assay is primer extension in which the RNA product of the in vitro transcription reaction is annealed with a ^{32}P-end labeled deoxyoligonucleotide complementary to a discrete sequence downstream of the transcription start site. Reverse transcription of the hybrid RNA product/DNA primer yields a labeled DNA fragment of discrete length that spans the sequence between the 5'-end of the primer and the transcription start site. This labeled cDNA fragment is resolved by denaturing PAGE followed by autoradiography. The product of both assays can be quantitated by scanning densitometry or by more sensitive phosphor image analysis.

In general, primer extension is more sensitive than run-off assay and permits a more accurate measurement of the transcription product from the authentic transcription start site. For many studies, a sequence that is specific for a common reporter gene, such as CAT or luciferase, allows the convenient use of a single primer downstream of a variety of promoters for primer extension analysis. This also permits the use of multiple DNA template promoters in a single transcription reaction because the 5'-transcribed sequence from each promoter is usually sufficiently different in size and thus gives rise to primer extension products of different lengths, which can be visualized on a single lane of a gel.

3.4.1. Preparation of DNA Templates for In Vitro Transcription

Plasmid templates containing the promoter of interest must be of high purity for in vitro transcription. Plasmids purified by double CsCl density gradient centrifugation are of high quality and can be used directly. Plasmid templates prepared by standard alkaline lysis procedures or by use of commercial plasmid preparation kits may require further purification. It is often important to further purify those templates by phenol-chloroform extraction followed by ethanol precipitation. For primer extension, superhelical plasmid templates are used. For run-off analysis, templates are linearized with a restriction enzyme in order to terminate transcription at a known site.

3.4.2. Preparation of Radiolabeled Deoxyoligonucleotides for Primer Extension

Deoxyoligonucleotide primers are designed to be about 30 nucleotides (nt) in length and to be complementary to a region from 100 to 200 nt downstream of the transcription initiation site. Primers with self-complementary sequences are avoided. To label a primer, incubate 10 pmol of deoxyoligonucleotide at 37°C for 30 min in a 10-µL reaction containing 1 µL of 10X forward exchange buffer, 6 µL of γ^{32}P-ATP (1000 Ci/mmole), and 1 µL of T4 polynucleotide kinase (10 U). Heat the reaction mixture at 100°C for 2 min and add 190 µL of water. Store at –20°C. The radiolabeled primer may be used as long as a suitable transcription signal is obtained, usually 4–6 wk.

3.4.3. In Vitro Transcription Analysis Using Primer Extension Analysis

3.4.3.1. TRANSCRIPTION REACTION

Reaction mixtures are assembled in RNase-free microfuge tubes containing 2.5 µL of 10X transcription buffer, 2.5 µL of rNTP mixture (5 m*M* each of ATP, GTP, CTP, and UTP), DNA template (typically 500 ng), nuclear extract (typically 10 µL), and water to give a final volume of 25 µL. The reaction is initiated by the addition of nuclear extract followed by incubation at 30°C for 60 min. For each new DNA template and for each batch of nuclear extract, preliminary titrations are done with different amounts of template (100–1000 ng) and nuclear extract (5.0–12.5 µL) to optimize the transcription signal (strength and authenticity, i.e., correct size). Compensate for volumes of nuclear extract less than 10 µL by the addition of an equivalent volume of buffer D.

3.4.3.2. ISOLATION OF THE RNA TRANSCRIPTION PRODUCT

When the transcription reaction is completed after 60 min, terminate the reaction by addition of 100 µL of stop mix. Then add 300 µL of 0.3 *M* sodium acetate and 300 µL of phenol/chloroform/isoamyl alcohol (50:50:2). Vortex the sample for 30 s and separate the phases by centrifugation at 10,000*g* for 2 min. Transfer the upper aqueous phase to a fresh tube containing 1 mL of ethanol, mix, and place on dry ice for 15 min. Centrifuge the sample at 10,000*g* for 10 min at 4°C, rinse the pellet carefully with 500 µL of 80% ethanol (–20°C), and briefly dry the pellet containing the RNA product in a vacuum dessicator.

3.4.3.3. PRIMER EXTENSION ANALYSIS

Resuspend the RNA pellet completely in 5 µL of 2X PE buffer, 5 µL of water, and 1 µL of the labeled primer (~0.5 pmol) by repeated vortexing and centrifugation. Denature the RNA by incubation at 65°C for 10 min and anneal the RNA with the primer by incubation at 42°C for 10 min. Next add 5 µL of 2X PE buffer, 1.6 µL of water, 1.4 µL of 40 m*M* sodium pyrophosphate, 1 µL of AMV reverse transcriptase (2.5 units, Promega) and incubate the mixture at 42°C for 30 min. Terminate the reverse transcriptase reaction by the addition of 20 µL of formamide loading mix.

3.4.3.4. Resolution of the Labeled Primer Extension Product by Denaturing Polyacrylamide Gel

Resolve the labeled product by electrophoresis on a 0.75-mm-thick/20-cm-long 6% urea polyacrylamide gel using a vertical gel chamber. The smaller plate (facing the apparatus) is siliconized with Rain-X to facilitate separation of the plates after electrophoresis. To prepare the gel, stir with a magnetic bar 12 g of urea (ultrapure, RNase-free) with 6.0 mL of 5X TBE, 4.5 mL of 40% acrylamide/*bis*-acrylamide (29:1), 300 µL of 10% ammonium persulfate, and water to a final volume of 30 mL. When the urea is completely dissolved, add 20 µL of TEMED and pipet or pour the mixture into an assembled gel sandwich held at about a 40° angle. Insert a 15-well comb and then allow the gel to polymerize in a horizontal position for 3 h or overnight. Prior to loading the samples, mount the gel on the electrophoresis chamber, add TBE to the upper and lower chambers, and clean the gel teeth by pipetting TBE up and down. Pre-electrophorese the gel for about 15 min at 400 V constant. Load the samples (10–20 µL) and a labeled size marker (e.g., a 50-bp ladder; *see* **Note 2**) into individual lanes on the gel and electrophorese at 400 V constant until the bromophenol blue marker just migrates off the gel.

The gel is pulled onto 3MM paper as follows. Briefly, separate the gel sandwich by gentle prying with a fine-bladed spatula. The gel will adhere to the larger, unsiliconized plate. Put the gel-containing plate flat on a bench top and carefully position a sheet of 3MM paper over the gel. Rub the paper gently but firmly and carefully pull off the paper with the adherent gel. The gel can then be dried on a gel dryer prior to autoradiography on X-ray film or simply covered with plastic wrap and autoradiographed with X-ray film and an intensifying screen at –70°C overnight. For quantitation and highly sensitive detection of signals, gels are dried and analyzed by phosphor image analysis as described in **Subheading 3.4.5.**

3.4.4. In Vitro Transcription Using the Run-Off Assay

The transcription reaction for the run-off assay is virtually the same as that described in **Subheading 3.4.3.1.** with the following modifications. First, the DNA template is cut at a convenient site with a restriction enzyme. Second, the 10X rNTP mixture consists of 5 m*M* ATP, GTP, and CTP but only 0.25 m*M* UTP. Third, ^{32}P-UTP (0.5 µL of 10 mCi/mL, 800–1000 Ci/mmol) is added to the reaction mixture. Finally, RNA is isolated as described in **Subheading 3.4.3.2.** and is directly analyzed by gel electrophoresis as described in **Subheading 3.4.3.4.**

3.4.5. Phosphor Image Analysis and Quantitation of Transcription Products From Primer Extension and Run-Off Assays

Regulatory proteins and mutants of those proteins may activate or repress transcription at varying rates as opposed to an all-or-none phenomenon. Multiple titrations with different levels of polypeptides followed by careful quantitation of transcription product are important to assess function. Phosphor image analysis provides for an efficient, accurate, and highly sensitive way to measure radiolabeled transcription products. Phosphor image analysis typically has a linear dynamic range of 5 orders of magnitude as compared to only about 1.5 for X-ray film. The pixels measured by the PhosphorImager can be quantitatively and statistically analyzed using software provided by the manufacturer, e.g., Image Quant for the Molecular Dynamics PhosphorImager system used here.

Briefly, cover the dried gel with plastic wrap and place a phosphor screen over the dry gel in a PhosphorImager cassette. Expose the phosphor screen from overnight to 2 d. Place the screen face-down on the PhosphorImager glass plate and select the scanning area using the appropriate software. The image of the gel is visualized and bands of the correct size are selected and quantified using the tools provided by the PhosphorImager software. These images can be printed or saved as digital files for permanent storage and for publication.

4. Notes

1. It is important that the transcription signal be optimized prior to assessing the effects of added regulatory protein. In the case of transcriptional activation, it is important to use conditions that are suboptimal in the absence of added activator so that transactivation can be easily measured over a wide dynamic range. By contrast, when analyzing transcription repression, it is desirable to use conditions that provide a moderate transcription signal, thus providing a sensitive assay for measuring repression by added protein. These goals can often be achieved empirically by one of two means. First, the amount of nuclear extract in the reaction mixture can often be titrated to provide the desired level of transcription in the absence of added protein factor. Second, it is possible to prepare extracts that possess innately high or low activity by altering the ionic strength of the salt used to prepare the nuclear extract. These extracts may contain greater or lesser amounts of transcription factors or cellular inhibitors of transcription, some of which may represent a cellular target of the added regulatory protein. The putative cellular target may not be rate-limiting in the nuclear transcription extract prepared under standard conditions. Specific modifications of the reaction mix-

ture can also achieve the desired results. For example, to demonstrate in vitro transactivation of the HIV LTR promoter by the HIV-1 Tat transactivator protein in run-off assays, it was necessary to reduce the basal transcription level of extracts by the addition of 6 m*M* sodium citrate *(15)*.

2. As a size marker for analysis of transcription products, it is convenient to end-label a 50- or 100-bp DNA ladder as described by the manufacturer (Invitrogen).

Acknowledgments

We thank Carolyn E. Mulhall for editorial assistance. The work from the authors' laboratories was supported by Research Career Award AI-04739 and Public Health Service grant CA-29561 to M.G. from the National Institutes of Health.

References

1. Shenk, T. (2001) *Adenoviridae*: the viruses and their replication, in *Fundamental Virology*, Lippincott, Williams and Wilkins, New York, pp. 1053–1088.
2. Chinnadurai, G. (2002) CtBP, an unconventional transcriptional corepressor in development and oncogenesis. *Mol. Cell* **9**, 213–224.
3. Song, C.-Z., Tierney, C. J., Loewenstein, P. M., et al. (1995) Transcriptional repression by human adenovirus E1A N-terminus/conserved domain 1 polypeptides in vivo and in vitro in the absence of protein synthesis. *J. Biol. Chem.* **40**, 23,263–23,267.
4. Roeder, R. G. (2005) Transcriptional regulation and the role of diverse coactivators in animal cells. *FEBS Lett.* **579**, 909–915.
5. Dignam, J. D., Lebovitz, R. M., and Roeder, R. G. (1983) Accurate transcription initiation by RNA polymerase II in a soluble extract from isolated mammalian nuclei. *Nucl. Acids Res.* **11**, 1475–1489.
6. Lee, K. A. W., and Green, M. R. (1990) Small-scale preparation of extracts from radiolabeled cells efficient in pre-mRNA splicing. *Methods Enzymol.* **181**, 20–30.
7. Lusser, A., and Kadonaga, J. T. (2004) Strategies for the reconstitution of chromatin. *Nat. Methods* **1**, 19–26.
8. Jenuwein, T., and Allis, C. D. (2001) Translating the histone code. *Science* **293**, 1074–1080.
9. Fyodorov, D. V., and Kadonaga, J. T. (2003) Chromatin assembly in vitro with purified recombinant ACF and NAP-1. *Methods Enzymol.* **371**, 499–515.
10. Green, M., Loewenstein, P. M., Pusztai, R., and Symington, J. S. (1988) An adenovirus E1A protein domain activates transcription in vivo and in vitro in the absence of protein synthesis. *Cell* **53**, 921–926.
11. Tang, H., Sun, X., Reinberg, D., and Ebright, R. H. (1996) Protein-protein interactions in eukaryotic transcription initiation: structure of the preinitiation complex. *Proc. Natl. Acad. Sci. USA* **93**, 1119–1124.
12. Boyd, J. M., Loewenstein, P. M., Tang, Q.-Q., Yu, L., and Green, M. (2002) Adenovirus E1A N-terminal amino acid sequence requirements for repression of

transcription in vitro and in vivo correlate with those required for E1A interference with TBP-TATA complex formation. *J. Virol.* **76,** 1461–1474.

13. Song, C.-Z., Loewenstein, P. M., Toth, K., and Green, M. (1995) TFIID is a direct functional target of the adenovirus E1A transcription-repression domain. *Proc. Natl. Acad. Sci. USA* **92,** 10,330–10,333.

14. Green, M. and Loewenstein, P. M. (2005) Human adenoviruses: propagation, purification, quantification, and storage, in *Current Protocols in Microbiology*, John Wiley and Sons, Inc, New York, pp. 14C.1.1.–14C.1.9.

15. Kato, H., Sumimoto, H., Pognonec, T., Chen, C.-H., Rosen, C. A., and Roeder, R. G. (1992) HIV-1 TaT acts as a processivity factor in vitro in conjunction with cellular elongation factors. *Gene Dev.* **6,** 655–666.

3

Preparation of Soluble Extracts From Adenovirus-Infected Cells for Studies of RNA Splicing

Oliver Mühlemann and Göran Akusjärvi

Summary

Here we describe a collection of methods that have been adapted to produce highly efficient nuclear and cytoplasmic extracts from adenovirus-infected HeLa cells. We describe how to produce extracts from virus-infected cells and how to analyze RNA splicing in vitro using T7 RNA polymerase-derived splicing substrate RNAs.

Key Words: Adenovirus; HeLa spinner cells; RNA splicing; nuclear extracts; S100 extract; polymerase chain reaction; T7 RNA polymerase.

1. Introduction

The interest in RNA splicing and alternative RNA splicing as important mechanisms controlling gene expression in health and disease has increased dramatically during the last decade. The fact that approx 15% of all human genetic disorders are a result of defects in RNA splicing highlights the significance of this regulatory step in eukaryotic gene expression. Also, in the wake of large genome-sequencing projects it has become abundantly clear that alternative RNA splicing is an important mechanism expanding our proteome (reviewed in **ref. 1**).

Adenovirus (Ad) has contributed significantly to our current understanding of the organization and expression of genes in eukaryotic cells. The most startling discovery was probably the discovery of the split gene and RNA-splicing concepts (reviewed in **ref. 2**). It was also shown early that the accumulation of alternatively spliced mRNAs was temporally regulated during virus infection. The shift from the early to late pattern of mRNA accumulation was shown to be dependent on viral late protein synthesis (for a recent review *see* **ref. 3**).

From: *Methods in Molecular Medicine, Vol. 131:*
Adenovirus Methods and Protocols, Second Edition, vol. 2:
Ad Proteins, RNA, Lifecycle, Host Interactions, and Phylogenetics
Edited by: W. S. M. Wold and A. E. Tollefson © Humana Press Inc., Totowa, NJ

Experiments aimed at reproducing RNA splicing in vitro initially progressed slowly. The first examples of successful in vitro RNA splicing were published in the early 1980s *(4–8)*. Success was, to a large extent, hampered by the difficulty of synthesizing good-quality substrate RNAs. A major step forward was therefore the development of the SP6 in vitro transcription system for pre-mRNA substrate synthesis *(9)*. With an easy method for production of large quantities of substrate RNA, the basic mechanisms of RNA splicing were rapidly established (reviewed in **ref. 10**).

An illustration summarizing the steps used to prepare splicing competent nuclear extracts is shown in **Fig. 1**. Nuclear extracts are prepared essentially as described by Dignam et al. *(11)*. This protocol was originally designed for in vitro transcription, but produces nuclear extracts that are also competent for in vitro RNA splicing. However, we (and most other groups) extract nuclei with 0.6 *M* KCl instead of 0.42 *M* NaCl, as described in the original protocol. The usage of a higher salt concentration is essential to obtain nuclear extracts that reproducibly splice the regulated Ad L1-IIIa mRNA *(12)*. Substrate RNA is synthesized using bacteriophage T7 or SP6 polymerase. The efficiency of in vitro splicing of pre-mRNAs with weak 3'-splice sites, such as L1-IIIa, is sometimes very low and therefore difficult to detect. Thus, in our earlier work we often included a U1 snRNA-binding site (which functions as a splicing enhancer) downstream of the 3'-splice site to improve the efficiency of weak splice site usage *(13)*. Such a U1 tag can increase the efficiency of splicing up to 20 times. However, the inclusion of a U1 tag may for some experiments be counterproductive, because it results in a disproportionate increase in the splicing efficiency in uninfected nuclear extracts, resulting in an underestimation of the actual difference in splicing efficiencies between infected and uninfected nuclear extracts *(14)*.

2. Materials

2.1. Infection of HeLa Spinner Cells With Ad

The protocol below is adapted for 4–6 L exponentially growing HeLa spinner cells, giving approx 10 mL of nuclear extract, but can be easily adjusted for cell cultures ranging between 500 mL and 30 L. To prepare small-scale nuclear extracts, *see* Lee et al. *(15)*.

1. Minimum essential medium (MEM) spinner cell medium (can be obtained from Gibco BRL).
2. Newborn calf serum.
3. Penicillin (10,000 IU/mL)/streptomycin (10,000 μg/mL) stock solution.
4. Purified Ad stock (wild-type or mutant of your choice) with a titer greater than 10^{10} PFU/mL.

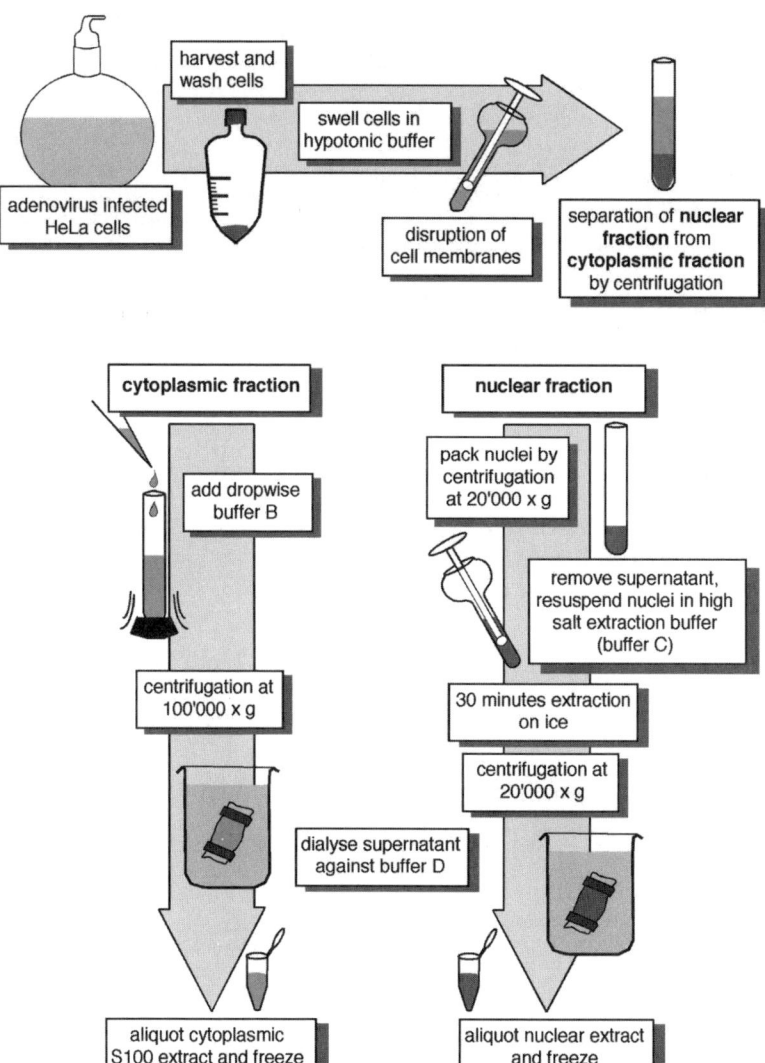

Fig. 1. Flowchart illustrating important steps in nuclear extract and cytoplasmic S100 extract preparation.

2.2. Preparation of Splicing Competent Nuclear Extract and Cytoplasmic S100 Extract

1. Dounce homogenizers with tight-fitting pestles, sizes 10 and 40 mL (Kontes).
2. Phosphate-buffered saline (PBS): 20 mM potassium phosphate, pH 7.4, 130 mM NaCl. Make up in autoclaved double-distilled H_2O (ddH_2O); store at 4°C.

3. Stock solutions for buffer preparation (make up with autoclaved ddH$_2$O): 0.5 M N-2-hydroxyethylpiperazine-N'-2-ethanesulfonate (HEPES), pH 7.9 (Sigma, St. Louis, MO; H 0891), adjusted to pH 7.9 with 5 M KOH, store at 4°C. 2 M KCl and 1 M MgCl$_2$, store at room temperature. 1 M dithiothreitol (DTT), store in 1-mL aliquots at –20°C.

4. Buffer A (hypotonic): 100 mL 10 mM HEPES, pH 7.9, 10 mM KCl, 1.5 mM MgCl$_2$, 0.5 mM DTT. Make fresh in ddH$_2$O from stock solutions, filter through a 0.2-μm membrane, keep on ice. Optional: add phenylmethylsulfonyl fluoride (PMSF) to a concentration of 0.2 mM just before use (see **Note 5**).

5. Buffer B (for S100): 100 mL 300 mM HEPES, pH 7.9, 1.4 M KCl, 30 mM MgCl$_2$. Make up in ddH$_2$O from stock solutions, filter through a 0.2-μm membrane. Can be stored at 4°C up to 6 mo.

6. Buffer C (hypertonic): 10 mL 20 mM HEPES, pH 7.9, 600 mM KCl, 25% glycerol, 0.2 mM ethylene diamine tetraacetic acid (EDTA), and 0.5 mM DTT. Make fresh in ddH$_2$O from stock solutions, filter through a 0.2-μm membrane, keep on ice.

7. Buffer D (dialysis): 2 L 20 mM HEPES, pH 7.9, 100 mM KCl, 20% glycerol, 0.2 mM EDTA, and 0.5 mM DTT. Make up with ddH$_2$O, filter through a 0.2-μm membrane, cool down to 4°C.

8. Dialysis tubing, molecular-weight cutoff of 12–14 kDa (Spectrum Medical Industries, Houston, TX).

2.3. Preparation of Radioactively Labeled Splicing Substrates

1. Template DNA (see **Notes 9** and **11**).

2. Nucleotide mix: 5 mM ATP, 5 mM UTP, 0.5 mM GTP, and 0.5 mM CTP. Stock solutions of nucleotides (Pharmacia Biosciences) are made by dissolving in autoclaved ddH$_2$O to a concentration of 100 mM, store at –70°C. From these stock solutions a nucleotide mix is prepared. The GTP concentration should be kept at 0.5 mM to favor incorporation of the cap nucleotide as the start nucleotide. In our example, the CTP concentration is also kept at 0.5 mM because we use α ^{32}P-CTP (Amersham Arlington Heights, IL; 20 mCi/mL, 800 Ci/mmol) to label the transcript. Mix the nucleotides in autoclaved ddH$_2$O, store at –20°C in 100-μL aliquots.

3. 10 mM mGpppG-cap nucleotide (Pharmacia Biosciences 27-4635): dissolve 5 U of lyophilized mGpppG in 24 μL autoclaved ddH$_2$O, store at –20°C.

4. 5X Transcription buffer: 200 mM Tris-HCl, pH 7.9, 30 mM MgCl$_2$, 10 mM spermidine, 50 mM NaCl, store at –20°C (this buffer is usually provided by the manufacturer of the RNA polymerase).

5. Porcine RNase Inhibitor (Amersham Biosciences): 39 U/μL.

6. T7 RNA polymerase (New England BioLabs): 50 U/μL.

7. RQ1 RNase-free DNase (Promega): 1 U/μL.

8. Loading buffer: 50 mM Tris-HCl, pH 7.9, 10 mM EDTA, 0.025% bromophenol blue/0.025% xylene cyanol, 80% formamide (Baker, proanalysis grade).

9. 40% Acrylamide stock solution (acrylamide:bis-acrylamide ratio 29:1).

10. 10X TBE stock solution: 108 g Tris base, 55 g boric acid, 7.44 g EDTA, add deionized H_2O to 1 L, let dissolve, autoclave. Use 1X TBE in gel and 0.5X TBE in the running buffer.

11. 10% Ammonium persulfate: for 10 mL dissolve 1 g ammonium persulfate in 10 mL ddH_2O. Aliquots should be stored at –20°C. Keep a working solution at 4°C no longer than 1 mo.

12. Elution buffer: 0.75 M NH$_4$-acetate, 0.1 % (w/v) sodium dodecyl sulfate, 10 mM Mg-acetate, 0.1 mM EDTA. Make 10 mL from sterile stock solutions and store at room temperature wrapped in aluminium foil. Store the 7.5 M NH$_4$-acetate stock protected from light.

2.4. In Vitro Splicing Assay

1. Nuclear extract (*see* **Subheading 3.2.**).
2. In vitro transcribed, radioactively labeled pre-mRNA (*see* **Subheading 3.3.**).
3. 13% (w/v) Polyvinyl alcohol (PVA): dissolve 1.3 g PVA (Sigma P8136) in 10 mL ddH_2O by adding the powder to the water on a magnetic stirrer. If the PVA does not dissolve completely, gentle heating helps; vortexing should be avoided. Store 1-mL aliquots at –20°C.
4. 62.5 mM MgCl$_2$: 1-mL aliquots are made up by dilution from a 1 M stock. Note that the MgCl$_2$ requirement has to be experimentally tested (*see* **Note 14**).
5. 0.5 M Creatine phosphate dissolved in autoclaved ddH_2O. Store 100-µL aliquots at –20°C.
6. 10 mg/mL Yeast tRNA (Sigma R5636), dissolved in autoclaved ddH_2O, store in aliquots at –20°C.
7. 2X Proteinase K buffer: 200 mM Tris-HCl, pH 7.5, 300 mM NaCl, 25 mM EDTA, 2% sodium dodecyl sulfate made up in autoclaved ddH_2O, store at room temperature.
8. 20 mg/mL Proteinase K: dissolve 20 mg lyophilized Proteinase K (Merck, Rahway, NJ) in 1 mL (50 mM Tris-HCl, pH 7.5, 10 mM CaCl$_2$), store in small (e.g., 30 µL) aliquots at –20°C. Do not freeze–thaw more than five times.
9. Proteinase K mix: consisting of 100 µL 2X Proteinase K buffer, 10 µg tRNA, 2 µL 20 mg/mL Proteinase K + 75 µL autoclaved ddH_2O per reaction. Do not store, make up fresh, use immediately.

3. Methods

3.1. Infection of HeLa Spinner Cells With Ad

HeLa spinner cells are grown in round cell-culture bottles on a magnetic stirrer at 37°C in MEM spinner cell medium, 5% newborn calf serum, optionally containing 1% penicillin/streptomycin. The cells must be kept in log phase (cell density 2–6 × 10^5 cells/mL), doubling time approx 24 h.

1. Start with 2–3 ×10^9 HeLa spinner cells, collect them by centrifugation in sterile 1-L plastic centrifuge bottles by spinning at 900g at room temperature for 20 min (Beckman J6M/E centrifuge, JS-4.2 rotor at 2000 rpm).

2. Decant medium back into the cell-culture bottle (handle under sterile conditions—the medium will be reused later), resuspend cells in 200–300 mL MEM without serum (*see* **Note 1**), transfer to a 1-L cell-culture bottle.

3. Infect cells with approx 10 PFU/cell of Ad from a high-titer virus preparation. Leave at 37°C on a magnetic stirrer for 1 h. Dilute cells to approx 4×10^5 cells/mL in a large cell-culture bottle with the old MEM saved during **step 2**. Add fresh medium if necessary.

4. Continue incubation at 37°C for 20–24 h for preparation of late-infected extracts.

3.2. Preparation of Splicing-Competent Nuclear Extracts and Splicing-Deficient Cytoplasmic S100 Extracts

This protocol describes a modification of the original Dignam protocol *(11)*. The procedure is our lab protocol adapted to yield high Ad major-late L1 splicing activity *(16)* (*see* **Note 3**). However, other modifications of the protocol might be better suited for splicing of other pre-mRNAs *(17,18)*.

3.2.1. Nuclear Extract Preparation

1. Check the cell density at the end of infection. It should be almost the same as at the start of infection. An Ad infection inhibits cell division. Thus, if cells continue to divide during the course of infection, the infection most probably did not work.

2. Collect cells by centrifugation at 1800g at 15°C for 20 min. (Beckman J6M/E centrifuge, JS-4.2 rotor, 3000 rpm). Decant medium (*see* **Note 2**) and resuspend cells gently in a small volume of the remaining medium. Transfer into two conical 250-mL centrifuge bottles.

3. Pellet cells at 1000g at 4°C for 10 min (Beckman GPKR centrifuge, 2200 rpm). Carefully remove all medium and resuspend cells in 20 mL ice-cold phosphate-buffered saline (PBS) and pool into one bottle, measure volume. From here on all steps should be performed on ice (*see* **Note 4**).

4. Determine the packed cell volume (pcv) according to the formula: total volume of suspension (**step 3**) – 20 mL = pcv.

5. Fill up with ice-cold PBS and repeat centrifugation as in **step 3**.

6. Decant PBS (carefully remove all remains of PBS with a Pasteur pipet) and resuspend cells in (5 × pcv) mL ice-cold buffer A (hypotonic buffer) by gently pipetting up and down. Leave on ice for 10 min.

7. Transfer cells into transparent 50-mL centrifuge tubes (note total volume) and centrifuge at 500g at 4°C for 10 min (Beckman JS 13.1 rotor, 2300 rpm).

8. Carefully remove the supernatant and measure its volume. The volume of the swollen cells is total volume (**step 7**) – supernatant and should be about 2–2.5 × pcv.

9. Add (2 × pcv) mL ice-cold buffer A and gently resuspend the cells with a pipet.

10. Transfer the solution to a 40-mL homogenizer, prechilled on ice. Disrupt cells with 10–12 strokes using the tight pestle. To check the efficiency of disruption, dilute a drop of the solution with buffer A, add 2 µL 0.4% Trypan blue (Sigma),

and check under the microscope. Nuclei stain blue; intact cells do not stain. If necessary, give a few additional strokes until approx 90% of the cells are disrupted (*see* **Note 6**).

11. Transfer the solution into a transparent centrifuge tube (note volume) and spin at 750*g* at 4°C for 10 min (Beckman JS 13.1 rotor, 2700 rpm).
12. Transfer the supernatant (cytoplasmic fraction) into a new tube (note volume; *see* **Note 7**). To prepare cytoplasmic S100 extract continue with this fraction at **Subheading 3.2.2.**
13. Spin the pellet (the nuclei fraction) at 20,000*g* at 4°C for 20 min (Beckman JA20 rotor, 16,000 rpm), then remove and discard the supernatant (note its volume). Calculate the packed nuclei volume (pnv) according to the formula: pnv = total volume after cell disruption (**step 11**) − [cytoplasmic supernatant (**step 12**) + supernatant from 20,000*g* spin (**step 13**)].
14. Add (1 × pnv) mL of ice-cold buffer C (high-salt buffer) to the nuclei pellet, transfer to the small (10-mL) homogenizer, and resuspend nuclei with several (6–12) gentle strokes using the tight-fitting pestle.
15. Transfer the homogenate back to the centrifuge tube, embed it in ice, and shake on a rocking table for 30 min.
16. Spin the homogenate at 20,000*g* at 4°C for 30 min (Beckman JA20 rotor, 16,000 rpm).
17. Collect the supernatant (nuclear extract) and dialyze it against ice-cold buffer D at 4°C, using a membrane with a cutoff at 12-14 kDa. Use 4 × 250 mL buffer D and change buffer at 1-h intervals.
18. Spin the extract at 20,000*g* at 4°C for 20 min (Beckman JA20 rotor, 16,000 rpm), discard the pellet, aliquot the supernatant into Eppendorf tubes, quick-freeze in liquid nitrogen, and store at −70°C (*see* **Note 8**).

3.2.2. S100 Extract Preparation

For convenience, these steps should be performed simultaneously with the nuclear extract preparation.

1. Add dropwise (0.11 × cytoplasmic supernatant from **Subheading 3.2.1.**, **step 12**) mL buffer B under gentle vortexing.
2. Spin at 100,000*g* at 4°C for 1 h (Beckman SW41 Ti rotor).
3. Dialyze the supernatant (S100 extract) against ice cold buffer D and freeze in aliquots as described in **steps 17** and **18** of **Subheading 3.2.1.**

3.3. Preparation of Radioactively Labeled Splicing Substrates

3.3.1. In Vitro Transcription With T7 RNA Polymerase

1. Make a master mix by combining at room temperature in an Eppendorf tube (*see* **Note 17**):
   a. 3.75 μL ddH$_2$O.
   b. 2.5 μL Nucleotide mix (5 m*M* ATP and UTP, 0.5 m*M* CTP and GTP).
   c. 1.25 μL 10 m*M* mGpppG cap-nucleotide.

   d. 5 µL 5X Transcription buffer.

   e. 2.5 µL 100 m*M* DTT.

   f. 0.5 µL RNase inhibitor.

   g. 2.5 µL (50 µCi) α^{32}P-CTP.

2. Mix 5 µL template DNA (concentration 0.2–0.3 µg DNA/µL; *see* **Notes 9** and **11**), 18 µL of the master mix, and 2 µL T7 RNA polymerase (20 U/µL) in a Safe-lock Eppendorf tube and incubate at 37°C for 2 h .

3. Optionally (*see* **Note 10**): dilute 2 µL of the remaining master mix with 48 µL H$_2$O, measure Cerenkov counts of 5 µL of the dilution. Recalculate the value to dpm based on the efficiency of your Cerenkov counter. Calculate the total amount of dpm per reaction according to the formula: dpm (total) = dpm (measured in aliquot) × 90.

4. Add 1 µL of RQ1 DNase to the reaction and continue incubation for 30 min at 37°C.

5. Add 15 µL loading buffer to stop the reaction. Store samples at –20°C or continue directly.

3.3.2. Recovery of Full-Length RNA Transcripts

1. Prepare a 4% acrylamide 8 *M* urea gel (acrylamide:*bis*-acrylamide = 29:1; 1X TBE) using a wide preparative comb.

2. Incubate samples at 100°C for 5 min, chill on ice, and load on gel. Run gel in 0.5X TBE at 50 W until the bromophenol blue dye reaches the middle of the gel (*see* **Note 12**).

3. Detach the gel from the electrophoresis equipment. Carefully remove the upper glass plate, wrap the gel with a plastic wrap, and cover with a clean glass plate. Move the whole setup to the darkroom, place an X-ray film on the upper half of the gel (covering the area from the wells to the bromophenol blue band), mark the position of the X-ray film with a waterproof pen on the plastic wrap, cover with a clean glass plate, and expose for 5 min at room temperature. Develop the X-ray film. For your own safety, work should be done behind a Plexiglass shield.

4. Copy the position of the band corresponding to your full-length transcript from the autoradiograph back to the gel using a needle and cut out the full-length band with a disposable scalpel or razor blade; place the gel piece (volume 100–150 µL) in an Eppendorf tube and add 500 µL elution buffer.

5. Elute the transcript at room temperature on a turning wheel for at least 4 h (or up to overnight).

6. Remove gel debris by a short spin in the microfuge, and transfer the solution to a new Eppendorf tube. A check with the hand monitor should give you roughly 80% of the counts in the solution. If necessary, the 20% remaining in the gel piece can be recovered by repeating the elution step.

7. Add an equal volume of phenol:chloroform:isoamyl alcohol (25:24:1) to each tube, vortex vigorously for 1 min, spin in the microfuge for 3 min, and carefully transfer the aqueous phase (upper) to a new Eppendorf tube.

8. Repeat **step 7** two times with chloroform:isoamyl alcohol (24:1).

9. Optionally: determine the volume of the samples by aspirating in a pipet tip (usually approx 400 µL) and measure dpm of a 5-µL aliquot. Calculate the total amounts of counts in each sample. The yield of transcript can then be determined according to the formula:

yield (mol) = total dpm in transcript × mol CTP in the reaction/total dpm
(**Subheading 3.3.1.**, **step 3**) × number of C residues in transcript

Following the protocol described here, CTP = 1.31×10^{-9} mol per reaction. (*see* **Note 13**).

10. Add 2.5 × volumes 95% ethanol and store the transcript in the ethanol in a lead or Plexiglass box at –20°C. For in vitro splicing reactions (**Subheading 3.4.**, **step 2**), transcripts up to 2 wk old will work, but fresh transcripts usually give a higher splicing efficiency.

3.4. In Vitro Splicing Assay

1. Spin aliquots with desired amount of transcript immediately before use, wash the pellet with 80% ethanol, remove all ethanol with an outdrawn, sterile Pasteur pipet, and dry under the bench lamp for 3–5 min. Dissolve the pellet in autoclaved ddH$_2$O to 20,000–25,000 dpm/µL.
2. For a 25-µL standard reaction, mix the following ingredients in an Eppendorf tube on ice (*see* **Notes 15–17**):
   a. 5 µL 13% PVA.
   b. 5 µL Buffer D.
   c. 10 µL Nuclear extract.
   d. 1.0 µL 80 m*M* MgCl$_2$ (*see* **Note 14**).
   e. 1.0 µL 0.5 *M* Creatine phosphate.
   f. 0.5 µL 100 m*M* ATP (*see* **Note 14**).
   g. 2.5 µL RNA transcript, approx 50,000–70,000 dpm (5–10 fmol).
   Collect material at the bottom of the tube by a short pulse in the microfuge, mix by pipetting up and down, *do not vortex*, and incubate for 90 min at 30°C.
3. Add 175 µL Proteinase K mix to each tube and incubate at 37°C for 30–45 min (*see* **Note 18**).
4. Add 200 µL ddH$_2$O, extract with 400 µL phenol:chloroform:isoamyl alcohol (25:24:1), and transfer upper phase (380 µL) into a new tube.
5. Repeat extraction with 380 µL chloroform/isoamyl alcohol, and transfer upper phase (360 µL) into a new tube.
6. Add 40 µL 3 *M* sodium acetate and 1 mL 95% ethanol, and precipitate at –20°C overnight or 1 h in dry ice.
7. Spin tubes at 14,000 rpm (16,000*g*) at 4°C for 30 min in an Eppendorf centrifuge and remove all ethanol with an outdrawn, sterile Pasteur pipet. Dry pellets under the bench lamp (3–5 min) and dissolve in 5–10 µL (depending on the well size in the gel) loading buffer by vortexing and pipetting up and down.
8. Boil the samples for 2 min, chill on ice, and collect all material at the bottom of the tube by a short pulse in the microfuge.

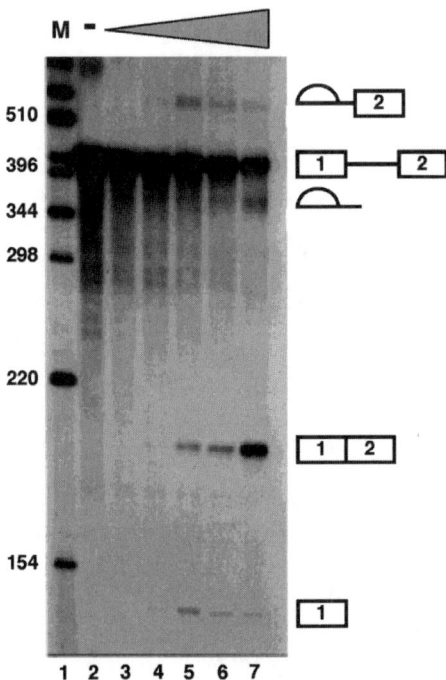

Fig. 2. Time-course experiment showing the change in pre-mRNA, intermediates, and product accumulation. In this experiment a penton base minigene pre-mRNA *(19)* was spliced in nuclear extracts prepared from uninfected HeLa cells. From a 160-µL splicing reaction, 25-µL aliquots were removed after 0 (lane 2), 15 (lane 3), 30 (lane 4), 45 (lane 5), 60 (lane 6), and 120 (lane 7) min. RNA was isolated as described and products resolved on a 6% polyacrylamide 8 *M* urea gel. Lane 1: Size markers.

9. Load samples on a 5–10% acrylamide 8 *M* urea gel (acrylamide:*bis*-acrylamide = 19:1; 1X TBE). The lengths of the expected splicing products determine the percentage of the gel. 8% Gels are suitable to resolve products ranging from 100 to 500 nucleotides; longer fragments should be run on a 6% gel.
10. Run the gel until desired separation is obtained, and dry it on a gel dryer.
11. Expose the gel to an X-ray film overnight and/or expose it on a PhosphorImager screen for subsequent quantitation of the result (*see* **Fig. 2** for an example).

4. Notes

1. Cells are infected in a small volume of medium (10^7 cells/mL) in order to achieve efficient adherence of virus to cells and a somewhat synchronous infection. The infection is carried out in MEM without serum to avoid a potential problem of neutralizing antibodies in the newborn calf serum. We use a 10-fold excess of virus over cells to ensure infection of every cell, but that number can probably be decreased if virus amounts are limiting.

2. During the whole extract preparation the material is potentially infectious, and everything must therefore be disinfected (glassware and solutions) or autoclaved (disposable material) after use.

3. In the standard Dignam protocol *(11)*, nuclei are extracted with 0.42 M NaCl. We have found that a higher salt concentration is essential to obtain nuclear extracts that reproducibly splice the L1-IIIa pre-mRNA *(12)*. Thus, we routinely use 0.6 M KCl to extract nuclei.

4. Low temperature during the whole procedure and total processing time are critical parameters for extract activity. Therefore, prepare all materials and precool solutions on ice before beginning the preparation. When started, proceed without prolonged incubations from step to step; the whole preparation including dialysis should be completed in one day.

5. We routinely include PMSF in buffers A, B, C, and D to minimize proteolytic degradation during extract preparation. However, we obtain satisfactory extracts also without protease inhibitors. From a 200 mM PMSF stock solution dissolved in 95% ethanol, add PMSF to a final concentration of 0.2 mM just before use; PMSF is unstable in aqueous solutions.

6. The number of strokes required to obtain disruption of 90% of cells can vary considerably between different homogenizers and should therefore be checked as described. Avoid too fast movement of the pestle and too many strokes because this tends to result in considerable leakage of splicing factors into the cytoplasm. This is particularly important if you are interested in producing splicing-deficient S100 extracts suitable for biochemical complementation. It is noteworthy that S100 extracts prepared from Ad-infected cells are, in our hands, always splicing-competent, probably because of leakage from more labile nuclei.

7. The nuclei fraction (lower) should be roughly half the volume of the cytoplasmic supernatant. The border of the two phases can be most easily seen when the tube is held up against a bright background.

8. The extracts can be stored for years at −70°C, but repeated freeze–thawing should be avoided (we observe decreased activity after more than five freeze–thaw cycles). If the extracts are used mainly for in vitro splicing assays, 100-μL aliquots are suitable. This protocol gives extracts with a concentration of total protein of 6–10 μg/μL.

9. The DNA template for in vitro transcribed RNA can be a plasmid with an SP6 or T7 promoter cut with the restriction endonuclease of your choice or a DNA fragment amplified by PCR using an oligo tagged with the T7 promoter (at**TAATACGACTCACTATAG**: lowercase = additional 5' nucleotides; uppercase = promoter sequence; bold = transcription start site) as the forward primer and a reverse primer of your choice. The latter strategy allows you to transcribe any specific part of any cloned gene and to modify sequences at the 5'- and 3'-ends of your transcript. However, this strategy usually gives a lower yield of ^{32}P-labeled transcript compared to phage polymerase transcription of plasmid DNA cut with a restriction endonuclease. In both cases the DNA template must be purified by phenol/chloroform extractions and ethanol precipitation or by using a commercially available purification kit. The quality and amount of

template DNA is critical to obtain high amounts of RNA transcript. Templates using the SP6 instead of T7 polymerase can be transcribed using the same protocol described here, except that the incubation temperature should be raised to 40°C.

10. Steps marked "Optionally" let you calculate the molar yield of transcribed RNA, which is especially useful if splicing efficiencies from different RNA transcripts will be compared to each other.

11. The efficiency of in vitro RNA splicing is length dependent; long introns reduce the splicing efficiency. Thus, try to design pre-mRNAs so that they do not exceed 500 nucleotides (nts) in length.

12. For transcripts fewer than 200 nts, a 6% polyacrylamide gel should be used instead of 4%. The gel can also be run longer if separation of the full-length transcript so requires. However, the free radioactive nts (about 80% of the input radioactivity) will then run out into the lower buffer chamber and contaminate the electrophoresis equipment.

13. Typically, about 10–20% of the initial radioactivity will be incorporated into the transcript, which corresponds to a yield of 1–3 pmol for a 200- to 500-nt-long transcript.

14. Optimal $MgCl_2$ and ATP concentrations are critical and vary for different splicing substrates and should be determined in a pilot experiment. Also, the amount of nuclear extract required varies between 20 and 60% (5–15 μL per 25-μL reaction) depending on the pre-mRNA and the source and activity of the nuclear extract.

15. For each new RNA substrate, it is wise to start with a time-course experiment to follow the accumulation of splicing products and intermediates and to determine the optimal incubation time (**Fig. 2**). The standard reaction is scaled up for this purpose, and aliquots are removed at different time points (e.g., 0, 30, 60, 90, and 180 min). It is recommended that splicing products be verified by reverse transcriptase-polymerase chain reaction sequencing of gel-purified bands. To facilitate identification of splicing products, a ^{32}P-labeled size marker should always be included on each gel. To help evaluate the result: linear RNA molecules (substrate RNA, its spliced product, and free exon 1) migrate according to their size and can be identified by comparison with the size marker. In a time-course experiment, accumulation of product and free exon 1 and decrease of the substrate RNA facilitates identification of these bands. The intermediates (lariat and lariat-exon 2) are branched molecules and migrate at unpredictable positions in the gel and can be most easily identified in a time-course experiment by their accumulation. When duplicate samples are run on two gels with different polyacrylamide concentrations, the intermediates are the bands that change their positions relative to the size marker.

16. A reaction omitting ATP and creatine phosphate is a valuable negative control.

17. For *n* reactions, always prepare master mixes for (*n* + 1) samples containing all common ingredients to minimize pipetting errors. When PVA is included in a master mix, viscosity of the solution changes, and therefore approx 20% more master mix than calculated should be prepared.

18. For most transcripts, phenol/chloroform and chloroform extractions (**Subheading 3.4.**, **steps 4** and **5** in the protocol) can be skipped without any negative effects on the result. Precipitate RNA by adding 160 μL ddH$_2$O, 40 μL 3 *M* sodium acetate, and 1 mL 95% ethanol after the Proteinase K treatment (**Subheading 3.4.**, **step 3**). However, note that some RNAs will give smeary bands and a large proportion of radioactivity retarded in the well when these extractions are omitted.

Acknowledgments

This work was supported by the Swedish Cancer Society.

References

1. Graveley, B. R. (2001) Alternative splicing: increasing diversity in the proteomic world. *Trends Genet.* **17**, 100–107.
2. Imperiale, M., Akusjärvi, G., and Leppard, K. (1995) Post-transcriptional control of adenovirus gene expression. *Curr. Topics Microbiol.* **199**, 139–171.
3. Akusjärvi, G. and Stevenin, J. (2003) Remodelling of the host cell RNA splicing machinery during an adenovirus infection. *Curr. Topics Microbiol.* **272**, 253–286.
4. Weingärtner, B. and Keller, W. (1981) Transcription and processing of adenoviral RNA by extracts from HeLa cells. *Proc. Natl. Acad. Sci. USA* **78**, 4092–4096.
5. Kole, R. and Weissman, S. M. (1982) Accurate in vitro splicing of human β-globin RNA. *Nucleic Acids Res.* **10**, 5429–5445.
6. Hernandez, N. and Keller, W. (1983) Splicing of in vitro synthesized messenger RNA precursors in HeLa cell extracts. *Cell* **35**, 89–99.
7. Padgett, R. A., Hardy, S. F., and Sharp, P. A. (1983) Splicing of adenovirus RNA in a cell-free transcription system. *Proc. Natl. Acad. Sci. USA* **80**, 5230–5234.
8. Krainer, A. R., Maniatis, T., Ruskin, B., and Green, M. R. (1984) Normal and mutant human β-globin pre-mRNAs are faithfully and efficiently spliced in vitro. *Cell* **36**, 993–1005.
9. Green, M. R., Maniatis, T., and Melton, D. A. (1983) Human β-globin pre-mRNA synthesized in vitro is accurately spliced in Xenopus oocyte nuclei. *Cell* **32**, 681–694.
10. Krämer, A. (1996) The structure and function of proteins involved in mammalian pre-mRNA splicing. *Ann. Rev. Biochem.* **65**, 367–409.
11. Dignam, J. D., Lebovitz, R. M., and Roeder, R. G. (1983) Accurate transcription initiation by RNA polymerase II in a soluble extract from isolated mammalian nuclei. *Nucleic Acids Res.* **11**, 1475–1489.
12. Zerivitz, K., Kreivi, J.-P., and Akusjärvi, G. (1992) Evidence for a HeLa cell splicing activity that is necessary for activation of a regulated adenovirus 3' splice site. *Nucleic Acids Res.* **20**, 3955–3961.
13. Kreivi, J.-P., Zerivitz, K., and Akusjärvi, G. (1991) A U1 snRNA binding site improves the efficiency of in vitro pre-mRNA splicing. *Nucleic Acids Res.* **19**, 6956.
14. Mühlemann, O., Yue, B. G., Petersen-Mahrt, S., and Akusjärvi, G. (2000) A novel type of splicing enhancer regulating adenovirus pre-mRNA splicing. *Mol. Cell. Biol.* **20**, 2317–2325.

15. Lee, K., Zerivitz, K., and Akusjärvi, G. (1995) Small-scale preparation of nuclear extracts from mammalian cells, in *Cell Biology. A Laboratory Handbook*, Vol.1 (Celis, J. E., ed.), Academic, London, pp. 668–673.
16. Kreivi, J.-P., Zerivitz, K., and Akusjärvi, G. (1991) Sequences involved in the control of adenovirus L1 alternative RNA splicing. *Nucleic Acids Res.* **19,** 2379–2386.
17. Eperon, I. C. and Krainer, A. R. (1993) Splicing of mRNA precursors in mammalian cells, in *RNA Processing, A Practical Approach*, Vol. 1 (Higgins, S. J. and Hames, B. D., eds.), IRL Press, Oxford, pp. 757–801.
18. Abmayr, S. M. and Workman, J. L. (1995) Preparation of nuclear and cytoplasmic extracts from mammalian cells, in *Current Protocols in Molecular Biology*. (Ausubel, F. A., Brent, R., Kingston, R. E., et al., eds.). John Wiley & Sons, New York, pp. 12.1.1–12.1.9.
19. Mühlemann, O., Kreivi, J.-P., and Akusjärvi, G. (1995) Enhanced splicing of nonconsensus 3' splice sites late during adenovirus infection. *J. Virol.* **69,** 7324–7327.

4

In Vitro Methods to Study RNA Interference During an Adenovirus Infection

Gunnar Andersson, Ning Xu, and Göran Akusjärvi

Summary

RNA interference (RNAi) has attracted a lot of interest during recent years as a method to knock-down gene expression and as a possible antiviral system. Here we present a collection of in vitro methods to study RNAi and the effect of an adenovirus infection on RNAi. We describe methods to measure the two key enzymatic complexes involved in RNAi: Dicer and RISC.

Key Words: Adenovirus; RNA interference; Dicer; RNA-induced silencing complex; RISC; dsRNA; siRNA; VA RNA; S15 extract; S100 extract; polymerase chain reaction; T7 RNA polymerase; SP6 RNA polymerase; HeLa cells; 293 cells.

1. Introduction

When double-stranded RNA (dsRNA) is introduced into a cell, it will induce a specific degradation of mRNA with the homologous sequence through a mechanism that is referred to as RNA interference (RNAi) *(1)*. The mechanism for RNAi involves an initial processing of the trigger dsRNA into short interfering RNAs (siRNAs) of 21–23 nucleotides (nt) by an RNase III type enzyme, called Dicer. The siRNAs are subsequently incorporated into the RISC enzyme complex, which targets and degrades mRNA with the homologous sequence (reviewed in **refs.** *2* and *3*). It is well documented that RNAi has an important function as an antiviral defense mechanism in plants (reviewed in **refs.** *4* and *5*). As a consequence of this, many plant viruses have evolved proteins that act as suppressors of RNAi (reviewed in **ref.** *6*). We have recently shown that human adenovirus (Ad) inhibits the key enzymes executing RNAi, Dicer and RNA-induced silencing complex (RISC) *(7)*. The Ad VA RNAs block the activity of Dicer by functioning as competitive substrates *(7)*, as well as by

From: *Methods in Molecular Medicine, Vol. 131:*
Adenovirus Methods and Protocols, Second Edition, vol. 2:
Ad Proteins, RNA, Lifecycle, Host Interactions, and Phylogenetics
Edited by: W. S. M. Wold and A. E. Tollefson © Humana Press Inc., Totowa, NJ

titrating out the Exportin 5 nuclear export factor *(8)*. The mechanism by which Ad blocks RISC is currently not known.

To be able to perform a biochemical analysis of RNAi in Ad-infected cells the development of simple in vitro assays is critical. Such assays have been described in other systems, and we present here an adaptation of these methods for work with virus-infected cells. It should be noted that large-scale S100 or nuclear extracts, prepared as described in Chapter 3, can be used for biochemical characterization of the RNAi machinery. However, processing extracts from multiple infections or mutant virus infections imposes practical as well as economical limitations. Thus, we present an alternative small-scale protocol for cytoplasmic extract preparation. With this protocol it is possible to reproducibly obtain high-quality extracts from only one or two 10-cm tissue culture plates. We routinely prepare up to 10 extracts in parallel in only 3 h using this small-scale method.

2. Materials

2.1. Infection of Adherent Cells With Ad

1. Dulbecco's modified Eagle's medium (DMEM).
2. Newborn calf serum or, alternatively, fetal calf serum.
3. Penicillin (10,000 IU/mL)/streptomycin (10,000 μg/mL) stock solution.
4. Purified Ad stock (wild-type or mutant) with a titer preferably higher than 10^{10} focus forming units (FFU)/mL.

2.2. Preparation of S15 Extracts for Dicer and RISC

1. Cell-lifters.
2. 1-mL Syringes with 23-gauge needles.
3. Phosphate-buffered saline (PBS): 20 mM potassium phosphate, pH 7.4, 130 mM NaCl. Make up in autoclaved double-distilled H_2O (ddH_2O), store at 4°C.
4. Buffer A: 10 mM *N*-2-hydroxyethylpiperazine-*N*'-2-ethanesulfonate (HEPES), pH 7.9, 10 mM KCl, 1.5 mM MgCl$_2$, 1 mM dithiothreitol (DTT). Make fresh in ddH_2O, filter through a 0.2-μm membrane, and keep on ice.
5. Buffer A + 50% glycerol: 500 μL 87% glycerol made up to 870 μL with Buffer A. Although this practice results in a dilution of salts it does not affect the performance of the extracts.

2.3. Preparation of Radioactively Labeled dsRNA for Dicer Assay

1. Template DNA: a 159-base-pair polymerase chain reaction (PCR) fragment generated from pGL2-Control (Promega) with T7 and SP6 promoter sequences at opposite ends *(7)* (*see* **Note 1**).
2. Nucleotide mix: 5 mM ATP, 5 mM UTP, 5 mM GTP, 0.2 mM CTP (*see* **Note 2**).
3. T7 (50 U/μL) and SP6 (20 U/μL) RNA polymerase with the buffers supplied (New England BioLabs) (*see* **Note 3**).

4. α^{32}P-CTP, 800 Ci/mmol: 20 µCi/µL.
5. Porcine RNase inhibitor (Amersham Biosciences): 39.2 U/µL.
6. RQ1 DNase (Promega): 1 U/µL.
7. Phenol: phenol:chloroform:isoamyl alcohol (25:24:1).
8. dsRNA annealing buffer: 30 mM Tris-HCl, pH 8.0, 25 mM NaCl.
9. Sample buffer: 80% formamide (Baker, proanalysis [p. a.] grade), 50 mM Tris-HCl, pH 7.9, 10 mM ethylene diamine tetraacetic acid (EDTA), 0.025% bromophenol blue/0.025% xylene cyanol.
10. 6% Acrylamide 8 M urea gel (acrylamide:*bis*-acrylamide = 29:1; 1X TBE).
11. dsRNA elution buffer: 0.5 M NH$_4$-acetate, 1 mM EDTA, 0.2% sodium dodecyl sulfate (SDS).
12. RNase mix: RNase A 0.5 U/µL, RNase T1 0.1U/µL in distilled water (*see* **Note 4**).

2.4. In Vitro Dicer Assay

1. Dicer 4X A buffer: 87 mM HEPES, pH 7.0, 8.7 mM Mg-acetate, 8.7 mM DTT, and 8 mM ATP (*see* **Note 5**). This is the 4X Dicer reaction buffer used for extracts in buffer A (i.e., S15 extracts).
2. Porcine RNase inhibitor (optional): 39.2 U/µL.
3. 2X Proteinase K buffer: 200 mM Tris-HCl, pH 7.5, 300 mM NaCl, 25 mM EDTA, 2% SDS, 25 µg/mL tRNA.
4. 20 mg/mL Proteinase K (Merck): dissolve in 50 mM Tris-HCl, pH 7.5, 10 mM CaCl$_2$, store at −20°C in small aliquots.
5. Phenol: phenol:chloroform:isoamyl alcohol (25:24:1).
6. 1 M MgCl$_2$: added to facilitate precipitation of small RNA species.
7. Sample buffer: 80% formamide (Baker, p. a. grade), 50 mM Tris-HCl, pH 7.9, 10 mM EDTA, 0.025% bromophenol blue/0.025% xylene cyanol.
8. 15% Acrylamide 7 M urea gel (acrylamide:*bis*-acrylamide = 19:1; 1X TBE).
9. Gel-fix solution: 5% methanol, 5% acetic acid (*see* **Note 6**).

2.5. Preparation of Radioactively Labeled p99 mRNA for the RISC Assay

1. Template DNA: plasmid pRSETA-p99 *(9)* cleaved with *Acc*I (*see* **Note 7**).
2. Nucleotide mix: 5 mM ATP, 5 mM UTP, 1 mM GTP, 1 mM CTP. The reaction contains a lower concentration of GTP to favor incorporation of the synthetic mGpppG-cap nucleotide.
3. 10 mM mGpppG-cap nucleotide (Pharmacia Biosciences, catalog number 27-4635): Dissolve 5 U of lyophilized mGpppG in 24 µL autoclaved ddH$_2$O, store at −20°C.
4. 300 mM DTT in ddH$_2$O, store at −20°C.
5. T7 RNA polymerase (New England BioLabs) at 50 U/µL in the corresponding buffer.
6. α^{32}P-CTP, 800 Ci/mmol: 20 µCi/µL.
7. RQI DNase (Promega): 1 U/µL.
8. Phenol: phenol:chloroform:isoamyl alcohol (25:24:1).

9. Sample buffer: 80% formamide (Baker, p. a. grade), 50 m*M* Tris-HCl, pH 7.9, 10 m*M* EDTA, 0.025% bromophenol blue/0.025% xylene cyanol.
10. 4% Acrylamide 8 *M* urea gel (acrylamide:*bis*-acrylamide = 29:1; 1X TBE).
11. mRNA elution buffer: 0.75 *M* NH$_4$-acetate, 10 m*M* Mg-acetate, 0.1 m*M* EDTA, 0.1% (w/v) SDS, 10 m*M* Mg-acetate, 0.1 m*M* EDTA.

2.6. In Vitro RISC Assay

1. RISC 4X A buffer: 60 m*M* HEPES-HCl, pH 7.4, 1 m*M* MgCl$_2$, 180 m*M* KCl, 2 m*M* DTT. This is the 4X RISC reaction buffer used for extracts in buffer A (i.e., S15 extracts).
2. 100 m*M* ATP (Amersham Biosciences), stored in aliquots at –20°C.
3. 20 m*M* GTP (Amersham Biosciences), stored in aliquots at –20°C.
4. Creatine phosphate (Sigma, St. Louis, MO): dissolve to 0.5 *M* in autoclaved ddH$_2$O. Store in 100-µL aliquots at –20°C.
5. Creatine phosphokinase (Sigma, St. Louis, MO): dissolve to a concentration of 1.2 mg/mL in autoclaved ddH$_2$O. It is important that this solution is made fresh. Dissolve immediately before use.
6. Porcine RNase inhibitor (Amersham Biosciences): 39.2 U/µL.
7. Proteinase K buffer: 100 m*M* Tris-HCl, pH 7.5, 150 m*M* NaCl, 12.5 m*M* EDTA, 1% SDS, 25 µg/mL tRNA.
8. Sample buffer: 80% formamide (Baker, p. a. grade), 50 m*M* Tris-HCl, pH 7.9, 10 m*M* EDTA, 0.025% bromophenol blue/0.025% xylene cyanol.
9. 8% Acrylamide 8 *M* urea gel (acrylamide:*bis*-acrylamide = 19:1; 1X TBE).

3. Methods

This section outlines the following:

1. Infection with Ad.
2. Procedures to prepare extracts from Ad-infected cells.
3. Strategies to synthesize radioactively labeled RNA.
4. In vitro assays to measure the activity of Dicer and RISC.

3.1. Infection of Adherent Cells With Ad

The protocol is adapted for small-scale cytoplasmic extract preparation from adherent cells, such as HeLa, C33A, or 293 cells. We routinely infect three 10-cm plates, but the protocol may be scaled up or down. The limiting factor is the total cell volume at the time of harvest, which should not be less than approx 50 µL if the cells are to be disrupted by passage through a syringe needle.

Monolayer cultures (we use 293, HeLa, or C33A cells primarily) are split the day before infection and seeded in 10-cm plates such that the monolayers are 60–80% confluent at the time of infection.

1. Calculate the amount of virus needed for the desired multiplicity of infection (MOI).
2. Thaw virus stock on ice. Mix stock briefly. Dilute virus in 3 mL DMEM containing 2% newborn calf serum per 10-cm plate. Quick-freeze the remaining virus stock in dry ice/ethanol.
3. Remove medium from the plate that you intend to infect.
4. Add 3 mL virus inoculum/plate.
5. Incubate for 60 min in CO_2 incubator (37°C, 7% CO_2).
6. Remove virus inoculum and dispose of, using an established and safe procedure.
7. Add 10 mL DMEM containing 10% newborn calf serum per plate (optionally containing 1% penicillin/streptomycin).
8. Return cells to CO_2 incubator.
9. Harvest at desired time point.

3.2. S15 Extract Preparation

To prepare cytoplasmic extracts from small volumes of cells requires special handling. To be able to recover reasonable amounts of extract using a syringe requires a volume of at least 300 µL (50 µL of cells and 250 µL of Buffer A). We typically produce S15 extracts from two to three 10-cm plates of HeLa or 293 cells. Also, it is possible to measure Dicer and RISC in S100 and splicing competent nuclear extracts prepared as described in Chapter 3. The protocol described below is adapted from a similar protocol designed for work with extracts prepared from *Drosophila* cells *(10,11)*.

1. The starting material is three 10-cm plates of 293 cells collected at a desired time point after Ad infection (*see* **Subheading 3.1.**).
2. Carefully remove the growth medium by gentle aspiration. If cells are loosely attached to the plate, it is better to resuspend all cells in the growth medium with a cell lifter and collect cells by centrifugation for a few min at 2000*g* in a 50-mL plastic Falcon tube.
3. Add 1 mL ice-cold PBS and resuspend cells with a cell-lifter or by pipetting. Transfer to three 1.5-mL Eppendorf tubes and collect cells by centrifugation at 3000*g* for 2 min.
4. Remove the supernatant by gentle aspiration.
5. Resuspend the pellets from the three tubes in a total volume of 1 mL ice-cold PBS. Transfer to a clean Eppendorf tube, centrifuge at 3000*g* for 2 min, and remove supernatant.
6. Estimate the packed cell volume (PCV). Adjust the volume to 6X PCV by addition of ice-cold Buffer A. A simple trick to estimate the PCV is to resuspend the pellet in a small volume of Buffer A (e.g., 100 µL) and measure the total volume of the solution using a 200-µL pipet. The PCV is total volume − 100 µL. This method gives extracts with a protein concentration of approx 3–4 µg/µL. Accurate measurement of PCV is important to obtain reproducible extracts.
7. Allow cells to swell on ice for 15 min.

8. Disrupt cells by passing the solution through a 23-gage syringe needle. Press the needle opening to the surface of the tube. Approximately 30–40 strokes are required. Be careful not to introduce bubbles. To check the efficiency of disruption, place a small drop of the solution on a plate, add 2 µL 0.4% Trypan blue, and check under the microscope. Nuclei stain blue; intact cells do not stain. Continue until more than 90% of the cells are disrupted.
9. Spin down the nuclei at 7000g for 5 min at 4°C.
10. Transfer the supernatant (cytoplasmic extract) to a new tube. Add 0.1 vol of Buffer A + 50% glycerol. It should be noted that S15 extracts without glycerol work well in the Dicer assay. However, glycerol is important for the RISC assay.
11. Clear the supernatant by centrifugation at 15,000g for 60 min at 4°C.
12 Aliquot the S15 extract into Eppendorf tubes, quick-freeze in liquid nitrogen, and store at –70°C. The extracts can be thawed a few times without losing activity if they are refrozen in liquid nitrogen.

3.3. Preparation of Radioactively Labeled Substrate dsRNA for the Dicer Assay

The Dicer reaction can be performed with any type of dsRNA that is long enough to be cleaved to an siRNA. Our standard Dicer substrate is a 159-bp dsRNA transcript derived from a Luciferase reporter construct *(7)*. Below we describe how this dsRNA is synthesized by simultaneous transcription of both strands of a PCR DNA template with a T7 and an SP6 promoter sequence at opposite ends. The two strands are annealed to form a dsRNA, excess ssRNA is removed by RNase treatment, and the final dsRNA is gel-purified on a polyacrylamide gel.

1. Combine in an Eppendorf tube at room temperature:
   a. 3 µL Template DNA (170 ng/µL for a 150 bp dsRNA).
   b. 2.5 µL 10X T7 RNA polymerase buffer.
   c. 9 µL ddH$_2$O.
   d. 2.5 µL Nucleotide mix.
   e. 1 µL 300 m*M* DTT.
   f. 0.5 µL Porcine RNase inhibitor.
   g. 0.75 µL T7 RNA polymerase.
   h. 0.75 µL SP6 RNA polymerase.
   Mix by pipeting up and down and briefly spin down the content. Add 5 µL α^{32}P-CTP and mix.
2. Incubate at 37°C for 2 h.
3. Add 1 µL RQ1 DNase to the reaction and continue incubation for 30 min at 37°C.
4. Add 400 µL of ddH$_2$O.
5. Purify by a single round of phenol extraction (400 µL).
6. Add Na-acetate to 0.3 *M* and 3 vol of ethanol; precipitate at –20°C overnight.
7. Spin tubes at 16,000g for 30 min at 4°C in an Eppendorf centrifuge, remove ethanol, and wash the pellet with 70% ethanol; spin again as above.

8. Dry the pellet and dissolve in 50 μL of dsRNA annealing buffer. Remove 2 μL for gel analysis.

9. Place tube in a beaker containing approx 200 mL of boiling water for 1 min, and let the water cool slowly to room temperature (this takes approx 2 h) (*see* **Note 8**). Remove 2 μL for gel analysis.

10. To the remaining 46 μL dsRNA add 13 μL of 5X T7 RNA polymerase buffer. Mix by vortexing and add 5 μL RNase mix (*see* **Note 9**).

11. Incubate 15 min at 37°C.

12. Purify dsRNA by a single round of phenol extraction (400 μL). Add Na-acetate to 0.3 *M* and 3 vol of ethanol; precipitate at –20°C overnight. Spin down, remove supernatant, dry pellet, and dissolve in 50 μL loading buffer. Take out 2 μL for control gel.

13. Prepare a 6% acrylamide 8 *M* urea gel. The gel does not have to be the large sequencing type of gel. A small minigel (Bio-Rad, cat. no. 165-3301) works well.

14. Load the sample on the gel without boiling. By this treatment ssRNA will be denatured while dsRNA will stay intact (*see* **Note 10**).

13. Expose to X-ray film and excise the dsRNA band (*see* **Note 11**).

14. Elute in 400 μL dsRNA elution buffer for 4 h at 55°C.

15. Transfer the eluted dsRNA into a new Eppendorf tube, add 3 volumes of ethanol and precipitate at –20°C overnight.

16. Spin tube at 16,000*g* at 4°C for 1 h in an Eppendorf centrifuge.

17. Wash the pellet twice with 75% ethanol.

18. Dissolve the dsRNA in a small volume of ddH$_2$O (*see* **Note 12**).

19. To verify the different steps in the generation of the dsRNA, the samples taken out at **steps 8**, **9**, and **12** may be analyzed on a polyacrylamide gel (*see* **Note 13**).

3.4. In Vitro Dicer Assay

In the standard Dicer reaction, we use 20 fmol substrate/10 μL = 2 n*M*. This corresponds to approx 50,000 dpm for a 150-bp dsRNA, labeled as described above. To see inhibition of Dicer in some virus-infected cell extracts, it is necessary to use a lower substrate concentration (*7*). Either label at higher specific activity or use fewer dpm in the reaction.

1. Prepare a reaction master mix. Each reaction should contain:
   a. 2.5 μL Dicer 4X A buffer (*see* **Note 14**)
   b. X μL ^{32}P-labeled dsRNA (10,000–50,000 dpm).
   c. 0.5 μL Porcine RNase inhibitor
   d. X μL ddH$_2$O (to a final volume of 5.5 μL).

2. Add 5.5 μL master mix to each Eppendorf tube.

3. Add 5 μL S15 extract, and mix by pipetting up and down. Be careful not to introduce bubbles.

4. Incubate in a cabinet incubator at 30°C for 1–4 h (*see* **Note 15**).

5. Add 10 μL 20 m*M* EDTA to terminate the reaction. Chill on ice (*see* **Note 16**).

6. Add 200 μL 2X Proteinase K buffer and 1 μL Proteinase K, and incubate at 65°C for 30 min.

7. Spin briefly to pellet any insoluble material.
8. Transfer the supernatant to a new Eppendorf tube, and extract twice with 200 μL phenol (*see* **Note 17**).
9. Add 4 μL 1 *M* MgCl$_2$ (facilitates precipitation of small RNAs) and 3 vol of ice-cold ethanol, precipitate overnight at –20°C.
10. Spin tubes at 16,000*g* for 1 h at 4°C in an Eppendorf centrifuge.
11. Remove supernatant, spin down briefly, and remove remaining ethanol with a fine pipet. Dry on the bench and dissolve the pellet in 20 μL sample buffer.
12. Heat samples at 100°C for 5 min, chill on ice, and separate 8–15 μL on a 15% acrylamide 7 *M* urea gel (*see* **Note 18**).
13. Elute salts from the gel by incubating the gel in gel-fix solution for 1 h. Transfer the gel to a sheet of 3MM paper and dry it on a gel dryer.
14. Expose the gel to X-ray film or a PhosphorImager screen. An example gel is shown in **Fig. 1**.

3.5. Preparation of Radioactively Labeled p99 Substrate RNA for the RISC Assay

In our assay system we use a capped transcript uniformly labeled with α^{32}P-CTP. To use a capped transcript is essential because the 3'-fragment produced after RISC cleavage often is rapidly degraded, and therefore lost, especially when S15 extracts are used. Our target RNA for the RISC reaction is a synthetic p99 mRNA (*9*) with a target sequence for a relevant siRNA located in the 5'-half of the mRNA (*see* **Note 7**). Remember that there is a strand bias for siRNA incorporation into the RISC complex (*12*). Thus, in unlucky situations it may become necessary to design alternative siRNAs or use two different templates, each transcribing one mRNA strand.

1. Combine in an Eppendorf tube at room temperature:
   a. 2.5 μL 10X T7 RNA polymerase buffer.
   b. 5 μL pRSETA-p99/*Acc*I cleaved template DNA (200–500 ng).
   c. 2.5 μL Nucleotide mix.
   d. 2.5 μL mGpppG-cap nucleotide.
   e. 1 μL 300 m*M* DTT.
   f. 5 μL α^{32}P-CTP (100 μCi).
   g. 0.5 μL Porcine RNase inhibitor.
   h. 1 μL T7 RNA polymerase.
   i. 5 μL ddH$_2$O.
2. Incubate 1–2 h at 37°C.
3. Add 1 μL RQ1 DNase, and incubate 30 min at 37°C.
4. Add 15 μL sample buffer. Store at –20°C or continue directly.
5. Heat samples at 100°C for 5 min, chill on ice, and separate on a 4% acrylamide 8 *M* urea gel (*see* **Note 11**).
6. Cut out the full-length band and place it in an Eppendorf tube; add 500 μL mRNA elution buffer.

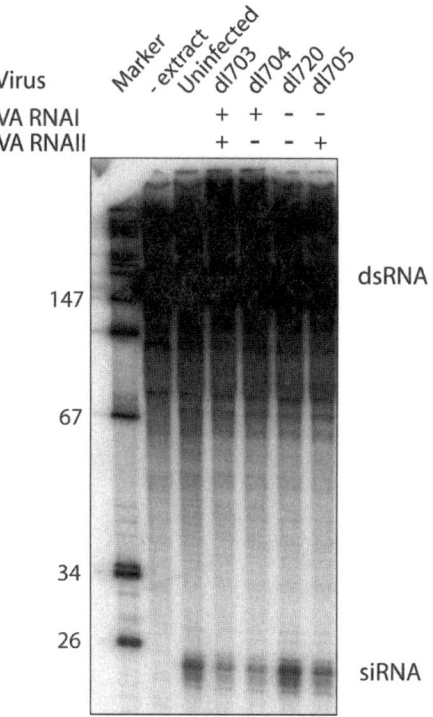

Fig. 1. Adenovirus VA RNAI blocks the activity of Dicer in virus-infected cells. 293 cells were infected with a wild-type adenovirus or mutant viruses defective in VA RNAI or VA RNAII expression *(15)*. The VA RNA status of the different viruses is shown at the top of the figure. S15 extracts were prepared from uninfected 293 cells or 20 h postinfection with the different mutant adenoviruses. The activity of Dicer was assayed using the 159-bp dsRNA described in the text. The lane labeled "-extract" is a control lane to show that the band of VA RNA size is generated during the incubation in the different extracts.

7. Elute the transcript at 37°C on a turning wheel for 2 h.
8. Remove insoluble material by a short spin and transfer the supernatant to a fresh tube.
9. Phenol-extract one time, add 3 vol of ethanol to the supernatant, and precipitate the RNA at –20°C overnight.
10. Dissolve in suitable volume of ddH$_2$O.

3.6. In Vitro RISC Assay

We have adapted a protocol described by Tuschl and collaborators *(11)* for use with our infected-cell extracts. This protocol can be used to analyze the endogenous activity of RISC in an extract or to assemble new RISC complexes by addition of synthetic siRNA to the extract. For the RISC reaction to be successful, it is important that the reaction master mix is made fresh immediately before mixing the reaction. The crucial part appears to be the creatine kinase, an enzyme needed to provide the RISC reaction with ATP. To obtain a reproducible and high performance, we recommend the following procedure. Plan every step of the process before starting the experiment.

1. Thaw S15 extracts and pellet insoluble components by centrifugation for 5 min at 16,000g in an Eppendorf centrifuge. Place extracts on ice.
2. Add 1 µL of 1 mM p99 siRNA to each reaction tube.
3. Prepare a reaction master mix. One RISC reaction should contain the following components:
   a. 2.5 µL RISC 4X A buffer (*see* **Note 19**).
   b. 0.1 µL 100 mM ATP.
   c. 0.1 µL 20 mM GTP.
   d. 0.5 µL 500 mM Creatine phosphate.
   e. 0.25 µL Creatine kinase (freshly dissolved).
   f. 0.05 µL Porcine RNase inhibitor.
   g. 0.3 µL p99 mRNA (50,000 dpm).
   h. 0.25 µL ddH$_2$O.
4. Add 4 µL of master mix into each reaction tube. Place the tubes at room temperature.
5. Add 5 µL S15 extract to each tube. Mix by gently pipetting up and down 20 times. Be careful not to introduce any bubbles.
6. Incubate in a cabinet incubator at 30°C for 1–4 h (*see* **Note 15**).
7. Add 1 µL 0.5 M EDTA to terminate reaction. Chill on ice.
8. Add 400 µL Proteinase K buffer and incubate at 65°C for 45 min.
9 Spin briefly to pellet any insoluble material.
10. Transfer the supernatant to a new Eppendorf tube and extract once with an equal volume of phenol (avoid transferring the interphase to the new tubes).
11. Add 3 vol of cold 100% ethanol, and precipitate at –20°C for 1 h to overnight.
12. Pellet RNA by centrifugation at 16,000g for 1 h at 4°C.
13. Remove supernatant, wash pellet with 80% ethanol, dry on the bench, and dissolve in 20 µL sample buffer.
14. Heat samples at 100°C for 5 min, chill on ice, and separate on an 8% acrylamide 8 M urea gel (*see* **Note 12**).
15. Expose to X-ray film or a PhosphorImager screen. For an example gel, *see* **Fig. 2**.

4. Notes

1. dsRNA is transcribed from template DNA with either T3, T7, or SP6 promoters, either the same at both ends or two different promoters. We use a forward primer with a 5'-extension encoding a minimal T7 promoter (ATTAATACGACTCA CTATAG; bold nucleotide = transcription start site) and a reverse primer with a 5'-extension encoding a minimal SP6 promoter (ATTTAGGTGACACTATAG;

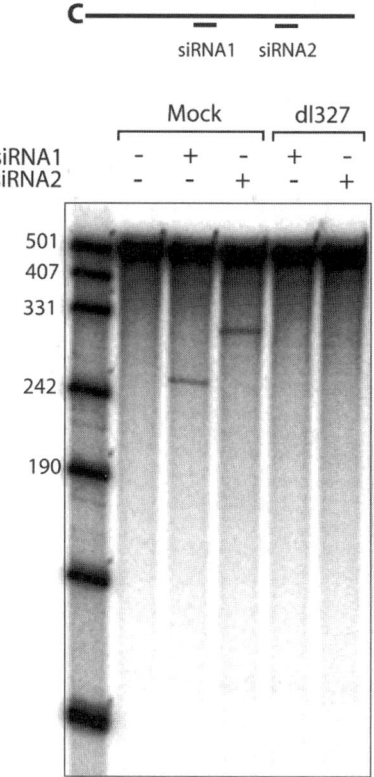

Fig. 2. Inhibition of RISC during an adenovirus infection. S15 extracts were prepared from Mock- or dl327-infected 293 cells at 20 h postinfection. The activity of RISC was assayed using two different p99 siRNAs (depicted schematically at the top). Note that only the capped 5'-fragment is stable in the extract. The downstream fragment generated by RISC cleavage will be uncapped and therefore rapidly degraded.

bold nucleotide = transcription start site) to generate our standard pGL2 control dsRNA. The length of the dsRNA is of your own choice. However, we have noted that relatively short transcripts (100–200 nt) form dsRNA more readily than longer species (500–1000 nt). Alternatively, RNA may be transcribed from a template encoding a hairpin-forming RNA placed after a single promotor. It is absolutely essential that the DNA is RNase-free and free from inhibitory contaminants.

2. This NTP mix is used in a reaction with 50 µCi α^{32}P-CTP. In order to obtain a higher yield we often use a master mix with 0.4 mM CTP and 100 µCi α ^{32}P-CTP. The standard reaction contains 500 pmol of cold CTP. 50 µCi of α ^{32}P-CTP at 800 Ci/mmol will add 62.5 pmol CTP (a total of 562 pmol CTP per reaction). An NTP mix without cold CTP may be used to get dsRNA with 10-fold

higher specific activity. In this case, no less than 200 µCi of hot CTP should be used in the reaction to give a sufficiently high concentration of CTP to drive the reaction. The high specific activity results in rapid radiolysis, and thus dsRNA prepared with this master mix should be used within a day after purification. Otherwise the degradation products will result in a smear that masks the reaction product.

3. Most suppliers provide enzymes of good quality. We have observed a poor performance of some provided buffers with the T7 RNA polymerase. The problem appears to be that spermidine precipitates in the buffer; this is detrimental for the activity of the spermidine-dependent T7 RNA polymerase.

4. The RNases must be of high quality, e.g., sequencing grade and absolutely free of dsRNA-degrading activities, helicases, etc. The crude RNase A used in plasmid preparation will also degrade dsRNA.

5. The reaction buffer includes ATP mainly for historical reasons. We have observed that ATP is not necessary in our system, and it has been shown *(13)* that the activity of recombinant Dicer is not ATP-dependent. By adding a complete ATP-regenerating system with creatine phosphate and creatine kinase, as described in the RISC reaction, it is possible to obtain a strong increase in the dsRNA-degrading activity (unpublished observations). However, the rapid disappearance of the dsRNA is not accompanied by an increased accumulation of siRNA. Thus, the disappearance of dsRNA in this case is most likely a combination of specific (for us Dicer) and nonspecific nuclease attack on the dsRNA. To be able to specifically study Dicer activity, we recommend omitting an ATP-regenerating system.

6. A 15% polyacrylamide gel is very prone to cracking unless the urea is removed by fixation prior to drying.

7. Plasmid pRSETA contain a T7 promoter. Thus, we generate our standard RISC substrate by T7 transcription of a linearized template. However, alternative templates can be designed for virtually any sequence by PCR amplification using one primer containing a 5'-extension encoding a minimal T7 promoter sequence (*see* **Note 1**). It should be remembered that there might be a strand bias for incorporation into RISC. Thus, for some siRNAs only one strand is efficiently incorporated into the RISC complex *(12)*. It may be necessary to make two different templates, each transcribing one mRNA strand. For the RISC reaction it is not necessary that the mRNA contain an open reading frame, polyadenylation signals, etc. However, it is essential that the transcript be capped. Otherwise it will be rapidly degraded by 5'-exonucleases in the extract.

8. This protocol gives almost perfect annealing of a 150-bp dsRNA. Thus, a transcription reaction with a short template like ours (approx 150 bp) will contain a large proportion of dsRNA so may proceed directly to the RNase treatment. Longer templates may require an annealing step (*see* **Subheading 3.3.**, **step 9**). It should be remembered that for long templates it is important that the transcribed RNA be dissolved in a low-salt buffer because high salt concentrations may give melting temperatures that exceed 100°C.

9. Selective degradation of ssRNA by RNase A and RNase T1 requires a minimum concentration of salt in the reaction. We routinely use the T7 RNA polymerase

buffer as the high-salt buffer because it fulfills our needs; i.e., is RNase-free and generally is supplied, in great excess, with the polymerase. If higher specific activity dsRNA was made by omitting the cold CTP from the transcription reaction (*see* **Note 2**), use 10 times less RNase mix to minimize degradation of the dsRNA.

10. The dsRNA is loaded on the 8 *M* urea polyacrylamide gel without boiling. The high melting temperature of the dsRNA prevents separation of the two strands. The conditions are, however, sufficient to denature the secondary structures of a single-stranded RNA. In this way it is possible to separate ssRNA and dsRNA on the same gel and still have the superior resolution of a denaturing urea acrylamide gel. Thus, the purified dsRNA will be of homogeneous size, trimmed of single-stranded overhangs, and essentially free from contaminant ssRNA (**Fig. 3**). This method gives much better separation than native polyacrylamide gel electrophoresis or agarose gel electrophoresis. Do not run the gel "too warm" because this may result in denaturation of the dsRNA. This protocol is adapted from a method previously described by Rotondo and Frendeway *(14)*.

11. For more detailed information on how to prepare the separation gel and the techniques used to identify and elute RNA from the polyacrylamide gel, *see* Chapter 3.

12. The supernatant frequently contains a small fraction of insoluble polyacrylamide residual, which binds some radioactivity. Spin down and transfer supernatant to a fresh tube before measuring activity.

13. The control gel is run in order to test the efficiency of the annealing and RNase treatment (*see* **Fig. 3**).

    a. Prepare a 4% acrylamide 8 *M* urea minigel (acrylamide:*bis*-acrylamide = 29:1; 1X TBE).

    b. Samples taken at **Subheading 3.3.**, **steps 8**, **9**, and **12** are divided into two tubes each and mixed with 1 µL of sample buffer.

    c. Boil one of the tubes from each sample.

    d. Load the gel with boiled and unboiled samples.

    e. Run the gel until the first dye marker reaches the front of the gel.

    f. Dry the gel and expose to X-ray film or PhosphorImager. Do not run the gel "too hot;" it may melt the native dsRNA. In the boiled sample, the denatured RNA will migrate as one single band. In the unboiled sample two bands can generally be seen; the faster migrating band will correspond to the dsRNA.

14. The final concentrations of components in the Dicer reaction mixture, including the salts present in the cell extracts, are: 20 m*M* HEPES-HCl, pH 7.0, 2 m*M* Mg-acetate, 2 m*M* DTT, 1 m*M* ATP, and 0.75 m*M* $MgCl_2$. If Dicer is assayed in S100 extracts prepared as described in Chapter 3, a different Dicer reaction buffer has to be used. Thus, extracts dialyzed against buffer D have a lower magnesium concentration than buffer A used to prepare S15 extracts. This is compensated in the Dicer 4X D buffer, which contains: 87 m*M* HEPES-HCl, pH 7.0, 8.7 m*M* Mg-acetate, 95.7 m*M* $MgCl_2$, 8.7 m*M* DTT, and 8 m*M* ATP.

15. Because the reactions are performed in a small volume (10 µL), it is important that the Eppendorf tubes be incubated in a cabinet incubator rather than in a water

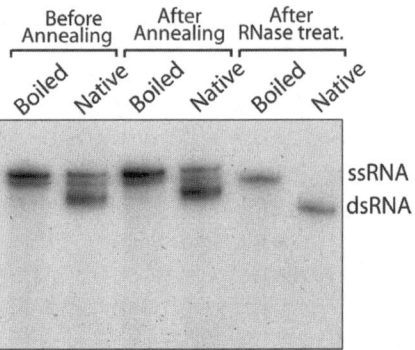

Fig. 3. A control gel to check the efficiency of annealing and RNase treatment during the synthesis of the Dicer substrate dsRNA. The small aliquots of dsRNA taken out at the different steps during the synthesis reaction (*see* **Subheading 3.3.**) were divided into two tubes and mixed with sample buffer. One tube from each step was boiled and both separated on an 8% denaturing gel. Under these conditions the unboiled dsRNA will not melt and can be distinguished from single-stranded RNA, which migrates with a different mobility in this gel system.

bath. Otherwise most of the reaction volume ends up as condensed water under the lid.

16. Some Dicer activity remains if samples are put on ice without EDTA. In negative controls, add EDTA at $t = 0$ and incubate at 30°C for the same time as other samples.

17. Do not dilute the sample with water. The Proteinase K buffer is used at double strength in order to have a higher salt concentration, which facilitates precipitation of siRNA. Do not be "greedy" during the phenol extraction. Trying to get all of the supernatant frequently results in contamination of the supernatant, something that subsequent results in smearing on the gel.

18. A 15% gel gives best resolution between 10 and 50 nt. Bromophenol blue migrates at approx 12 nt, whereas xylene cyanol migrates at 55 nt, and siRNA right in between. It is sufficient to run the gel until the bromophenol blue has migrated approx 15–20 cm. Use a 5'-end labeled synthetic siRNA as a marker.

19. The final concentrations of components in the RISC reaction mixture, including the salts present in the cell extracts are: 15 mM HEPES-KOH, pH 7.4, 50 mM KCl, 1 mM MgCl$_2$, 1 mM ATP, 0.2 mM GTP, 10 µg/mL RNasin, 30 µg/mL creatine kinase, 25 mM creatine phosphate, 0.5 mM DTT, 2.5% glycerol. If RISC is assayed in S100 extracts prepared as described in Chapter 3, a different 4X reaction buffer should be used. Extracts dialyzed against buffer D have a lower magnesium concentration than if buffer A is used to prepare S15 extracts. This is compensated for by the RISC 4X D buffer, which contains 60 mM HEPES-HCl, pH 7.0, 4.4 mM MgCl$_2$, and 2 mM DTT.

Acknowledgments

This work was supported by the Swedish Cancer Society and the Wallenberg Consortium North.

References

1. Fire, A., Xu, S., Montgomery, M. K., Kostas, S. A., Driver, S. E., and Mello, C. C. (1998) Potent and specific genetic interference by double-stranded RNA in Caenorhabditis elegans. *Nature* **391,** 806–811.
2. Hutvagner, G. and Zamore, P. D. (2002) RNAi: nature abhors a double-strand. *Curr. Opin. Genet. Dev.* **12,** 225–232.
3. Denli, A. M. and Hannon, G. J. (2003) RNAi: an ever-growing puzzle. *Trends Biochem. Sci.* **28,** 196–201.
4. Baulcombe, D. (2002) Viral suppression of systemic silencing. *Trends Microbiol.* **10,** 306–308.
5. Lecellier, C. H. and Voinnet, O. (2004) RNA silencing: no mercy for viruses? *Immunol. Rev.* **198,** 285–303.
6. Li, W. X. and Ding, S. W. (2001) Viral suppressors of RNA silencing. *Curr. Opin. Biotechnol.* **12,** 150–154.
7. Andersson, M. G., Haasnoot, P. C., Xu, N., Berenjian, S., Berkhout, B., and Akusjärvi, G. (2005) Suppression of RNA interference by adenovirus virus-associated RNA. *J. Virol.* **79,** 9556–9565.
8. Lu, S. and Cullen, B. R. (2004) Adenovirus VA1 noncoding RNA can inhibit small interfering RNA and MicroRNA biogenesis. *J. Virol.* **78,** 12,868–12,876.
9. Kreivi, J. P., Trinkle-Mulcahy, L., Lyon, C. E., Morrice, N. A., Cohen, P., and Lamond, A. I. (1997) Purification and characterisation of p99, a nuclear modulator of protein phosphatase 1 activity. *FEBS Lett.* **420,** 57–62.
10. Zamore, P. D., Tuschl, T., Sharp, P. A., and Bartel, D. P. (2000) RNAi: double-stranded RNA directs the ATP-dependent cleavage of mRNA at 21 to 23 nucleotide intervals. *Cell* **101,** 25–33.
11. Tuschl, T., Zamore, P. D., Lehmann, R., Bartel, D. P., and Sharp, P. A. (1999) Targeted mRNA degradation by double-stranded RNA in vitro. *Genes Dev.* **13,** 3191–3197.
12. Schwarz, D. S., Hutvagner, G., Du, T., Xu, Z., Aronin, N., and Zamore, P. D. (2003) Asymmetry in the assembly of the RNAi enzyme complex. *Cell* **115,** 199–208.
13. Provost, P., Dishart, D., Doucet, J., Frendewey, D., Samuelsson, B., and Radmark, O. (2002) Ribonuclease activity and RNA binding of recombinant human Dicer. *EMBO J.* **21,** 5864–5874.
14. Rotondo, G. and Frendewey, D. (2001) Pac1 ribonuclease of Schizosaccharomyces pombe. *Methods Enzymol.* **342,** 168–193.
15. Bhat, R. A. and Thimmappaya, B. (1984) Adenovirus mutants with DNA sequence perturbations in the intragenic promoter of VAI RNA gene allow the enhanced transcription of VAII RNA gene in HeLa cells. *Nucleic Acids Res.* **12,** 7377–7388.

5

Simultaneous Detection of Adenovirus RNA and Cellular Proteins by Fluorescent Labeling *In Situ*

Eileen Bridge

Summary

Investigating the cell biology of gene expression requires methodologies for localizing RNA relative to proteins involved in RNA transcription, processing, and export. Adenovirus is an important model system for the analysis of eukaryotic gene expression and is also being used to investigate the organization of gene expression within the nucleus. Here are described the combined *in situ* hybridization and immunofluorescence staining techniques that have been used to study the localization of viral RNA relative to nuclear structures that contain splicing factors.

Key Words: Adenovirus; *in situ* hybridization; RNA localization; RNA processing; gene expression; nuclear organization.

1. Introduction

Eukaryotic mRNA production is a complex process that involves capping, splicing, and polyadenylation of the primary transcript. Nuclear-processing activities remodel the protein composition of the primary RNP transcript and play important roles in the export, translation, and stability of the mRNA *(1)*. Adenovirus (Ad) mRNA production has been extensively studied as a model system for eukaryotic gene expression and was instrumental in the discovery of RNA processing *(2,3)*. A major goal of current research in cell biology is to understand how gene expression activities are organized relative to each other and to the structural framework of the nucleus. There is considerable evidence linking transcription with subsequent RNA processing events *(4,5)*. Ad late mRNA transcription occurs at the periphery of nuclear inclusions that contain viral DNA and high concentrations of the Ad 72-kDa DNA binding protein *(6)*. Ad transcription sites have been visualized by fluorescence detection of

From: *Methods in Molecular Medicine, Vol. 131:*
Adenovirus Methods and Protocols, Second Edition, vol. 2:
Ad Proteins, RNA, Lifecycle, Host Interactions, and Phylogenetics
Edited by: W. S. M. Wold and A. E. Tollefson © Humana Press Inc., Totowa, NJ

modified nucleotides incorporated into elongating transcripts in either permeabilized *(7)* or intact *(8)* cells. Colocalization studies indicate that factors involved in splicing *(7,9)* and RNA export *(8)* are present at transcription sites containing Ad late RNA. Spliced viral RNA also accumulates posttranscriptionally in nuclear interchromatin granules, which contain splicing factors *(9,10)*. The nuclear organization of Ad gene expression activities has been investigated using immunostaining techniques to analyze the distribution of cellular factors involved in RNA processing and export and *in situ* hybridization (ISH) to determine the localization of Ad RNA *(7,11)*. A combination of ISH and immunostaining to simultaneously determine the localization of both specific nucleic acids and specific proteins can thus be a powerful method for studying their relationship within the organizational framework of the cell. We have used these methods to study the localization of viral RNA relative to nuclear structures that contain splicing factors *(9–12)*. Such localization studies require careful attention to a number of different parameters. These include fixation and pretreatment of cells, preparation of suitable probes, and hybridization conditions. Each of these parameters will be discussed separately in the sections below.

1.1. Fixation and Pretreatment of Cells

We have routinely grown and infected cells on glass slides or coverslips. At appropriate times after infection, the cells must be treated with a fixative to preserve structure and permeabilized to allow access of probes and antibodies to their target molecules. Common fixatives include cross-linking agents such as paraformaldehyde (PFA), formaldehyde, or glutaraldehyde. Fixed cells can be permeabilized by extraction with buffers containing detergents such as Triton X-100, NP-40, saponin, and digitonin *(13)* or mild treatments with sodium dodecyl sulfate (SDS) *(7)*. Alternatively, fixing with agents such as methanol or acetone can simultaneously fix cellular structure and permeabilize the cells to allow access to probes. The choice of treatments for fixing and extracting cells varies greatly and can critically affect the outcome of the experiment. Unfortunately, there is no one protocol that works reliably for all antibodies, cell types, and probes. We have found it necessary to titrate both fixation and extraction conditions in order to obtain the best staining for each combination of antibody and *in situ* probes. In every case it is necessary to balance the need for good accessibility of target molecules with the best preservation of cellular structure.

1.2. Probes for ISH

Nucleic acid probes for ISH can be obtained by a variety of methods. We have used synthetic oligonucleotide probes coupled to biotin or digoxigenin for detection using fluorescence-based methods *(11)*. It is also possible to use oligonucleotides directly coupled to fluorochromes, such as fluorescein isothiocyanate (FITC) *(14,15)*. Oligonucleotide probes are small enough to easily penetrate fixed and permeabilized cells and are frequently the probe of choice. However, it should be noted that not every oligonucleotide probe will give a good ISH signal; if bound proteins mask the target sequence, it will be difficult to get a signal above the background. When possible, it can be useful to test several different probes to the RNA of interest in order to identify one that has good access to its target sequence in fixed cells. Oligonucleotide probes have been particularly useful for identifying spliced forms of RNA *in situ* because they can be designed to hybridize specifically to splice junctions of processed RNA *(11,14)*. Probes can also be produced by nick-translation, by polymerase chain reaction (PCR) amplification, and by in vitro transcription reactions performed in the presence of nucleotides coupled to biotin or digoxigenin *(16)*. Probe length is an important consideration for developing a useful reagent for ISH. Ideally, probes should not be much longer than about 200 nucleotides in order to efficiently enter fixed and permeabilized cells *(17)*.

1.3. Hybridization to Nucleic Acids in Ad-Infected Cells

Hybridizations are normally done in buffers containing formamide since the presence of formamide reduces the melting temperature of nucleic acid duplexes. This property allows the hybridizations to be done at lower temperatures and thus helps to preserve cellular structure during the procedure *(16)*. Other parameters important for efficient hybridization include temperature, pH, and the concentration of monovalent cations. Many hybridization solutions also contain dextran sulfate, which is strongly hydrated in aqueous solutions and can thus increase the effective concentration of probes.

Hybridization to double-stranded DNA will require a denaturation step. This can be done by heat-treating the cells or by treatment with NaOH *(17)*. In most studies of cellular genes, hybridization to the gene and hybridization to the RNA produced from the gene are easily distinguished by doing the hybridization under denaturing or nondenaturing conditions. In Ad-infected cells, differentiating between hybridization to DNA and hybridization to RNA is considerably complicated by the fact that the virus produces a large amount of single-stranded DNA during its replication *(18)*. These sequences are available for hybridizing to many probes even under nondenaturing conditions. Thus, extra enzymatic treatments with RNase and DNase are required as controls to determine what part of the signal is the result of hybridizing to RNA and what part of the signal is the result of hybridizing to DNA. In Ad-infected cells, the localization of double-stranded DNA (detected in denatured samples) is separate from that of single-stranded DNA (detected in nondenatured samples) during the viral late phase *(19,20)*. During the late phase single-stranded DNA accumulates in nuclear inclusions that also contain the viral 72-kDa DNA-binding protein *(20,21)*. Antibodies against the 72-kDa protein can thus serve as a useful marker for the location of single-stranded DNA during the late phase. This chapter will focus on the methodology we have used for localizing viral RNA and cellular or viral proteins in double-labeling experiments.

2. Materials

Note: All solutions and materials must be kept RNase-free for the detection of RNA.

1. Biotin-16-dUTP (Boehringer Mannheim; Indianapolis, IN).
2. G-50 sephadex spin chromatography. G-50 sephadex (Pharmacia; Piscataway, NJ) is prepared for use according to the manufacturer's instructions and using precautions to prevent contamination with RNases. We routinely autoclave the hydrated G-50 sephadex and store it at 4°C.
3. Glass cover slips. We use 12 × 12-mm glass cover slips that are prepared by washing in 70% ethanol and then distributing the cover slips between layers of Whatman 3MM paper in a glass Petri dish. The cover slips in the Petri dish are autoclaved and are then ready to use.
4. 20X Salt sodium citrate (SSC) (1 L): 175.3 g NaCl, 88.2 g sodium citrate. Adjust to pH 7.4 with NaOH. Autoclave.
5. Phosphate-buffered saline (PBS) (1 L): 8.0 g NaCl, 0.2 g KCl, 1.44 g Na_2HPO_4, 0.24 g KH_2PO_4. Adjust to pH 7.4 with HCl. Autoclave.
6. 4.0% (w/v) PFA in PBS: mix PFA with PBS. With stirring, add NaOH until the PFA goes into solution. Add $MgCl_2$ to a final concentration of 5 mM and adjust to pH 7.4 with HCl. Filter through a 0.2-μm filter. PFA can be frozen in aliquots at –20°C for long-term storage or kept at 4°C and used within 2–3 d.
7. 0.5% Triton X-100 in PBS. This solution is filtered through a 0.2-μm filter and stored at 4°C.
8. Hybridization solution: 50% deionized formamide, 2X SSC, 5% dextran sulfate, 50 mM phosphate buffer, pH 7.0 *(22)*, 1 mg/mL *E. coli* t-RNA (Sigma; St. Louis, MO). Store at –20°C.
9. Hybridization chamber. A moist chamber is required to prevent the cells from drying out during hybridization. Our homemade chambers consist of a plastic box that contains paper towels wetted with 50% formamide 2X SSC. Slides containing the cover slips are placed on stands within the box and the lid sealed with parafilm before placing at the appropriate temperature.
10. Blocking solution: 0.5% blocking reagent (Boehringer Mannheim, Indianapolis IN) in 100 mM Tris-HCl, pH 7.5, 150 mM NaCl. Autoclave the solution to dissolve the blocking reagent. Aliquot and freeze at –20°C.
11. Tween-20.
12. ExtrAvidin-FITC (Sigma, St. Louis, MO).
13. Primary antibody against the protein of interest.

14. Secondary antibody conjugated to Texas Red or other appropriate fluorochrome.
15. Mounting medium: Vecta-Shield (Vector Laboratories, Burlingame, CA). This contains antifading reagents to preserve the fluorescence during microscopy.
16. DNase I digestion buffer: 20 mM Tris-HCl, pH 7.5, 6 mM MgCl$_2$.
17. RNase H buffer: 40 mM Tris-HCl, pH 7.5, 4.0 mM MgCl$_2$, 1.0 mM dithiothreitol, 4% glycerol, 30 µg/mL BSA, 100 mM KCl.

3. Methods

3.1. Preparation of Probes

Oligonucleotide probes with nucleotides modified for fluorescence-based detection can be purchased (e.g., from GeneDetect.com). It should be noted that it is possible to prepare probes by a variety of standard techniques such as random priming, in vitro transcription, or nick-translation *(16)*. Probes can also be made with digoxigenin *(16)*. The procedure that we have used for making biotin-labeled probes by PCR is given here. Appropriate primers are chosen to amplify a target sequence of approx 200 nucleotides. PCR is then performed under standard conditions except that the concentration of dTTP is 134 µM and biotin-16-dUTP is present at 66 µM. Amplification is done with 200 µM dATP, dCTP, and dGTP in a 100-µL reaction volume. Following amplification, the PCR product is purified by spun-column chromatography using G-50 sephadex *(22)* to remove unincorporated nucleotides, precipitated in ethanol, dissolved in a small volume of RNase-free water, and then stored at –20°C. The concentration of the PCR product is estimated by gel electrophoresis with standards of known concentration.

3.2. Preparation and Fixation of Cells

1. Cells growing in monolayers are transferred to tissue culture dishes containing sterile glass cover slips and allowed to grow on these cover slips for 1–2 d to form a subconfluent monolayer. Cells can be infected with virus or mock-infected as desired.
2. At the desired time after infection, remove cover slips with attached cells with small forceps and transfer to a separate dish and rinse two times with 1 mL of PBS. I find it convenient to process three 12 × 12-mm cover slips in a 3.5-cm Petri dish and buffer volumes given here are for this size container. If other containers are used, buffer volumes should be sufficient to adequately cover the cells.
3. Fix the cells with 3–4% PFA in PBS. First, rinse cells on cover slips with 1 mL of this solution and then incubate with 1 mL for 10 min at room temperature. Rinse cover slips with 1 mL of 0.5% Triton X-100 in PBS and then incubate with 1 mL of this buffer for 15 min at room temperature (*see* **Note 1**).
4. Wash cover slips three times for 5 min with 2.0 mL of PBS at room temperature (*see* **Note 2**).
5. Wash cells twice for 5 min at room temperature with 2.0 mL 2X SSC; they are then ready for ISH.

3.3. ISH to Ad-RNA and/or Single-Stranded Ad-DNA

1. For each 12 × 12-mm cover slip, dry down approximately 50 ng of labeled PCR product or oligonucleotide probe in a speed vac (the optimal amount of probe can be determined by titration). Dissolve the probe in hybridization solution (8 μL/ cover slip).
2. Denature the probe by heating to 70°C for 5 min. Chill on ice.
3. Pipet 8 μL of denatured probe onto a clean glass slide. Lift out a cover slip using a needle and forceps, blot the back of the cover slip dry with a tissue, and then carefully layer the cover slip cell-side down onto the drop of hybridization solution containing the probe. Incubate the slide and the cover slip in a hybridization chamber equilibrated with 50% formamide, 2X SSC at 37°C for 1–3 h.
4. Loosen the cover slip by pipeting 2X SSC around the edges. Gently pry up the cover slip using a needle and forceps and transfer it cell-side up to a 3.5-cm dish containing 2 mL 2X SSC. Wash with 2 mL 2X SSC three times for 15 min at 37°C. Wash with 2 mL 1X SSC for 15 min at room temperature (*see* **Note 3**).
5. Rinse cover slips with 4X SSC containing 0.05% Tween-20. Aspirate this buffer taking care to dry around the edges of the cover slip and then add 8 μL of blocking solution to the top of each cover slip. Incubate for 30 min at room temperature.
6. Rinse cover slips with 1 mL 4X SSC containing 0.05% Tween-20. Aspirate, add 8 μL of FITC-coupled ExtrAvidin diluted 1/200 in blocking solution (12.5 μg/mL final concentration) to the top of each cover slip, and incubate for 30 min at room temperature. When performing double stainings in which ISH is performed together with immunostaining, it is convenient to incubate the primary antibody together with the ExtrAvidin at an appropriate dilution (*see* **Note 4**).
7. Wash cover slips three times for 5 min with 2.0 mL 4X SSC at room temperature.
8. Rinse cover slips with 4X SSC containing 0.05% Tween-20. Aspirate, and add 8 μL of an appropriate secondary antibody coupled to a suitable fluorochrome that has been diluted into blocking solution to a final concentration of 0.02 mg/mL. Incubate the cover slip at room temperature for 30 min.
9. Wash cover slips three times for 5 min with 2 mL 4X SSC at room temperature.
10. Postfix the cells with 1 mL 4% PFA in PBS for 5 min at room temperature.
11. Wash the cells three times for 5 min with 2 mL PBS at room temperature (*see* **Note 5**).
12. Mount the cover slips onto a slide containing a 3.5-μL spot of Vecta-shield mounting medium, seal with nail polish, and then observe the staining by fluorescence microscopy using appropriate filters for the two different fluorochromes.

3.4. RNase and DNase Controls

Enzymatic digestions with RNase A, RNase H, and DNase I may be performed to determine whether probes are hybridizing to DNA or RNA (*17*). These controls are essential in Ad-infected cells because the large amount of ssDNA produced during viral replication can be available for hybridization even under nondenaturing conditions. When possible, the use of splice junction oligonucleotide probes of 22–24 nucleotides for detecting spliced RNA can greatly simplify the analysis, since these probes do not hybridize efficiently to single-stranded Ad DNA *(12,14)*.

3.4.1. RNase A Digestion

1. Following the fixation procedure (**Subheading 3.2.**, **step 4**), transfer the cover slips to be treated with RNase to separate 3.5-cm dishes (cell-side up).
2. Rinse the cover slips with 1.0 mL of 2X SSC containing 0.05% Tween-20. Aspirate the buffer and then add 8 µL of 100 µg/mL RNAse A in 2X SSC to the top of the cover slip. Incubate the dish in a moist chamber (a covered plastic box containing paper towels soaked in water) at 37°C for 45 min.
3. Wash the cover slips three times for 10 min with 2X SSC at room temperature.
4. Proceed with the hybridization protocol (**Subheading 3.3.**, **step 1**).

3.4.2. DNase I Digestion

1. Following the fixation procedure (**Subheading 3.2.**, **step 4**), transfer the cover slips to be treated with DNase to separate 3.5-cm dishes (cell-side up).
2. Rinse the cells with DNase I digestion buffer containing 0.05% Tween-20. Aspirate the buffer and then add 8 µL of 100 U/mL RNase-free DNase in DNase I digestion buffer to the top of the cover slip. Incubate the dishes in a moist chamber (*see* **Subheading 3.4.1.**, **step 2**) at 37°C for 45 min.
3. Wash the cells three times for 10 min with 2X SSC at room temperature.
4. Proceed with the hybridization protocol (**Subheading 3.3.**, **step 1.**).

3.4.3. Digestion With RNase H

RNase H degrades RNA present in RNA–DNA hybrids. Thus, if DNA-based probes are used, the signal resulting from hybridization to RNA should be sensitive to RNase H treatment. Conversely, if RNA-based probes are used, the signal resulting from hybridization to DNA will be sensitive to RNase H. This can be a powerful tool for determining the location of RNA or DNA *(11,17,23)*.

1. Following the ISH washes (**Subheading 3.3.**, **step 4**), rinse cells with RNase H buffer containing 0.05% Tween-20. Aspirate the buffer and then add 8 µL of 75 U/mL RNase H in RNase H buffer to the top of the cover slip. Incubate the cover slip at 37°C in a moist chamber (*see* **Subheading 3.4.1.**, **step 2**) for 45 min.
2. Wash the cover slips three times for 10 min in 4X SSC at room temperature.
3. Proceed with the biotin detection (*see* **Subheading 3.3.**, **steps 5–12**).

4. Notes

1. The procedure for fixing and permeabilizing the cells to both preserve structure and allow access to macromolecular reagents is highly variable. We have found it necessary to titrate the amount of PFA used for fixation and usually get good results with 2–4% PFA solutions. We try to use as close to 4% PFA as possible in

order to maintain the best preservation of structure. Preextraction of the cells with 0.5% Triton X-100 in PBS for 30 s to 3 min on ice prior to fixation with PFA can help to increase permeability of the fixed cells, but can also result in loss of cytoplasmic RNA *(17)*. We have used a mild treatment with SDS-containing buffers *(7,11)* to allow greater access of probes to the cytoplasmic RNA. In this procedure, incubate cells for 5 min at room temperature in buffer containing 0.1% SDS, 100 mM Tris-HCl, pH 7.5, 150 mM NaCl, 12.5 mM ethylene diamine tetraacetic acid (EDTA) following fixation in 4% PFA. This incubation replaces the extraction with 0.5% Triton X-100 containing buffer (during **Subheading 3.2.**, **step 3**). It is advisable to titrate the amount of SDS in the solution. We have used a range between 0.05 and 0.2%. Detection of cytoplasmic RNA may require additional treatments with protease to remove the proteins and allow access of the probes to the RNA *(24)*. We have tried to avoid this step because we are specifically interested in looking at the localization of both proteins and RNA, but there may be circumstances in which protease treatments are desirable.

2. Regarding storage of cover slips following fixation: we have observed the best staining of RNA using cells that are fixed and permeabilized and then used directly for the ISH protocol. However, several laboratories obtain good results when fixed cells on cover slips are stored in 70% ethanol and then used for ISH within several months *(14,15,17,25)*.

3. Hybridization and washing conditions may require titration. We have done the hybridization at temperatures ranging from 37 to 55°C, and washing temperatures have ranged from room temperature to 42°C. Where viral RNA is being detected, the goal is to optimize the signal-to-noise ratio between infected and uninfected cells.

4. ISH is a relatively harsh procedure; we have encountered several instances in which an antibody no longer detects its antigen following the ISH protocol. Presumably, this is because ISH affects the epitope recognized by the antibody. It can be useful to test several immunological reagents against the protein of interest to find one that still efficiently recognizes the protein following ISH. It may also be helpful to perform the antibody immunodetection first followed by ISH *(26)*.

5. **Subheading 3.3.**, **steps 10** and **11** are optional, but I have found that the ISH signal remains stable for a longer period of time if the cells are postfixed following the ISH protocol.

Acknowledgments

I thank the members of my laboratory for their suggestions and support. This work was supported by the National Cancer Institute (grant CA82111).

References

1. Kuersten, S. and Goodwin, E. B. (2005) Linking nuclear mRNP assembly and cytoplasmic destiny. *Biol. Cell* **97,** 469–478.

2. Berget, S. M., Moore, C., and Sharp, P. A. (1977) Spliced segments at the 5' terminus of adenovirus 2 late mRNA. *Proc. Natl. Acad. Sci. USA* **74,** 3171–3175.
3. Chow, L. T., Gelinas, R. E., Broker. T. R., and Roberts, R. J. (1977) An amazing sequence arrangement at the 5' ends of adenovirus 2 messenger RNA. *Cell* **12,** 1–8.
4. Proudfoot, N. J., Furger, A., and Dye, M. J. (2002) Integrating mRNA processing with transcription. *Cell* **108,** 501–512.
5. Proudfoot, N. (2004) New perspectives on connecting messenger RNA 3' end formation to transcription. *Curr. Opin. Cell Biol.* **16,** 272–278.
6. Bridge, E. and Pettersson, U. (1996) Nuclear organization of adenovirus RNA biogenesis. *Exp. Cell Res.* **229,** 233–239.
7. Pombo, A., Ferreira, J., Bridge, E., and Carmo-Fonseca, M. (1994) Adenovirus replication and transcription sites are spatially separated in the nucleus of infected cells. *EMBO J.* **13,** 5075–5085.
8. Custodio, N., Carvalho, C., Condado, I., Antoniou, M., Blencowe, B. J., and Carmo-Fonseca, M. (2004) In vivo recruitment of exon junction complex proteins to transcription sites in mammalian cell nuclei. *RNA* **10,** 622–633.
9. Aspegren, A. and Bridge, E. (2002) Release of snRNP and RNA from transcription sites in adenovirus-infected cells. *Exp. Cell Res.* **276,** 273–283.
10. Aspegren, A., Rabino, C., and Bridge, E. (1998) Organization of splicing factors in adenovirus-infected cells reflects changes in gene expression during the early to late phase transition. *Exp. Cell Res.* **245,** 203–213.
11. Bridge, E., Riedel, K. U., Johansson, B. M., and Pettersson, U. (1996) Spliced exons of adenovirus late RNAs colocalize with snRNP in a specific nuclear domain. *J. Cell Biol.* **135,** 303–314.
12. Bridge, E., Mattsson, K., Aspegren, A., and Sengupta, A. (2003) Adenovirus early region 4 promotes the localization of splicing factors and viral RNA in late-phase interchromatin granule clusters. *Virology* **311,** 40–50.
13. Earnshaw, W. C. and Rattner, J. B. (1991) The use of autoantibodies in the study of nuclear and chromosomal organization. *Methods in Cell Biol.* **35,** 135–175.
14. Zhang, G., Taneja, K. L., Singer, R. H., and Green, M. R. (1994) Localization of pre-mRNA splicing in mammalian nuclei. *Nature* **372,** 809–812.
15. Zhang, G., Zapp, M. L., Yan, G., and Green, M. R. (1996) Localization of HIV-1 RNA in mammalian nuclei. *J. Cell Biol.* **135,** 9–18.
16. Gruenewald-Jahno, S., Keesey, J., Leous, M., van Miltonberg, R., and Schroeder, C., eds. (1996) *Nonradioactive In Situ Hybridization Application Manual,* 2nd ed. Boehringer Mannheim GmbH Biochemica, Mannheim, Germany, pp. 8–56.
17. Johnson, C. V., Singer, R. H., and Lawrence, J. B. (1991) Fluorescent detection of nuclear RNA and DNA: Implications for genome organization. *Methods Cell Biol.* **35,** 73–99.
18. Challberg, M. and Kelly, T. (1989) Animal virus DNA replication. *Ann. Rev. Biochem.* **58,** 671–717.
19. Puvion-Dutilleul, F. and Pichard, E. (1992) Segregation of viral double-stranded and single-stranded DNA molecules in nuclei of adenovirus infected cells as revealed by electron microscope *in situ* hybridization. *Biol. Cell* **76,** 139–150.

20. Puvion-Dutilleul, F. and Puvion, E. (1990) Analysis by in situ hybridization and autoradiography of sites of replication and storage of single- and double-stranded adenovirus type 5 DNA in lytically infected HeLa cells. *J. Struct. Biol.* **103,** 280–289.

21. Puvion-Dutilleul, F., Pédron, J., and Cajean-Feroldi, C. (1984) Identification of intranuclear structures containing the 72K DNA-binding protein of human adenovirus type 5. *Eur. J. Cell Biol.* **34,** 313–322.

22. Sambrook, J., Fritsch, E. F., and Maniatis, T. (1989) *Molecular Cloning. A Laboratory Manual,* 2nd ed. Cold Spring Harbor Laboratory Press, Cold Spring Harbor, NY.

23. Dirks, R. W., van de Rijke, F. M., Fujishita, S., van der Ploeg, M., and Raap, A. K. (1993) Methodologies for specific intron and exon RNA localization in cultured cells by haptenized and fluorochromized probes. *J. Cell Sci.* **104,** 1187–1197.

24. Jiménez-Garcia, L. F. and Spector, D. L. (1993) In vivo evidence that transcription and splicing are coordinated by a recruiting mechanism. *Cell* **73,** 47–59.

25. Xing, Y., Johnson, C. V., Dobner, P. R., and Lawrence, J. B. (1993) Higher level organization of individual gene transcription and RNA splicing. *Science* **259,** 1326–1329.

26. Jiménez-Garcia, L. F., Green, S. R., Mathews, M. B., and Spector, D. L. (1993) Organization of the double-stranded RNA-activated protein kinase DAI and virus-associated VA RNAI in adenovirus-2-infected HeLa cells. *J. Cell Sci.* **106,** 11–22.

6

Study of Nucleolar Localization of Adenovirus Core Proteins

David A. Matthews

Summary

This chapter describes the techniques used to study nucleolar-localized proteins. The chapter starts with cloning of viral proteins for expression in mammalian cells as fusion proteins to well-characterized tags such as enhanced green fluorescence protein (EGFP). This follows on to techniques for transient expression in mammalian cells and immunofluorescence techniques used to examine subcellular localization. Finally there is guidance on the types of antigens and metabolic features of the nucleolus that can be used as markers to confirm that the protein in question is indeed localized in the nucleolus and determine whether it affects gross rRNA synthesis.

Key Words: Adenovirus; nucleolus; enhanced green fluorescence protein; core; nucleus; expression; microscopy; rRNA.

1. Introduction

In recent years two things have become increasingly clear. First, the nucleolus is not merely a site for ribosomal biogenesis. Numerous reports are emerging that indicate the wide role of the nucleolus in the cell (*1,2*). Second, many viruses from a wide range of virus families either make proteins that are targeted to the nucleolus and/or viruses sequester nucleolar antigens and functions for their own purposes (*3*). With this in mind, this chapter will focus on some basic *in situ* techniques used to determine nucleolar localization signals and a simple method to examine rRNA synthesis *in situ*.

The use of adenovirus (Ad) as a model system for examining the role of the nucleolus has many advantages. First, the virus is safe to handle and is well characterized with an established system for introducing foreign genes into the viral genome (*4,5*). Second, the virus has been known for some time to affect

From: *Methods in Molecular Medicine, Vol. 131:*
Adenovirus Methods and Protocols, Second Edition, vol. 2:
Ad Proteins, RNA, Lifecycle, Host Interactions, and Phylogenetics
Edited by: W. S. M. Wold and A. E. Tollefson © Humana Press Inc., Totowa, NJ

rRNA biogenesis *(6)*. Third, the virus makes several proteins that are targeted to the nucleolus *(7–12)*. Finally, the virus affects nucleolar antigen distribution and utilizes nucleolar antigens to facilitate viral replication *(9,12–17)*. Thus, this virus is a very versatile model with which to examine the role of the nucleolus in infected cells.

2. Materials

1. Mammalian expression plasmid such as pcJMA2egfp *(18)*. This plasmid utilizes a cytomegalovirus (CMV) promoter to direct expression and allows proteins of interest to be cloned N-terminal of the enhanced green fluorescent protein (EGFP) open reading frame (ORF). However, any commercially available plasmid that performs a similar function can be used. Examples of such plasmids are pAcGFP1-N1 and pEGFP-N1, by Clontech (www.clontech.com).
2. Chemically competent *E. coli* strain DH5α; from Invitrogen (www.invitrogen.com), LB media, and kanamycin (final strength of 50 μg/mL).
3. Viral DNA. In this case the plasmid pFG140 was used, which can be bought from Microbix (www.microbix.com).
4. Restriction enzymes from New England Biolabs (www.neb.com) and T4 DNA ligase from Roche (www.roche-applied-science.com).
5. Agarose gel electrophoresis equipment, low-melting-point agarose from Invitrogen, and TBE buffer.
6. Gel extraction kits to purify DNA from agarose gels (e.g., Qiaquick gel extraction kit from Qiagen; www.qiagen.com).
7. Oligonucleotide primers, thermostable DNA polymerase such as *pfx* polymerase, and dNTP mixtures, all available from Invitrogen.
8. Midi prep DNA extraction kit from Qiagen.
9. HeLa cells, available from ECACC (www.ecacc.org.uk) or ATCC (www.atcc.org). Cell culture medium Dulbecco's modified Eagle's medium (DMEM) with Glutamax, fetal calf serum (FCS), penicillin/streptomycin solution, trypsin/ethylene diamine tetraacetic acid (EDTA) solution—all available from GibcoBRL (www.invitrogen.com). You will also need tissue culture plasticware and access to tissue culture class II cabinets and CO_2-gassed 37°C incubators.
10. Mammalian cell transfection reagents such as Lipofectamine 2000, available from Invitrogen.
11. Access to a fluorescent microscope, preferably a laser con focal microscope.
12. Jeweller's forceps or other fine-point forceps. Glass cover slips (round, diameter 9–16 mm) and glass slides.
13. Monoclonal antibody to nucleolar antigen Nucleolin/C23 (cat. no. sc-8031) or to B23 (sc-32256) from Santa Cruz Biotechnology (www.scbt.com).
14. Secondary antiserum such as AlexaFluor 568 goat anti-mouse IgG antibody from Invitrogen.
15. Phosphate-buffered saline (PBS) supplemented with formaldehyde (final concentration 4% v/v); store refrigerated. PBS supplemented with Triton X-100 (final concentration 1% v/v).

16. 5-Fluorouridine (5-FU) available from Sigma (cat. no. F 5130) and mouse mono-clonal anti-BrdU antibody, also from Sigma (cat. no. B 2531).

3. Methods

The methods outlined here will cover four main steps:

1. The PCR amplification and cloning of the viral gene of interest.
2. Transfection into human cells.
3. Examination of the location of the expressed protein using fluorescent techniques.
4. Assessment of the viral protein location with respect to rRNA synthesis and key nucleolar antigens.

While the approaches mainly concentrate on EGFP-tagged proteins, it is normal to also make fusion proteins with other tags such as MYC, FLAG, or HA. This is to confirm that the presence of EGFP does not influence the target-ing in some manner (*see* **Note 1**).

3.1. Cloning of Viral ORFs Into pAcGFP1-N1

This plasmid is typical of many commercially available plasmids for expression of proteins in mammalian cells. It contains a bacterial origin of replication for maintaining the plasmid at high copy number in bacterial cells and an antibiotic resistance marker for selection of transformed bacteria. When cloning ORFs into this vector it is important to keep the start codon, ATG, remove the stop codon from the ORF, and ensure that the cloning procedure maintains the reading frame between the ORF and the fluorescent protein. This is usually achieved by careful design of the polymerase chain reaction-amplifi-cation primers (*see* **Note 2**). For the cloning of Ad proteins V, pre-VII, and preMu, the plasmid pFG140 is used as a source of viral DNA.

1. While the target DNA is being amplified by PCR, digest approx 5 µg of vector DNA using appropriate enzymes.
2. Once the target gene has been amplified, separate the product on a low-melting-point agarose gel alongside the digested vector DNA using standard protocols *(19)*. Excise the two DNA samples from the gel and purify them separately using the gel extraction kit.
3. Digest the entire PCR product with appropriate restriction enzymes and purify the DNA from the enzymes using the gel extraction kit (there is no need to run the DNA on a second gel).
4. Ligate the digested product and vector DNA using a ratio of 3:1 and incubate on the bench for 2 h at room temperature (*see* **Note 3**). Next transform into compe-tent DH5α using the manufacturer's recommendations.
5. Select colonies for overnight growth and miniprep the plasmid DNA using stan-dard protocols *(19)*.
6. Screen potential minipreps for recombinant plasmids using appropriate restric-tion enzymes, again using standard protocols *(19)*.

7. Once a suitable clone has been identified, grow a medium-scale culture and isolate high-quality DNA using a commercially available miniprep kit from Qiagen.

3.2. Transfection of Plasmid DNA Into Mammalian Cells

Several commercial formulations are available. All rely on high-quality DNA, so it is important to check the quality and quantity of the DNA isolated by the Qiagen Miniprep kit by measuring the A_{260}/A_{280} ratio *(19)*. In the experiments with proteins V, pre-VII/VII, and preMu/Mu, we also infected the cells with Ad. We found this improved the transfection rate somewhat *(7–9)*.

1. Harvest all the HeLa cells from a 75-cm² culture dish with trypsin.
2. Resuspend the cells in 36 mL of growth medium.
3. Place a glass cover slip (diameter 9–16 mm) into ethanol and dab off excess ethanol onto a tissue; place the cover slip into a well on a 6-well dish. Repeat until you have enough cover slips for your needs, using only one cover slip per well.
4. Add 2 mL of the cells to each well. Thus, one 75-cm² culture vessel can seed three 6-well dishes. Shake well, ensuring that the cover slip does not float on top of the medium, and incubate overnight in a cell culture incubator at 37°C.
5. The next day, transfect 0.5 µg of plasmid DNA according to manufacturer's guidelines (*see* **Note 4**).
6. Once the transfection reagent has been added, Ad can be added at a multiplicity of infection (MOI) of about 0.5 PFU/cell, if required.
7. After 16–24 h the fluorescent fusion proteins should be clearly visible using an inverted fluorescent microscope.

3.3. Fluorescence Microscopy and Immunofluorescence Technique

There are many approaches to immunofluorescence; the one we use routinely is formaldehyde to fix the proteins followed by Triton X-100 to permeabilize the membranes. At this stage it is important to examine whether the nucleolus is indeed the site of localization rather than the protein forming an inclusion body in the nucleus of the cell. There are two complementary approaches: first, using phase-contrast to determine if the proteins are localized in bodies that resemble the nucleolus, and second, using antiserum to known nucleolar antigens. In addition, it is a good idea to include a reliable marker for the nucleus such as diamidino phenylindole (DAPI), which intercalates with dsDNA and fluoresces blue when exposed to UV irradiation. It is essential to remember here that HeLa cells often have multiple nucleoli; as such your protein of interest should be present in all the nucleoli you see in an individual cell with phase-contrast or immunofluorescence. These points are illustrated in **Fig. 1** and in papers we have published on this subject *(7–9,20)*.

1. Remove growth medium and replace with 2 mL formaldehyde/PBS (4% v/v, 4°C) per well, and leave for 5 min.

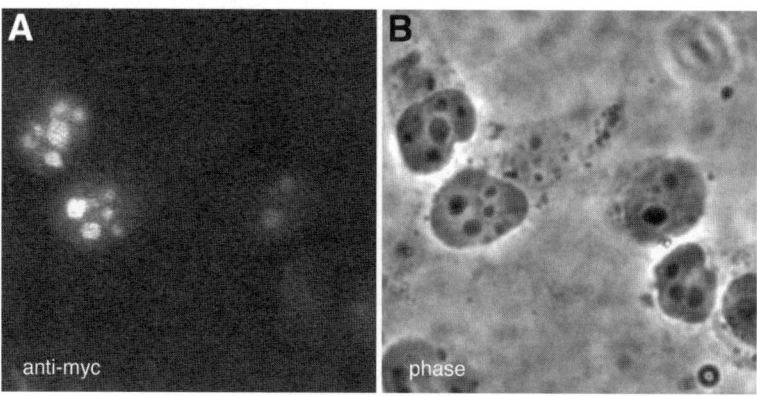

Fig. 1. Nucleolar localization of adenovirus preMu as shown by comparison with phase contrast. In (**A**), the localization of preMu is detected via a myc tag and anti-myc antiserum. In (**B**), the corresponding phase-contrast image is shown, revealing the nucleoli as characteristically dark round occlusions within the cell nucleus. Also note that the viral protein is present in every nucleolus within that nucleus.

2. Wash once with PBS, remove the PBS, and replace with 2 mL Triton X-100 (1% v/v in PBS) per well. Again, this should be left for 5 min before removing and washing the cells once with PBS.
3. Finally incubate the cells in a blocking solution. We normally use FCS/PBS (10% v/v), but dried skim milk in PBS can be used as well (10% w/v). The cells can be left overnight at 4°C or incubated at room temperature on a platform shaker with gentle agitation for at least 30 min.
4. Make up the primary antibody solution in the FCS/PBS blocking solution. We normally use antibodies from commercial suppliers at a concentration of ~4 µg/mL (*see* **Note 5**).
5. When the cover slips have been blocked, remove them from the well using watchmaker's forceps, keeping the side with the cells uppermost. Carefully dry the underside of the cover slip with a clean tissue and place the slide, cell-side up, onto a clean Eppendorf rack (*see* **Note 6**). Pipet 100 µL of primary antibody solution onto the surface of the cover slip, and leave it at room temperature for about 30 min.
6. After incubation, pick up the cover slip and tip off the antibody solution onto a tissue. Place the cover slip (cell-side up) back into the 6-well dish with the blocking solution and incubate at room temperature with gentle shaking for about 20 min. After this, repeat the procedure using the appropriate secondary antibody. Remember to use fresh blocking solution to wash off the secondary antibody (*see* **Note 7**).
7. While the secondary antibody is being washed off, spot 10 µL of mounting medium onto a glass slide roughly in the middle. Vectashield with DAPI is one

we use routinely in the laboratory because it will clearly mark the nuclei as well as preventing bleaching (*see* **Note 8**).

8. Once the secondary is washed off, carefully remove the slide and dry the back of the cover slip; invert it so that the side with the cells on it is placed face-down onto the spot of mounting medium.
9. View immediately on a suitable microscope. The slides can be stored at 4°C overnight and viewed first thing in the morning. However, the sooner images can be taken, the better quality they will be. Storage for longer than 2 d is not recommended.
10. It is also important to assess the integrity of the EGFP fusion protein by harvesting a sample of transfected cells and assessing the expression by Western blotting using standard techniques *(19)* and EGFP-specific antiserum. This ensures that the fluorescent signal is coming primarily from the full-length protein.

3.4. rRNA Synthesis Assessment

The method outlined here is simple and reliable *(21)*. All RNA polymerase activity is detected, but in a normal cell the bulk of RNA synthesis is rRNA.

1. Prepare cells for immunofluorescence as described in **Subheading 3.2.**
2. Prepare a stock of 10 m*M* 5-FU in PBS, store at –20°C in aliquots.
3. On the day you are ready to fix the cells for immunofluorescence add 5-FU to the culture medium to a final concentration of 2 m*M* and incubate the cells at 37°C in the cell culture incubator for a range of times (*see* **Note 9**).
4. After the desired time, fix and permeabilize the cells for immunofluorescence as in **Subheading 3.3.** In this case you will use the anti-BrdU monoclonal antibody instead of the anti-nucleolin antibody. Unlike detection of Br-dUTP incorporation into DNA, the cells do not require any special treatment in order to detect the Br-UTP incorporated into RNA in the cell.

4. Notes

1. The choice to use N-terminal or C-terminal fusions can be a difficult one. However, in the case of Ad, it was simplified by the fact that protein VII is made as a precursor that is cleaved at the N-terminus. Because it was important to compare the localization properties of pre-VII and mature VII side by side, I decided to make all my core protein fusions the same way, i.e., placing the EGFP motif after the C terminus of the viral protein. In this manner a more straightforward comparison between pre-VII and VII was possible, a fact made all the more important by the discovery that processing from pre-VII to VII made a marked difference to the localization of this protein *(7)*. However, it should be always borne in mind that any tag at any location could, theoretically, have dramatic implications for subcellular localization.
2. The PCR-amplification stage can be difficult. Problems include mispriming of one or both primers and low or no yield. In particular, GC-rich targets such as the protein VII open reading frame can be a real problem. When PCR-amplifying

protein VII, we found that redesigning the primers was often necessary, and we relied heavily on the PCR enhancer buffer supplied by the manufacturer. This enhancer buffer is designed to reduce the melting temperature of DNA, and we found it particularly useful.

3. You should aliquot the reaction buffer into 10-µL amounts and freeze the aliquots. For each ligation, thaw one aliquot of ligation buffer and throw away any you do not use, thus ensuring the buffer is only ever thawed twice before use.

4. We frequently set up the transfection mixture in PBS and added the mixture to cells growing in normal medium. Although this reduces the efficiency somewhat, very high levels of transfection were not needed, and we were still able to transfect more than 20% of the HeLa cells.

5. For other antibodies we normally start at a dilution of 1:50; alternatively, for antibodies we have used in Western blots, a good rule of thumb is to use the antibody at 10 times the normal working concentration for Western blotting (i.e., 1:50 for immunofluorescence corresponds to 1:500 for Western blotting).

6. Placing the cover slip so that it is in between the Eppendorf holes makes it easy to subsequently grip the cover slip. Ideally the rack will also have a lid so that you can put the lid over the cover slips, reducing evaporation and preventing unwanted objects from falling onto the cover slips.

7. It is a good idea to incubate secondary antibodies in the dark. During the washing stage you can wrap the 6-well dish in aluminum baking foil.

8. Another good reason for using DAPI routinely is that it guards against mycoplasma contamination *(22)*. Mycoplasma DNA will also stain, but the mycoplasma resides in the cytoplasm. Thus, contaminated cells show a clearly stained nucleus with numerous speckles in the cytoplasm. If the slide is a little dirty, then there will be speckles all over the surface of the slide. However, in mycoplasma-contaminated cells the speckles are clearly confined to the cytoplasm. While this technique is not the most sensitive method of detecting mycoplasma, it is simple and will, at least, detect gross contamination.

9. The length of time will vary from cell line to cell line. However, 10–20 min should be enough in most cases, and labeling for up to 2 h has been tried, although the level of signal is too high to be useful.

References

1. Pederson, T. (1998) The plurifunctional nucleolus. *Nucleic Acids Res.* **26,** 3871–3876.
2. Rubbi, C. P. and Milner, J. (2003) Disruption of the nucleolus mediates stabilization of p53 in response to DNA damage and other stresses. *EMBO J.* **22,** 6068–6077.
3. Hiscox, J. A. (2002) The nucleolus—a gateway to viral infection? *Arch. Virol.* **147,** 1077–1089.
4. Shenk, T. (2001) Adenoviridae: the viruses and their replication, in *Fields Virology* (Knipe, D. M., Howley, P. M. eds.), Vol. 2, Lippincott-Raven, Philadelphia, pp. 2265–2299.

5. Russell, W. C. (2000) Update on adenovirus and its vectors. *J. Gen. Virol.* **81,** 2573–2604.

6. Castiglia, C. L. and Flint, S. J. (1983) Effects of adenovirus infection on rRNA synthesis and maturation in HeLa cells. *Mol. Cell. Biol.* **3,** 662–671.

7. Lee, T. W., Blair, G. E., and Matthews, D. A. (2003) Adenovirus core protein VII contains distinct sequences that mediate targeting to the nucleus and nucleolus, and colocalization with human chromosomes. *J. Gen. Virol.* **84,** 3423–3428.

8. Lee, T. W., Lawrence, F. J., Dauksaite, V., Akusjarvi, G., Blair, G. E., and Matthews, D. A. (2004) Precursor of human adenovirus core polypeptide Mu targets the nucleolus and modulates the expression of E2 proteins. *J. Gen. Virol.* **85,** 185–196.

9. Matthews, D. A. (2001) Adenovirus protein V induces redistribution of nucleolin and B23 from nucleolus to cytoplasm. *J. Virol.* **75,** 1031–1038.

10. Matthews, D. A. and Russell, W. C. (1998) Adenovirus core protein V is delivered by the invading virus to the nucleus of the infected cell and later in infection is associated with nucleoli. *J. Gen. Virol.* **79 ,** 1671–1675.

11. Lutz, P., Puvion-Dutilleul, F., Lutz, Y., and Kedinger, C. (1996) Nucleoplasmic and nucleolar distribution of the adenovirus IVa2 gene product. *J. Virol.* **70,** 3449–3460.

12. Miron, M. J., Gallouzi, I. E., Lavoie, J. N., and Branton, P. E. (2004) Nuclear localization of the adenovirus E4orf4 protein is mediated through an arginine-rich motif and correlates with cell death. *Oncogene* **23,** 7458–7468.

13. Puvion-Dutilleul, F. and Christensen, M. E. (1993) Alterations of fibrillarin distribution and nucleolar ultrastructure induced by adenovirus infection. *Eur. J. Cell Biol.* **61,** 168–176.

14. Russell, W. C. and Matthews, D. A. (2003) Nuclear perturbations following adenovirus infection. *Curr. Top. Microbiol. Immunol.* **272,** 399–413.

15. Okuwaki, M., Iwamatsu, A., Tsujimoto, M., and Nagata, K. (2001) Identification of nucleophosmin/B23, an acidic nucleolar protein, as a stimulatory factor for in vitro replication of adenovirus DNA complexed with viral basic core proteins. *J. Mol. Biol.* **311,** 41–55.

16. Walton, T. H., Moen, P. T., Jr., Fox, E., and Bodnar, J. W. (1989) Interactions of minute virus of mice and adenovirus with host nucleoli. *J. Virol.* **63,** 3651–3660.

17. Rodrigues, S. H., Silva, N. P., Delicio, L. R., Granato, C., and Andrade, L. E. (1996) The behavior of the coiled body in cells infected with adenovirus in vitro. *Mol. Biol. Rep.* **23,** 183–189.

18. Askham, J. M., Moncur, P., Markham, A. F., and Morrison, E. E. (2000) Regulation and function of the interaction between the APC tumour suppressor protein and EB1. *Oncogene* **19,** 1950–1958.

19. Sambrook, J. and Russell, D. W. (2001) *Molecular Cloning: A Laboratory Manual,* 3rd ed., Cold Spring Harbor Laboratory Press, Cold Spring Harbor, New York.

20. Lee, T. W., Matthews, D. A., and Blair, G. E. (2005) Novel molecular approaches to cystic fibrosis gene therapy. *Biochem. J.* **387,** 1–15.

21. Boisvert, F. M., Hendzel, M. J., and Bazett-Jones, D. P. (2000) Promyelocytic leukemia (PML) nuclear bodies are protein structures that do not accumulate RNA. *J. Cell Biol.* **148,** 283–292.
22. Russell, W. C., Newman, C., and Williamson, D. H. (1975) A simple cytochemical technique for demonstration of DNA in cells infected with mycoplasmas and viruses. *Nature* **253,** 461–462.

7

Analysis of Adenovirus Infections in Synchronized Cells

David A. Ornelles, Robin N. Broughton-Shepard, and Felicia D. Goodrum

Summary

Adenoviruses (Ads) are small DNA tumor viruses that have played a pivotal role in under-standing eukaryotic cell biology and viral oncogenesis. Among other cellular pathways, Ad usurps cell cycle progression following infection. Likewise, progression of the viral infection is influenced by the host cell cycle. We describe here methods developed for synchronizing dividing cell populations and for analysis of cell cycle synchrony by flow cytometry. Further-more, three methods used to evaluate the outcome of Ad infection in synchronized cell popula-tions are described. These include two assays for infectious centers and an assay for analyzing production of progeny virus by transmission electron microscopy. These methods have been used to demonstrate that Ads that fail to direct synthesis of the E1B 55-kDa or E4orf6 proteins replicate most effectively upon infecting cells in S phase.

Key Words: Adenovirus; cell cycle; synchronization; mitotic shake; S phase; G1, infec-tious centers assay.

1. Introduction

Adenovirus (Ad) is well-known for its ability to manipulate progression of the cell cycle (reviewed in **refs. *1–3***). However, the cell cycle can influence the outcome of an Ad infection. Evidence for this relationship was obtained from studies of Ad mutants deleted of the genes for E1B 55-kDa (E1B-55K) protein or E4 open reading frame 6 (E4orf6). When individual cells infected with these mutant viruses were evaluated by electron microscopy or by assays for infec-tious centers, approximately one in five cells contained progeny virus *(4–6)*. Because approximately the same fraction of cells were in S phase at any time, the hypothesis that the E1B-55K- and E4orf6-mutant viruses replicated prefer-entially in S-phase cells was developed and confirmed. For these experiments it was necessary to obtain synchronously dividing cells that could be infected

From: *Methods in Molecular Medicine, Vol. 131:*
Adenovirus Methods and Protocols, Second Edition, vol. 2:
Ad Proteins, RNA, Lifecycle, Host Interactions, and Phylogenetics
Edited by: W. S. M. Wold and A. E. Tollefson © Humana Press Inc., Totowa, NJ

as the cells traversed the cell cycle as a cohort. The method that was developed to generate synchronously dividing cells is described here.

In addition to the method of synchrony, methods to determine the stage of the cell cycle by DNA content using propidium iodide staining and flow cytometry will be described. Three methods to evaluate the outcome of an Ad infection at the individual cell level will be described; these include a limiting dilution assay for productively infected cells (infectious centers), an assay for infectious centers based on agarose overlays, and a basic method to prepare cells for analysis by transmission electron microscopy. These assays have application to asynchronously infected cells as well and can be adapted for other lytic viruses.

Synchronously dividing cells are generated by a modification of the method of Cao et al. *(7)*. Naturally occurring mitotic cells are selected by mechanical agitation (i.e., mitotic shake) and further enriched by their ability to rapidly reattach to the substrate. The mitotic cells are allowed to cycle through G1 in the presence of hydroxyurea until reaching the border between G1 and S phase. The amount of hydroxyurea is adjusted for each cell line to kill any cells that were initially in S phase while allowing G1 cells to progress to the G1/S border. The drug is then removed, and the cells immediately enter the cell cycle at S phase in a synchronous fashion. This produces a population of cells that divide as a synchronized cohort for two or more cycles. These are precisely the requirements advocated by Cooper as being the most appropriate for the study of the cell cycle *(8)*. The reader is encouraged to consider the exchange on this subject between Cooper *(8,9)* and Spellman and Sherlock *(10,11)*.

An important criterion during the development of this method was to generate minimally perturbed cells with respect to Ad infection. The critical requirement was that the yield of Ad from nonsynchronized cells was the same as that from a reconstituted population of synchronized cells composed of comparable proportions of G1 and S-phase cells. This can be seen in the representative results in **Fig. 1** in which the yield of the E1B-55K-mutant virus *dl*338 from cells that were infected as an asynchronously dividing culture is the same as it would be from a mixture of 70% G1 cells and 25% S-phase cells. Many procedures have been described that can generate highly synchronized populations of cells. Alternative methods evaluated included the use of mitotic blockers (nocadazole, colcemid, and colchicine), inhibitors of DNA synthesis (aphidocolin, single- and double-thymidine arrest), and nutrient deprivation (leucine and serum starvation). In addition, several combinations of these treatments were applied, such as the use of mitotic blockers in combination with mitotic shake. Although these approaches produced highly synchronized cultures of viable cells, the ability of these cells to support an Ad infection was severely compromised. For example, the yield of virus from cells synchro-

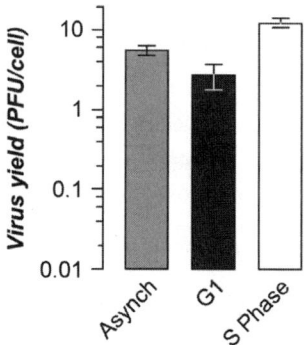

Fig. 1. Synchronized HeLa cells are not compromised in their ability to support an adenovirus infection. HeLa cells were synchronized by the method described in the text and then infected with the E1B-55K-mutant virus *dl*338 as the cells passed through either G1 or S phase. A comparable number of asynchronously dividing HeLa cells (Asynch) were also infected with *dl*338 and the yield of virus determined by plaque assay after 3 d of infection. The results are expressed as the average PFU per infected cell obtained by three to five independent infections.

nized by a double-thymidine block was more than 2 orders of magnitude less than that from untreated cells.

2. Materials

2.1. Cell Synchronization

1. Exponentially growing adherent cells.
2. Tissue culture plasticware and equipment to maintain adherent human cultured cells.
3. Cell culture medium, serum, and supplements.
4. 75-cm^2 plug-seal T-flasks (such as BD Falcon product no. 353135).
5. Phosphate-buffered saline (PBS): 137 mM NaCl, 2 mM KCl, 1.76 mM KH$_2$PO$_4$, 10 mM Na$_2$HPO$_4$. This solution can be prepared in large quantities and then sterilized by autoclaving and stored at 4°C indefinitely. Alternatively, a 10-fold concentrated stock can be prepared, sterilized by autoclaving, and diluted with sterile water as needed.
6. Hydroxyurea: prepare 1 M stock in sterile water. Store in single-used aliquots at –20°C for up to 1 yr.
7. Polypropylene 15- and 50-mL conical centrifuge tubes.
8. Hemacytometer.

2.2. Ad Infection

1. Infection medium: sterile PBS supplemented with 0.2 mM MgCl$_2$, 0.02 mM CaCl$_2$, and 2% normal serum. Store at 4°C for up to 4 mo.

2.3. DNA Profile Measured by Flow Cytometry

1. Flow cytometer equipped with an excitation light source and filters appropriate for analyzing propidium iodide fluorescence.
2. Ethanol, ice-cold 70% (v/v) solution in water.
3. 10X Propidium iodide stock solution: 1 M NaCl, 0.36 M sodium citrate, 0.5 mg per mL propidium iodide, 6% (w/v) Nonidet P-40 (or Polydet P-40). Filter through a 0.22-μm filter and store protected from light at 4°C for up to 1 yr. Propidium iodide is a mutagen and probably a carcinogen. Propidium iodide should be handled with care and disposed of according to local regulations for carcinogens.
4. 50X RNase (DNase-free): 2 mg per mL pancreatic RNase in water. Dissolve RNase at 2 mg per mL and incubate in a boiling water bath for 5 min to denature contaminating enzymes such as DNase. Remove the RNase from the heat and allow the RNase to refold by slowly cooling to room temperature. Dispense into small aliquots (0.2 mL) and store at –20°C. The RNase is stable for many years stored in this manner. Aliquots of RNase A can be thawed carefully and refrozen several times with no loss of potency.
5. Propidium iodide and RNase working solution: immediately before use, mix 44 vol of water, 5 vol of 10X propidium iodide stock solution, 1 vol of 2 mg per mL DNase-free RNase A. Keep the solution at 4°C protected from direct light.
6. Disposable syringe fitted with a 27.5-gage needle or a fine-mesh filter.

2.4. Limiting Dilution Assay for Infectious Centers

1. Sterile 0.2-μm filter apparatus suitable for serum-containing tissue culture medium.
2. Conditioned medium: establish a dense culture of cells in fresh cell growth medium and allow the cells to grow for 24 h. Harvest the medium and centrifuge at 1000g for 10 min at room temperature to remove cells and debris. Filter the supernatant fluid through a 0.2-μm filter. Store protected from light at 4°C for up to 2 wk before use.
3. Sterile 5-mL round bottom culture tubes to prepare serial dilutions of infected cells.
4. Sterile 96-well cell culture plates.
5. Eight- or 12-channel pipet device.

2.5. Agarose Overlay Assay for Infectious Centers

1. Tissue-culture-grade sterile water.
2. Low-melting-temperature agarose for cell suspension. Prepare a 1.8% (w/v) suspension of cell-culture-qualified agarose such as NuSieve GTG Agarose (Cambrex) or Type VII agarose (Sigma) in tissue-culture-grade water, heat to dissolve, dispense into 36-mL working aliquots, sterilize by autoclaving, and store at room temperature.

3. Agarose for plaque assays such as SeakemME Agar (Cambrex). Prepare a 1.8% (w/v) suspension in tissue-culture-grade water, autoclave to sterilize, and store in working aliquots at room temperature.
4. 5X Concentrated serum-free cell culture medium without antibiotics, without $NaHCO_3$.
5. Neutral Red: 0.33% (w/v) Neutral Red dye in tissue culture-grade sterile water.
6. Cell dilution agar: immediately before use, melt 1.8% low-melting-temperature agarose for cell suspension. Combine 4 parts molten agarose with 1 part 5X concentrated serum-free medium. Maintain in a molten state at 42°C.
7. First agar overlay: melt 36 mL 1.8% agarose for plaque assays and maintain at 45–55°C. In a separate sterile container, combine 30 mL sterile tissue-culture-grade water, 20 mL 5X concentrated serum-free medium, 10 mL 7.5% (w/v) $NaHCO_3$, and 4 mL serum. Maintain the media components at 37°C. Immediately before use, combine the molten agarose and media components.
8. Second agar overlay: melt 36 mL 1.8% agarose for plaque assays and maintain at 45–55°C. In a separate sterile container, combine 37 mL tissue culture grade water, 20 mL 5X concentrated serum-free medium, 5 mL 7.5% (w/v) $NaHCO_3$, and 2 mL serum. Maintain the media components at 37°C. Immediately before use, combine the molten agar and media components.
9. Neutral Red agar overlay: melt 36 mL 1.8% agarose for plaque assays and maintain at 45–55°C. In a separate sterile container, combine 35 mL sterile tissue-culture-grade water, 20 mL 5X concentrated serum-free medium, 5 mL 7.5% (w/v) $NaHCO_3$, 2 mL 0.33% Neutral Red solution, and 2 mL serum. Maintain the media components with Neutral Red at 37°C. Immediately before use, combine the molten agar and media components with Neutral Red.

2.6. Transmission Electron Microscopy for Viral Particles

1. Plastic transfer pipets, cell-scraping devices.
2. Sodium cacodylate, 0.2 M solution in deionized water. Sodium cacodylate, $(CH_3)_2AsO_2Na\cdot3H_2O$, is an arsenical compound that is an environmental hazard and is moderately toxic.
3. Glutaraldehyde, 4–25% (w/v) stock solution. Glutaraldehyde can be purchased in small aliquots stored under an inert gas. Prepare a 2.5% (w/v) solution in 0.1 M sodium cacodylate immediately before use. Glutaraldehyde is a hazardous compound that must be used with proper ventilation in a fume hood. Discard glutaraldehyde waste in accordance with local regulations.

3. Methods

3.1. Cell Synchronization

This procedure begins with exponentially growing adherent cells in 10-cm tissue culture plates. The cells are harvested and transferred to sufficient 75-cm^2 T-flasks in order to establish monolayers of cells that are approximately

Table 1
Sample Schedule to Concurrently Synchronize Cells to G1 and S Phase for Infection With Adenovirus

Procedure	Time G1	Time S phase	Day
Place cells in T-flasks for G1	5 PM		0
Mitotic shake for G1 cells	6 AM		1
Attach G1 cells in HU	7 AM		1
Replace HU medium (G1), cycle to G1/S	8 AM		1
Place cells in T-flasks for S phase		9 AM	1
Release HU block (G1), cycle to G1	7 PM		1
Mitotic shake for S phase		9 PM	1
Attach S-phase cells in HU		10 PM	1
Replace HU medium (S), cycle to G1/S		11 PM	1
Release S-phase HU block		10 AM	2
Infect G1 and S-phase cells	10 AM	10 AM	2

two-thirds confluent at the time of the mitotic shake. For HeLa cells, approx 10^7 cells can be placed into two 75-cm^2 T-flasks, each of which will generate one 60-mm plate of cells containing 4×10^5 cells following mitotic shake. A typical schedule showing two concurrent synchronizations in order to prepare S-phase and G1 cells that are infected at the same time is shown in **Table 1**.

3.1.1. Place Cells in T-Flasks for Mitotic Shake

1. Prepare approximately two 75-cm^2 plug-seal T-flasks with 10 mL of growth medium for each 10-cm plate of cells. Cells to be seeded for mitotic shake should be exponentially growing.
2. Replace the growth medium in each 10-cm dish with warm PBS to rinse the cells and then replace the PBS with 1 mL Trypsin/ethylene diamine tetraacetic acid (EDTA) solution. Incubate at room temperature with periodic agitation to detach the cells.
3. Once the cells begin to detach freely, stream a small volume of growth medium over the surface to dislodge cells. Collect the detached cells in 50-mL conical centrifuge tubes.
4. Recover the cells by centrifugation at 100*g* for 5 min at room temperature, and resuspend in sufficient growth medium to provide 1 mL of suspended cells per T-flask.
5. Add 1 mL of the cell suspension to each T-flask containing 10 mL of medium. Distribute the cells uniformly throughout the T-flask. Ensure that the cap is sufficiently loose to permit free gas exchange, and return the T-flasks to appropriate culture conditions.

6. Allow the cells sufficient time to attach to the substrate, spread, and resume progression through the cell cycle. For HeLa cells, 12–14 h is an appropriate length of time. Once attached, the cells should be approximately two-thirds confluent, with no cell–cell contact (*see* **Note 1**).

3.1.2. Mitotic Shake

1. Prepare and warm complete growth medium with 2 m*M* hydroxyurea (*see* **Note 2**).
2. In sets of five to eight, remove and discard 7 mL of medium from each T-flask prepared the previous day. This step leaves 4 mL of medium with which to dislodge and recover the mitotic cells.
3. Tighten the cap on each T-flask, and then crisply tap the T-flask on a hard surface three times on each long side to dislodge the mitotic cells. Monitor the efficacy of the mitotic shake with a microscope and adjust the strength of the tap so as to dislodge only mitotic cells. Reduce the vigor of each tap if many interphase cells are dislodged. Increase the vigor only if *many* mitotic cells remain attached (*see* **Note 3**).
4. Collect the fragile mitotic cells in 50-mL conical centrifuge tubes.
5. Repeat **steps 2** through **4** as required to process the remaining T-flasks.
6. Gently recover the mitotic cells by centrifugation at 100*g* for 5 min at room temperature.
7. Resuspend the mitotic cells in a small volume of growth medium with 2 m*M* hydroxyurea, and determine the number of cells with a hemacytometer.
8. Dilute the mitotic cells to the desired final concentration in normal growth medium with 2 m*M* hydroxyurea. Prepare 1.5×10^5 cells per mL for G1 studies, 2.5×10^5 cells per mL for S-phase studies (*see* **Note 4**).
9. Distribute the mitotic cells in normal growth medium with 2 m*M* hydroxyurea into the desired culture vessels and allow the mitotic cells to attach. If possible, save a portion of the mitotic cells to analyze by flow cytometry as described in **Subheading 3.3.**
10. Allow the mitotic cells to attach under normal growth conditions for exactly 1 h. Only mitotic cells should be able to reattach during this brief incubation. Avoid stacking plates of cells to ensure that each plate is exposed to an identical local environment during this incubation.
11. After 1 h, gently swirl the 2 m*M* hydroxyurea medium and then remove the medium by aspiration to discard the nonattached cells. Although the mitotic cells will not have spread, they will be difficult to dislodge. Take care to treat each plate identically.
12. Replace the discarded medium with fresh medium containing 2 m*M* hydroxyurea and return the cells to normal growth conditions for an additional 11–13 h. The length of this incubation should be slightly longer (by 2–3 h) than the time required for the mitotic cells to reach S phase (*see* **Note 5**).

3.1.3. Release From Hydroxyurea

1. Remove the 2 m*M* hydroxyurea medium and replace with warm PBS. Rinse the cells once or twice more with warm PBS to remove hydroxyurea from the culture dish.
2. Replace the final PBS rinse with warm complete growth medium lacking hydroxyurea.
3. The cells begin to enter S phase and will likely progress through at least two synchronous divisions. They can be infected at appropriate points in the cell cycle as described in **Subheading 3.2.** The progression through the cell cycle can be measured by DNA content using propidium iodide and flow cytometry, as described below in **Subheading 3.3.**

3.2. Ad Infection

A simple procedure for infecting adherent cells with Ad is described. The standard volume is the volume of growth medium that maintains a constant ratio of volume to surface area among culture vessels. For the experiments described here, 1.70 mL of growth medium is used for each 10-cm^2 of surface area.

1. Harvest a culture of cells prepared in parallel with the cells that are to be infected. Determine the number of cells to be infected, and process these cells for propidium iodide staining to determine the stage of the cell cycle as described in **Subheading 3.3.**
2. Dilute the requisite amount of virus to achieve the desired multiplicity of infection in one-tenth the standard volume of infection medium.
3. Thoroughly aspirate the growth medium on each plate to be infected and then replace with the diluted virus in one-fifth to one-tenth of the standard volume.
4. Incubate for exactly 1 h at 37°C with continuous rocking (*see* **Note 6**).
5. Remove the virus and infection medium, replace with the standard volume of normal growth medium, and return to normal growth conditions.

3.3. DNA Profile Measured by Flow Cytometry

The position in the cell cycle is determined by measuring the relative DNA content of individual cells by fluorescence-activated cytometry (FACS) of cells that have been permeabilized, depleted of RNA, and stained with the fluorescent DNA-intercalating dye, propidium iodide. The method originally described by Krishan can be completed in a very short amount of time *(12,13)*.

3.3.1. Collect Cells in 70% Ethanol

1. Recover the cells from a single-cell suspension by centrifugation at 400*g* for 5 min at room temperature.
2. Rinse the cells by resuspending them in ice-cold PBS. If necessary, determine the cell number at this time.

3. Recover the cells by centrifugation as before and resuspend in 2 mL ice-cold 70% ethanol in order to fix the cells and permeabilize the cell membrane. The ethanol can be added dropwise with continuous mixing or by forcefully dislodging the pellet with the addition of ethanol.

4. Store the cells in 70% ethanol at 4°C for at least 3 h to fix the cells. The cells can be stored at 4°C for several days or stored at −20°C for several weeks before analysis.

3.3.2. Stain Cells With Propidium Iodide

1. Prepare 1 mL of a working solution of propidium iodide and RNase per 10^6 cells.
2. Recover the cells stored in 70% ethanol by centrifugation at 400g for 5 min at 4°C in a conical centrifuge tube.
3. Wash the cells with ice-cold PBS and recover by centrifugation.
4. Resuspend the cell pellet in the working solution of propidium iodide and RNase solution at a density of $1–2 \times 10^6$ cells per mL and incubate in the dark for 30 min at room temperature.
5. Triturate the cells through a 27.5-gage needle or pass the cell suspension through a fine-mesh filter to achieve single cells.
6. Transfer the propidium iodide-stained cells to an appropriate tube for FACS analysis and store the samples at 4°C until analyzed by flow cytometry.

3.3.3. Flow Cytometry

The following procedures assume familiarity with the operation of a flow cytometer with the capacity to analyze the red fluorescent signal of propidium iodide. A flow cytometer equipped with a light source providing excitation in the UV to green range is required. The argon laser excitation wavelength of 488 nm (green) is ideal. Propidium iodide emits fluorescence primarily at wavelengths above 610 nm, and fluorescence can be collected through a red or orange long-pass filter. For the BD FACScan™ and FACSCaliburflow™ cytometers, the FL2 parameter is appropriate if no other fluorochrome is being used. A representative series of DNA profiles obtained from HeLa cells that were synchronized and analyzed at various times after release from hydroxyurea is seen in **Fig. 2**.

1. Establish an acquisition method on the instrument that includes time-based collection to acquire FL2 data. The time parameter allows one to confirm that the cells remained uniformly suspended during the data acquisition.
2. Gate the cells on the basis of forward and side scatter to limit the analysis to intact cells. Gate the results further on the basis of FL2 peak-width (FL2-W) and FL2 peak-height (FL2-H) to exclude cell doublets that have a greater FL2-W.
3. Collect FL2-H values in the linear mode for cell cycle analysis.
4. Collect FL2-H values in logarithmic mode to identify apoptotic cells (below-diploid content of DNA).

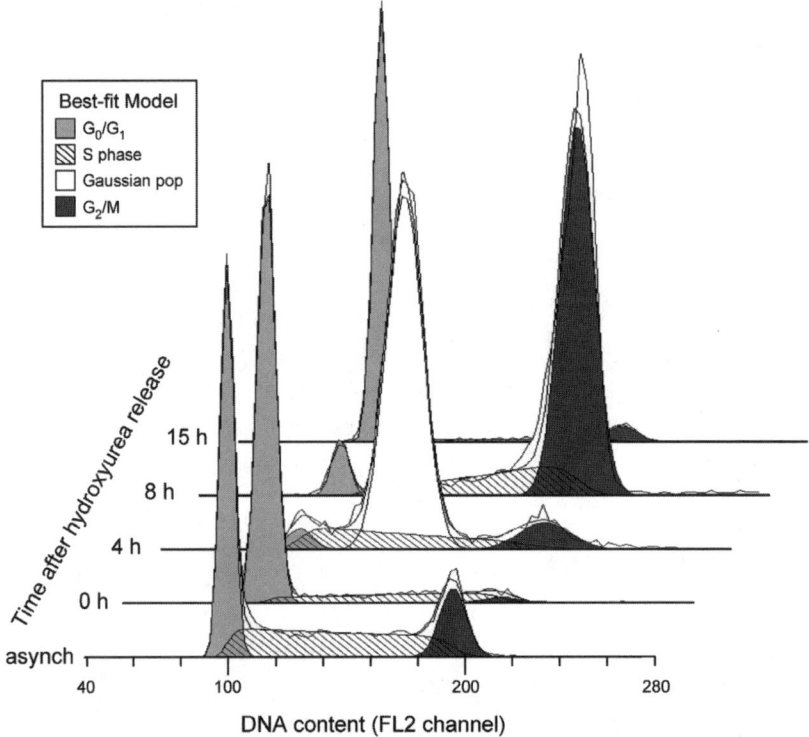

Fig. 2. DNA content of synchronized HeLa cells revealed by propidium iodide staining and flow cytometry. HeLa cells were synchronized as described in the text. Cells were harvested at the times indicated to the left of each axis. Asynchronously dividing cells are labeled as "asynch." The DNA content in the cells was determined by the intensity of propidium iodide fluorescence as described in the text. The data are presented as a histogram with the DNA content represented by the linear scale on the *x*-axis and the relative number of cells with the corresponding amount of DNA represented as the height of curve. The solid shaded portions represent the best-fit Gaussian curve for either G_0/G_1, G_2/M or a population of synchronized S phase cells seen at 4 h after release from hydroxyurea. The hatched shading represents the predicted population of S-phase cells as determined by the standard model for diploid, dividing cells, which is provided with the ModFit software program.

5. Analyze the results with appropriate software such as ModFIT (Verity Software House) to quantify the populations of cells with varying DNA content.

3.4. Limiting Dilution Assay for Infectious Centers

Uniformly infected cells are harvested and recovered as a single-cell suspension before progeny virus is generated. The infected single cells are diluted in serial fashion into a 96-well microtiter plate and returned to normal growth

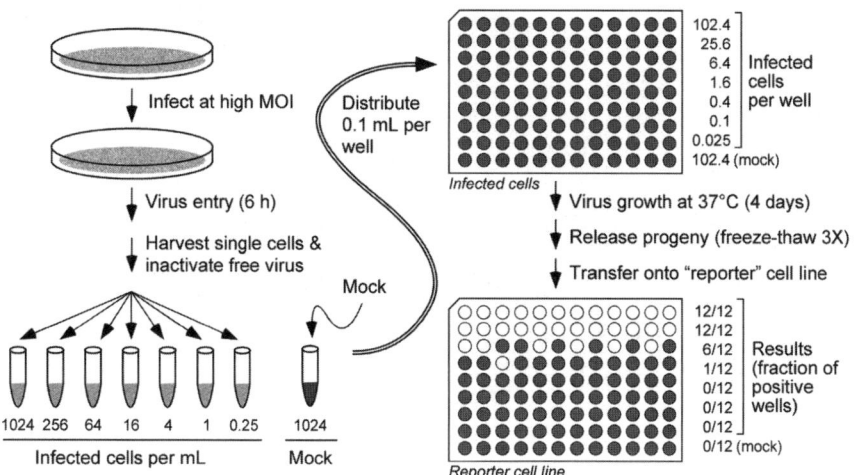

Fig. 3. Schematic representation of the procedure to determine infectious centers by limiting dilution. The procedural details are described in the text. MOI, multiplicity of infection.

conditions, where the infection is allowed to proceed to completion. Virus production is then measured as an all-or-none event using a reporter cell line by standard means such as the cytopathic effect. The number of infected cells that were required to yield a productively infected cell is determined by application of the Poisson distribution. A graphic summary of the method is shown in **Fig. 3**.

3.4.1. Conditioned Medium

For each infected culture to be evaluated, prepare approx 20 mL of conditioned medium from a dense culture of cells that have been allowed to grow for 24 h. Centrifuge the medium (1000g for 10 min at room temperature) to remove cells and debris. Filter the supernatant fluid through a 0.2-μm filter before use. The medium can be collected in advance and stored at 4°C.

3.4.2. Subculture Cells to Be Infected

If asynchronously growing cells are to be infected, subculture the cells under standard conditions. Include replicate cultures that will be used to determine cell numbers.

3.4.3. Infection and Single-Cell Harvest (Day 0)

1. Infect the cells as described in **Subheading 3.2.** at a sufficiently high multiplicity to ensure that all cells are infected. It is essential that all cells be successfully infected.

2. Allow the infection to proceed for a total of 6 h at 37°C. This is the optimal time after which infected HeLa cells can be harvested without compromising the viability of the infected cells. This time should be determined for each cell line (*see* **Note 7**).

3. Aspirate the growth medium, and thoroughly rinse the infected cells twice with warm PBS.

4. Recover the infected cells by extensive treatment with Trypsin-EDTA. For example, incubate a 60-mm plate with 1 mL Trypsin-EDTA solution for 10 min at 37°C with rocking. This step releases the cells and inactivates any extracellular virus.

5. Add an equal volume of normal growth medium with serum to overwhelm the Trypsin and to facilitate recovery of the infected cells.

6. Using a hemacytometer, count the number of cells that were recovered. If the cells have not been successfully dispersed to single cells, repeat the exposure to Trypsin-EDTA and use mechanical means to achieve a single-cell suspension.

3.4.4. Serial Dilution of Infected Cells (Day 0)

1. Prepare a serial dilution of the infected cells in the conditioned medium that was prepared as described in **Subheading 3.4.1.**

2. Prepare seven dilutions of the cells in fourfold increments to yield 1024, 256, 64, 16, 4, 1, and 0.25 cells per mL in a final volume of 2 mL.

3. Prepare 2 mL of mock-infected cells at 1024 cells per mL in the conditioned medium.

4. Distribute 0.1 mL of each dilution of infected cells as well as the mock-infected cells into each well of a row in the 96-well microtiter plate. Ensure that the cells remain uniformly suspended during this process (*see* **Note 8**).

5. Return the cells to normal growth conditions, and allow the infection to proceed to completion. In this example, the virus infectious cycle is complete on day 4.

3.4.5. Prepare Reporter Cells (Day 3)

1. For each microtiter plate of diluted, infected cells, prepare one 96-well microtiter plate with appropriate reporter cells such as 293 cells.

2. The reporter cells should form a sparse monolayer at the time of use on day 4.

3.4.6. Harvest Virus and Infect the Reporter Cells (Day 4)

1. Four days following the infection, transfer the entire 96-well plate of infected cells to a freezer. The virus is harvested after sufficient time to allow a complete replicative cycle. If the infection needs to be extended beyond 4 d, add fresh medium to the cultures every 3–4 d.

2. Subject the infected cells to three freeze–thaw cycles at either –20 or –80°C to release the progeny virus. If necessary, add a small volume of concentrated Tris or HEPES buffer to each well in order to keep the pH close to 7.6 throughout the freeze–thaw process.

3. Remove the growth medium from a microtiter plate of reporter cells that were prepared the previous day according to **Subheading 3.4.5.**

4. Using an 8- or 12-channel pipet device, transfer half of the entire freeze–thaw lysate potentially containing progeny virus onto the subconfluent monolayer of reporter cells. Use new pipet tips for each transfer.

3.4.7. Score for Infection (Days 7–9)

1. Determine the fraction of wells in each row that exhibit a productive infection by visual inspection. A productive infection can be recognized by a decrease in the pH of the medium or by examining the reporter cells for cytopathic effects with an inverted microscope.

2. To ensure an accurate assessment, reevaluate the reporter cells again after several days. Keep the culture viable by adding 0.05 mL of growth medium to each well before returning the plates to the incubator for an additional 3–4 d. Reassess the number of infected wells as in **step 1**.

3. Analyze the results by methods appropriate for limiting dilution ($TCID_{50}$) assays. Prism™ by GraphPad Software is an excellent program for this analysis.

3.5. Agarose Overlay Assay for Infectious Centers

Uniformly infected cells are harvested and recovered as a single-cell suspension before progeny virus can be generated. The infected single cells are diluted in serial fashion into growth medium, which is further diluted with an equal volume of molten agarose. A monolayer of reporter cells is overlaid with a small volume of the infected cells in the molten agarose. Once gelled, the cells are overlaid with an additional volume of nutrient agarose. Virus-producing cells, or infectious centers, are enumerated as plaques, and the fraction of productive cells is determined by adjusting for the efficiency of plaque formation.

3.5.1. Subculture Reporter Cells and Cells to Be Infected

Prepare 60-mm culture plates with reporter cells suitable for use in plaque assays such as 293 cells. A single infected sample typically requires 15 60-mm plates of reporter cells in order to evaluate five dilutions of infected cells in triplicate. If asynchronously growing cells are to be infected, prepare the cells to be infected under standard conditions at this time. Include replicate cultures of cells that will be used to determine cell numbers.

3.5.2. Infection and Single-Cell Harvest

1. Infect the cells as described in **Subheading 3.2.** at a sufficiently high multiplicity to ensure that all cells are infected. Note that it is essential that all cells be successfully infected.

2. After 60–90 min of virus adsorption, aspirate the virus infection medium and thoroughly rinse the infected cells twice with warm PBS (*see* **Note 7**).

3. Recover the infected cells by extensive treatment with 0.25% Trypsin-EDTA to release the cells and inactivate any noninternalized virus. For example, incubate a 60-mm plate with 1 mL Trypsin-EDTA solution for 10 min at 37°C with rocking.
4. Add an equal volume of normal growth medium with serum proteins in order to overwhelm the trypsin and to facilitate recovery of the infected cells.
5. Wash the cells twice more with warm normal growth medium and determine the number of infected cells that were recovered.

3.5.3. Serial Dilution of Infected Cells

1. Prepare a dilution series of the infected cells in normal growth medium from 4000 to 250 cells per mL in two steps (*see* **Note 9**).
2. Add 0.1 mL of each cell dilution to 0.1 mL of molten 1.4% type VII agarose in serum-free medium held at 42°C.
3. Thoroughly aspirate the medium from a 60-mm plate of reporter cells prepared according to **Subheading 3.5.1.** and then gently overlay the reporter cells with the entire 0.2 mL suspension of infected cells in molten agarose. Ensure that the agarose is uniformly distributed across the plate.
4. After the agarose has gelled, overlay with 3.5 mL of 0.7% SeaKem agarose in normal growth medium (first agarose overlay).
5. Every 3–4 d, supplement the cultures by adding 2–4 mL of nutrient agar (second agarose overlay). Maintain the plates at normal growth conditions to allow development of plaques.
6. Overlay the cultures 10–12 d after the initial infection with 4 mL of 0.07% Neutral Red in nutrient agar. Return the cultures to normal growth conditions to allow the dye to penetrate the agarose. Living cells will stain red, and clear plaques will become evident 8–16 h after adding the stain. With extensive cell death and cytolysis, plaques will appear clear; otherwise, the plaques can be recognized by the rounded cells within the plaque.

3.5.4. Analysis

Analyze the results by plotting the number of infectious centers (plaques) versus the number of infected cells plated. The efficiency of plaque formation must be determined in each experiment using an infected population of cells that are known to produce virus in every cell. For HeLa cells infected with the wild-type virus; approx 20% of the infected cells analyzed by this method register as infectious centers. A representative experiment showing the number of plaques recovered as a function of the number of infected cells for the wild-type and E1B-55K-mutant virus is shown in **Fig. 4**. In this experiment, the efficiency of plaque formation for the wild-type virus, which by other measures is known to produce virus in every infected cells, was 22.8%. By contrast, the frequency of infectious centers for the E1B-55K-mutant virus-infected cells

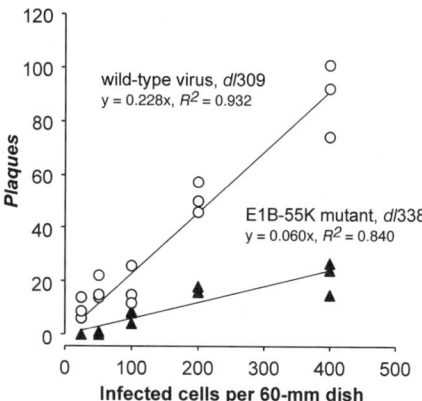

Fig. 4. Fewer HeLa cells infected with the E1B-55K-mutant virus produce infectious centers than cells infected with the wild-type virus. The results of a representative experiment to measure infectious centers by plaque assay are shown where each value from three replicates is shown. The experimental details are described in the text.

was 6.0%. The ratio of these values (6.0:22.8) indicates that the relative frequency of infectious centers measured in mutant virus-infected cells was 26%.

3.6. Transmission Electron Microscopy for Viral Particles

Individual cells are examined by transmission electron microscopy to determine the fraction of cells that contain progeny virus particles. The following is a basic method to collect and fix infected cells. The samples prepared by this method are appropriate for further processing by postfixation with osmium tetroxide, embedment, and thin sectioning. Only plasticware is used throughout to avoid introducing glass shards into the sample that can damage microtome knife edges.

3.6.1. Cell Fixation

1. At an appropriate time after infection, collect both detached and adherent cells by harvesting the medium from the infected cells and by scraping the adherent cells into a small volume of PBS. Rinse the culture dish and cell scraper with additional PBS and then pool the PBS rinse with the previously collected medium with detached cells (*see* **Note 10**).
2. Recover the cells by centrifugation at 400*g* for 5 min at room temperature.
3. Rinse the cell pellet by suspending in approximately 10 mL of 0.1 *M* sodium cacodylate. Sodium cacodylate is used in place of PBS to avoid precipitation artifacts that can occur with the use of phosphate and osmium.

4. Transfer the cells to a 15-mL conical centrifuge tube and recover the cells by gentle centrifugation as in **step 2**. A conical tube is used in order to form a compact pellet of fixed cells upon centrifugation in **step 7**.

5. Resuspend the cells in a small volume (0.2 mL) of 0.1 M sodium cacodylate.

6. With gentle mixing, add 4 volumes (0.8 mL) of 2.5% gluteraldehyde in 0.1 M sodium cacodylate to the cell suspension to achieve a final concentration of 2% gluteraldehyde.

7. Immediately centrifuge the cells with moderate force (1000g for 10 min at room temperature) to form a cohesive pellet.

8. Allow the cells to fix as a pellet for 1–3 h at room temperature.

9. Gently remove the gluteraldehyde fixative and discard as hazardous waste in accordance with local requirements.

10. Dislodge the cell pellet from centrifuge tube and rinse the fixed cell pellet twice with 0.1 M sodium cacodylate, taking care not to disturb the integrity of the pellet.

11. Store the fixed cell pellet in a small volume of 0.1 M sodium cacodylate at 4°C until it can be processed further.

12. The gluteraldehyde-fixed cells are further fixed with osmium tetroxide and embedded in resin in order to prepare thin sections (100 nm) by standard means.

3.6.2. Microscopic Evaluation

Thin sections of infected cells are stained with uranyl acetate and lead citrate and analyzed at 80 keV with a transmission electron microscope. Progeny virus can be identified in the nuclei of cells by their characteristic morphology (regular 75 nm particles) and staining properties. An experienced operator can identify viral particles while working at a magnification of ×3300 although it is frequently necessary to work at higher magnifications in order to differentiate progeny virus from dense-staining inclusions. Cells that contain any progeny viral particles are scored as positive. It is important to evaluate sections taken from different portions of the cell pellet to obtain statistically meaningful data.

4. Notes

1. Two-thirds confluence is recommended because cells that are in contact with each other tend to be more easily dislodged during the mitotic shake.

2. The concentration of hydroxyurea should be sufficiently high to kill cells in S phase and arrest cells in G1 at the G1/S border but should be nontoxic to cells in G1, G2, or M phase. If necessary, concentrations between 0.2 and 7.5 mM can be tested.

3. The ability to *selectively* dislodge only mitotic cells is a skill that requires practice. It is recommended that this step be performed with frequent feedback by examining the flask with an inverted microscope between strikes. Immediately before striking the flask, ensure that the medium is uniformly distributed over the culture surface. Rapidly raise the flask approx 12 in. (30 cm) above the hard surface such that the plane of the adherent cells is orthogonal to the hard surface.

Strike the flask with slightly more force than would be required to crush a raw chicken egg. In a typical asynchronously growing culture of cells, approx 4% of the cells are in mitosis at any time. This should be the largest number of cells that can be recovered. If significantly more cells are recovered, it is likely that many interphase cells were dislodged. These cells should not survive the subsequent hydroxyurea treatment but will lead to an overestimate of the number of mitotic cells. Cao et al. *(7)* suggested using a low concentration of trypsin (0.05%) to facilitate the recovery of mitotic HeLa cells with less force. However, this may be impractical for processing the large numbers of flasks used in a typical synchronization.

4. Cells that will be used in G1 will have passed through mitosis and thus will have doubled in number, whereas the cells used in S phase will not pass through mitosis and will not increase in number. In principle, therefore, twice as many mitotic cells should be placed in culture for S-phase samples as for G1 samples. However, some nonadherent cells in the culture medium will be collected during the mitotic shake. These nonadherent cells may comprise as much as 20–30% of the mitotic cells. Because these nonadherent cells fail to attach and divide, we have empirically determined that by plating 1.67-fold more HeLa cells for S phase than for G1, an identical number of cells are obtained at the time of release from hydroxyurea. It would probably be necessary to determine this factor for each cell line and set of culture conditions.

5. The length of time for each phase of the cell cycle can be determined by measuring the fraction of randomly cycling cells in G1, S, and G2/M phases and by measuring the generation time from growth curves.

6. The apparent particle-to-infectious unit ratio will vary with changes to the time of virus adsorption, the volume of infection medium, and the nature of agitation, as discussed by Nyberg-Hoffman et al. *(14)*. For this reason, it is important to apply a standard practice for each infection in order to maintain consistency between experiments.

7. This procedure requires that Ad-infected cells be detached and then reattached with minimal perturbation of the progression of the virus infection. We have found that HeLa cells can be harvested as late as 6 h after infection with no apparent impact on the progression of the infection. However, it would be prudent to determine the appropriate time after infection for each particular cell line and culture condition.

8. The concentration of cells used in the dilution series should be adjusted such that approximately one-half of the wells receiving the central dilution will contain an infectious center. The example dilution series includes values that place less than one cell per well (4, 1, and 0.25 cells per mL) to accommodate possible errors in counting and to ensure that there was no free virus in the dilution medium. The possibility of free virus in the medium used to suspend the infected cells can be evaluated by standard methods after removing the cells from the medium.

9. The concentration of cells used in the dilution series analyzed by plaque assay should be adjusted to yield approx 20–60 plaques per 60-mm dish. Once the

expected values are determined, a finer dilution series can be prepared to increase the precision of this assay. The possibility of free virus in the medium used to suspend the infected cells can be evaluated by standard methods after removing the cells from the medium.

10. The number of cells required to form a pellet suitable for further infiltration and embedding is dependent on the particular methods used to process samples for electron microscopy and should be determined in consultation with the laboratory or service that processes the specimens. We have found that two 10-cm plates provide enough cells (4×10^6) to prepare several blocks for sectioning.

Acknowledgments

This work was supported in part by Public Health Service grants AI 35589 from the National Institute of Allergy and Infectious Disease and grants CA 77342 and CA 77342S1 from the from the National Cancer Institute to DO. Tissue culture reagents and services were provided by the Tissue Culture Core Laboratory, and electron microscopy was performed through the Micromed facility, both services of the Comprehensive Cancer Center of Wake Forest University, which is supported in part by the National Cancer Institute grant CA 12197. We especially wish to acknowledge Ken Grant and Nora Zbieranski of Wake Forest University for invaluable assistance with electron microscopy and FACS analysis, respectively.

References

1. Vousden, K. H. (1995) Regulation of the cell cycle by viral oncoproteins. *Semin. Cancer Biol.* **6,** 109–116.
2. Op de Beeck, A. and Caillet-Fauquet, P. (1997) Viruses and the cell cycle. *Prog. Cell Cycle Res.* **3,** 1–19.
3. Ben-Israel, H. and Kleinberger, T. (2002) Adenovirus and cell cycle control. *Front. Biosci.* **7,** 369–1395.
4. Goodrum, F. D. and Ornelles, D. A. (1997) The early region 1B 55-kilodalton oncoprotein of adenovirus relieves growth restrictions imposed on viral replication by the cell cycle. *J. Virol.* **71,** 548–561.
5. Goodrum, F. D. and Ornelles, D. A. (1999) Roles for the E4 orf6, orf3, and E1B 55-kilodalton proteins in cell cycle-independent adenovirus replication. *J. Virol.* **73,** 7474–7488.
6. Shepard, R. N. and Ornelles, D. A. (2003) E4orf3 is necessary for enhanced S-phase replication of cell cycle-restricted subgroup C adenoviruses. *J. Virol.* **77,** 8593–8595.
7. Cao, G., Liu, L. M., and Cleary, S. F. (1991) Modified method of mammalian cell synchronization improves yield and degree of synchronization. *Exp. Cell Res.* **193,** 405–410.
8. Cooper, S. (2004) Is whole-culture synchronization biology's "perpetual-motion machine"? *Trends Biotechnol.* **22,** 266–269.

9. Cooper, S. (2004) Rejoinder: whole-culture synchronization cannot, and does not, synchronize cells. *Trends Biotechnol.* **22,** 274–276.

10. Spellman, P. T. and Sherlock, G. (2004) Reply: whole-culture synchronization—effective tools for cell cycle studies. *Trends Biotechnol.* **22,** 270–273.

11. Spellman, P. T. and Sherlock, G. (2004) Final words: cell age and cell cycle are unlinked. *Trends Biotechnol.* **22,** 277–278.

12. Krishan, A. (1975) Rapid flow cytofluorometric analysis of mammalian cell cycle by propidium iodide staining. *J. Cell Biol.* **66,** 188–193.

13. Krishan, A. (1990) Rapid DNA content analysis by the propidium iodide-hypotonic citrate method. *Methods Cell Biol.* **33,** 121–125.

14. Nyberg-Hoffman, C., Shabram, P., Li, W., Giroux, D., and Aguilar-Cordova, E. (1997) Sensitivity and reproducibility in adenoviral infectious titer determination. *Nat. Med.* **3,** 808–811.

8

Co-Immunoprecipitation of Protein Complexes

Peter Yaciuk

Summary

Co-immunoprecipitation is a common method used to determine protein–protein interactions. This method typically exploits the highly specific interaction between a monoclonal antibody and a protein of interest. Providing that the antibody–protein interaction does not interfere with the ability of the protein to interact with other proteins in a cell lysate, a protein complex containing the antibody, the protein of interest, and its interacting proteins could be isolated on protein A Sepharose beads and purified from the cell lysate. This chapter details many of the techniques and associated procedures of co-immunoprecipitation and highlights many of the concerns and considerations associated with each step of this method.

Key Words: Co-immunoprecipitation; protein–protein interactions; monoclonal antibody; protein complexes; radioactive labeling of proteins; nonionic detergents; protease inhibitors; phosphatase inhibitors; kinase inhibitors.

1. Introduction

Important insights into cell growth and transcriptional regulations have been realized by correlating the adenovirus E1A-induced cell-growth-control mechanism with E1A's specific interactions with host-cell proteins *(1)*. A key method for directly demonstrating these E1A–host-cell protein interactions is by immunoprecipitation. This assay is performed under conditions that efficiently lyse cells but preserve protein–protein interactions. This method combines the elegant specificities of monoclonal antibodies and the fortuitous properties of nonionic detergents. These highly efficient and specific means of isolating protein complexes from cell lysates, although detailed here for E1A-containing protein complexes, have general application for the study of other cellular protein complexes.

From: *Methods in Molecular Medicine, Vol. 131:*
Adenovirus Methods and Protocols, Second Edition, vol. 2:
Ad Proteins, RNA, Lifecycle, Host Interactions, and Phylogenetics
Edited by: W. S. M. Wold and A. E. Tollefson © Humana Press Inc., Totowa, NJ

When investigating potential protein–protein interactions of a specific protein, it is important that the investigator optimize assay conditions that both minimize the amounts of nonspecific-interacting proteins and maximize the amount of specific-interacting protein. Procedures to minimize the amounts of nonspecific-interacting protein are detailed below. These steps, along with immunoprecipitations done with control antibodies, should help to identify candidate proteins that specifically interact with the protein of interest. Once a candidate specific-interacting protein (or proteins) is (or are) identified, the investigator needs to demonstrate that the candidate specific-interacting protein is co-immunoprecipitated through protein–protein interaction and not through a direct recognition by the antibody used. To address this issue one should demonstrate that the antibody used to immunoprecipitate the protein complex will not recognize the candidate specific-interacting protein by Western blot analysis or that the antibody will not reimmunoprecipitate the candidate specific-interacting protein after the purified protein complex is denatured. One should also rigorously address this issue by using any additional relevant assays specific to a research field. Once it is clearly established that the candidate specific-interacting protein is co-immunoprecipitated by a protein–protein interaction, one can then determine the identity of this protein using various strategies; such as by mass spectrometry analysis, partial amino acid sequence, or developing specific antibodies. At this point the investigator can also start preliminary characterization of the protein–protein interaction, such as with the cell-cycle phase dependency or the cell-type specificity of this interaction.

An immunoprecipitation assay starts by radiolabeling an experimental cultured cell system. The cells are lysed under conditions in which protein–protein interactions are not disrupted. Cell debris and nonspecific-binding proteins are removed from the lysate. Then an antibody that has the properties of binding to a specific protein in a protein complex, without disrupting the binding of the other protein associations, is used for immunoprecipitation. The protein complex and antibody are then adsorbed onto Sepharose protein A beads and purified from the lysate. The protein complex components are then separated by standard sodium dodecyl sulfate (SDS)–polyacrylamide gel electrophoresis and analyzed.

2. Materials

2.1. Radiolabeling of Cultured Cells

1. Tissue culture medium, tissue culture cells, tissue culture plates (*see* **Note 1**).
2. Mammalian cells that constitutively express *E1A* gene products, such as 293 cells (ATCC, Manassas, VA, cat. no. CRL-1573), or other specific cell systems where E1A expression can be introduced (*see* **Note 2**).

3. Methionine/cysteine-free tissue culture medium (Invitrogen/Gibco, Grand Island, NY, cat. no. 21013-016).
4. ^{35}S-labeled methionine and cysteine (Tran^{35}S-label, MP Biomedicals, Irvine, CA, cat. no. 51006 or Express^{TM} Methionine/Cysteine Protein Labeling Mix, PerkinElmer Life And Analytical Services, Boston, MA, cat. no. NEG072; *see* **Note 3**).

2.2. Lysis of Radiolabeled Cells and Immunoprecipitation

1. Lysis buffer: 0.1% Nonidet P-40 (NP40) (*see* **Note 4**), 20 mM sodium phosphate, 250 mM sodium chloride, 30 mM sodium pyrophosphate, 5 mM ethylenediamine tetraacetate (EDTA), 10 mM sodium fluoride (*see* **Note 5**). Add dithiothreitol to 5 mM final concentration from 1 M dithiothreitol stock (*see* **Note 6**).
2. Protease-inhibitor stock solutions: aprotinin (5 mg/mL; add 1 µL per mL of lysis buffer), pepstatin (5 mg/mL; add 1 µL per 5 mL lysis buffer), phenylmethylsulfonyl fluoride (PMSF; *see* **Note 7**) (75 mg/mL in dimethyl sulfoxide [DMSO]; add 5 µL per mL of lysis buffer), or use commercially available mixtures of protease inhibitors (Protease Inhibitor Cocktail, Sigma, St. Louis, MO, cat. no. P8340; or Protease Inhibitor Set, Roche Applied Science, Indianapolis, IN, cat. no. 11 836 153 001) according to manufacturer's recommendations (*see* **Note 8**).
3. Phosphotyrosine-specific phosphatase inhibitor stocks solution: 100 mM sodium orthovanadate (Na$_3$VO$_4$); add 1 µL per mL of lysis buffer (*see* **Note 9**).
4. Stock 10% *Staphylococcus aureus* Cowan A strain slurry (Pansorbin, EMD Bioscience/Calbiochem, San Diego, CA, cat. no. 507858; *see* **Note 10**).
5. Anti-E1A monoclonal antibody, M73 (EMD Biosciences/Calbiochem, San Diego, CA, cat. no. DP11; *see* **Note 11**) or stock antigen-specific antibody solutions (*see* **Note 12**) and secondary antibody, such as polyclonal rabbit anti-mouse antibodies (EMD Biosciences/Calbiochem, San Diego, CA, cat. no. 402334, or various venders; *see* **Note 13**).
6. Stock 3% Protein A Sepharose CL-4B beads (Amersham Biosciences, cat. no. 17-0780-01; *see* **Note 14**).
7. 2X Sample loading buffer: 100 mM Tris, pH 6.8, 200 mM dithiothreitol, 4% SDS, 0.2% bromophenol blue, 10% glycerol.

2.3. Normalization of Cell Lysates

1. Bio-Rad Protein Assay kit (Bio-Rad, Hercules, CA, cat. no. 500-0001; *see* **Note 15**).
2. Glass-microfiber filter disks (Fisher Scientific, Pittsburgh, PA) (Whatman GF/C, cat. no. 1822-024, or various vendors).
3. 20% Trichloroacetic acid (TCA).

3. Methods
3.1. Radiolabeling of Cultured Cells

1. Infect or transfect cells with appropriate wild-type or mutant adenoviruses (*see* **Note 16**) or wild-type or mutant E1A-expression plasmids at appropriate times prior to the in vivo labeling of cells.

2. Remove medium from infected or transfected cell cultures or cell lines that constitutively express E1A.
3. Rinse tissue culture cells with 3 mL methionine/cysteine-free Dulbecco's modified Eagle's medium (DMEM).
4. Add 3 mL methionine/cysteine-free DMEM containing 100 μCi ^{35}S-labeled methionine/cysteine (*see* **Note 17**).
5. Incubate at 37°C in a tissue culture incubator for the appropriate labeling times (*see* **Note 18**).

3.2. Immunoprecipitation

1. After the incubation period, remove radioactive medium into designated radioactive waste containers.
2. Add 1 mL lysis buffer to each plate. Distribute lysis buffer over entire surface of plate, manually shake, and then rock plates in cold room for 10 min.
3. Transfer cell lysates to microfuge tubes.
4. Preclear lysates by adding 100 μL 10% slurry of the *S. aureus* Cowan A strain (*see* **Note 19**). Incubate on ice for 5 min.
5. Spin tubes in refrigerated microfuge at high speed for 10 min (or in the ultracentrifuge for 20 min at 70,000*g*) to pellet cell debris and *S. aureus* bacteria.
6. Normalize lysate volumes to TCA precipitable radioactive counts per min (cpm) or total cell protein (*see* **Subheadings 3.3.1.** and **3.3.2.** and **Note 20**), if required.
7. Transfer lysates to a new set of microfuge tubes.
8. Add titered amount of specified antibody and mix.
9. Add titered amount of specified secondary antibody (if necessary, *see* **Note 13**) and mix.
10. Add 100 μL of 3% slurry (w/v) protein A Sepharose beads and mix (*see* **Note 14**).
11. Rotate sample tubes for 1 h at 4°C.
12. Spin immunocomplex-containing beads down in a variable-speed microfuge at a low-speed setting for 10 s in the microfuge (*see* **Note 21**).
13. Remove supernatant into appropriate radioactive waste container.
14. Wash beads five times by adding 1 mL lysis buffer, mix, and spin as in **step 12**, and discard each wash solution into radioactive waste container (*see* **Note 22**).
15. Add 40 μL 2X sample buffer, mix, and boil tubes 3 min.
16. Mix and then spin tubes at high speed for 1 min.
17. Run samples on a SDS-polyacrylamide gel (*see* **ref. 2** and **Note 23**). Dry gel and expose to film or phosphoimaging plate and analyze results.

3.3. Normalization of Radioactive Lysates

3.3.1. Normalization to Amount of Total Cell Protein

In experiments in which cell lysates are normalized to amount of cell protein in the lysate, the Bio-Rad protein assay kit works well with these assay conditions.

1. Distribute 800 µL dH$_2$O to a series of six microfuge tubes for standards and duplicate tubes for each cell-lysate sample.
2. Add 5 µL of lysis buffer to the six protein-standard tubes.
3. Add 0, 2, 5, 10, and 20 µg of the IgG protein standard to the six protein standard tubes, respectively.
4. Add 5 µL of cell lysate of each sample to the corresponding sample tubes.
5. Add 200 µL of the dye reagent concentrate to each tube and mix well. Incubate at room temperature for at least 5 min.
6. Measure absorbance at 595 nm and calculate total cell protein per cell lysate sample. Use these calculated amounts to transfer equal amounts of total cell protein per sample to a new set of tubes, and adjust each sample to the same volume.

3.3.2. Normalization by Radioactive Incorporation Into Total Cellular Protein

In some experiments in which the relative levels of specific proteins need to be compared in different cell populations, lysates should be normalized to a total number of radioactive counts.

1. Spot 5 µL of radioactive lysate onto a glass microfiber filter disk and place in filter disk filtration apparatus with suction.
2. Wash the filter disks with 10 mL ice-cold 20% TCA, followed by 10 mL ice-cold 10% TCA and a final wash with 10 mL 100% ice-cold ethanol. Air-dry filter disks briefly.
3. Transfer the filter disks to scintillation vials, add scintillation fluid, and measure the amounts of incorporated radioactivity with a scintillation counter. Calculate and portion the lysate volumes to contain equal amounts of radioactivity. Adjust samples to the same volume.

4. Notes

1. A wide variety of tissue culture media and serum concentrations and types are used for growing specific cell types in culture. For 293 cells, grow in 10% calf serum in DMEM containing streptomycin and penicillin.
2. The E1A proteins have been introduced into numerous types of cells by transfection, such as the primary human embryonal kidney cell line called 293, in which sheared human adenovirus DNA was transfected and a transformed cell line that constitutively expresses the *E1A* gene products was isolated *(3)*. Because many cell types are permissive for adenovirus infection, the E1A products (and their cell growth control mechanisms) can be introduced into a wide variety of cells to address specific scientific questions about particular biological mechanisms.
3. All labeling and handling of radioactive materials should be done in compliance with Nuclear Regulatory Commission Radiation Safety Regulations. Tran[35]S-label and Express™Methionine/Cysteine Protein Labeling Mix are cellular hydrosylates of *E. coli* grown in the presence of $^{35}SO_4$. Typically, they contain

≥70% L-methionine, ≤15% L-cysteine, plus other labeled compounds, but serve as an economical source of ^{35}S-labeled amino acids and are good for general in vivo ^{35}S-labeling of cellular proteins. Please note that ^{35}S-labeled compounds have volatile components that can be inhaled or contaminate the working areas. It is important to initially open the ^{35}S-reagents stock vial in a fume hood and check for and clean up any working surfaces, including CO_2 incubator shelves! It is a good idea to equip tissue culture incubators, which circulate the incubator air through a sterilization filter, with a charcoal prefilter to remove any volatile ^{35}S-compounds. To maximize the in vivo incorporation of these radioactive amino acids, it is best to use them in conjunction with methionine/cysteine-free DMEM. Increases in incorporation have been noted by preincubating cells for up to 1 h in methionine/cysteine-free DMEM to deplete cellular methionine/cysteine concentrations prior to labeling cells. *See* **ref. 4** for a thorough discussion of labeling with other amino acids.

4. The nonionic detergents, such as Nonidet P-40 (NP40), tend to be much less effective at disrupting protein–protein interactions than the ionic detergents and therefore are better suited for the purification of protein complexes. Tween-20 and Triton X-100 have also has been used with success. Whereas E1A-containing protein complexes are purified using NP40 concentrations in the range of 0.1–0.5%, it is recommended that the investigators working with other cellular protein complexes optimize for the amount of their specific-interacting protein by varying the concentration of detergent and other lysis buffer components. *See* **ref. 5** for an overview of detergents.

5. Upon cell lysis, various enzyme activities, such as protease, kinase, and phosphatase activities, are released from their subcellular compartments and can potentially use the other proteins in the cell lysate as substrates. To minimize these effects, keep cell lysates and reagents ice-cold at all times. Keeping cell lysates cold will also enhance general protein stability. Also, to eliminate or minimize these effects, several components of this lysis buffer act as inhibitors of these activities. Whereas sodium phosphate is a good buffer at pH 7.0, it also functions as a phosphatase inhibitor. EDTA chelates divalent metal ions and therefore will inhibit the activity of enzymes that are metal ion-dependent, such as kinases and metalloproteases. Sodium pyrophosphate will inhibit protein phosphorylation. Sodium fluoride should inhibit some serine/theonine-specific protein phosphatases.

6. Dithiothreitol is a reducing agent that is added to inhibit protein aggregation through disulfide bond formation of cysteines. It is also highly volatile and therefore should be added fresh. If unacceptably high backgrounds of nonspecific-binding proteins are obtained, addition of dithiotreitol at a few steps in this procedure can reduce these backgrounds.

7. PMSF is highly toxic and should not be inhaled or come into contact with skin. It should be stored in a double-sealed container and only opened in the hood.

8. Aprotinin and PMSF are general serine protease inhibitors, leupeptin inhibits thioproteases, and pepstatin A inhibits acid proteases.

9. Add sodium orthovandate to eliminate potential phosphotyrosine-specific phosphatases, since this activity may affect protein–protein interactions and migration position in SDS-polyacrylamide gels. Add this reagent fresh.

10. Resuspend the freeze-dried bacteria to the recommended volume. It is strongly recommended that this slurry be sonicated for at least 5 min to ensure that it is homogeneous. Store each preparation at –20°C in 1-mL portions. Thaw one or more portions per experiment as needed. Mix well before use.

11. All antigen-specific monoclonal or polyclonal antibodies are not useful in demonstrating protein–protein interactions, because it is suspected that protein-binding sites serve as good immunogens and the antibody competes away the associated proteins. It is the rare antibody that binds to a region other than a protein-binding site and preserves the binding of all associated proteins.

12. Whereas antibodies are rather stable proteins, the buffers are ideal media for growth of contaminating microbes. Antibody solutions should be filter-sterilized and treated with strict aseptic technique. Store each antibody solution in multiple portions to prevent loss of the entire preparation by a single contamination event. It is recommended that multiple portions of antibody solution be frozen, while keeping one portion at 4°C as the current working stock. This will avoid loss of antibody affinity because of the detrimental effect of multiple freeze–thaw cycles and indefinitely preserve the initial preparation. Addition of bactericidal agents, such as sodium azide (at 0.02% [w/v]) or mercury-[(*o*-carboxyphenyl)thio]-ethyl sodium salt (at 0.01% w/v) (thimerosal, Sigma, T-5125) work well in preventing microbial contamination in antibody solutions without affecting most antibody–antigen interactions, but it should be strongly noted that these bactericidal agents are highly reactive molecules that can affect antibody affinity and in vitro biochemical reactions and are highly toxic in vivo. Also, antibody preparations should be titered to assure that the investigator is in antibody excess and that the amount of protein A Sepharose CL-4B beads is sufficient to bind that amount of antibody.

13. Protein A has a range of affinities for different types of antibodies. If protein A does not recognize or has low affinity for your specific antibody, you can add a secondary antibody to improve isolation of your protein of interest. For example, some mouse monoclonal antibodies of subclass IgG_1 are not recognized by protein A. In this case, by adding a titered amount of a rabbit anti-mouse secondary antibody, the IgG_1 is recognized well by the secondary antibody and the secondary antibody is recognized well by protein A, and your protein of interest can be isolated.

14. Sepharose protein A CL-4B beads are commercially supplied and are freeze-dried in the presence of additives. When the beads are rehydrated, they must be washed well to remove the additives. It is also important to mix these beads thoroughly to disperse any clumps of beads. These clumps have clogged Pipetman tips without preventing buffer from being drawn up into the pipet tip, resulting in the correct volume of solution transferred but with reduced or no bead transfer and therefore reduced immunocomplex isolation. To eliminate this problem, cut

off the end of the pipet tip (approx 2 mm) with a clean scissor or razor blade. This step has led to more consistent final bead volumes and immunocomplex isolation. Also, these protein A Sepharose bead preparations will start to settle within 1 min if left standing. It is critical to keep the slurry well mixed during transfer to multiple sample immunoprecipitation tubes to maintain consistent bead transfer.

15. Each commercially available protein assay kit has its advantages and disadvantages, depending on the specific contents of the protein solution buffers and amounts or type of proteins being measured. To minimize the effects caused by buffer contents, it is recommended that you add a volume of lysis buffer to your protein standards that equals the cell lysate volume that you use to do your protein measurements.

16. If using adenovirus as a vector to introduce your protein of interest into cultured cells and polyclonal antigen-specific serum in your analysis, it is important to note that most serum contains a high titer of antibodies to various adenovirus proteins. Thus, it is critical to compare your immune serum results with results using pre-immune serum from the same animal.

17. Adding 3 mL methionine/cysteine-free DMEM containing 100 µCi/mL ^{35}S-labeled amino acids is a good starting radioactivity concentration for several exponentially growing tissue culture systems. Different tissue culture cell systems may require as little as 10 µCi/mL or as much as 300 µCi/mL ^{35}S-labeled amino acids. The 3-mL volume is sufficient to cover a 10-cm plate (or 2 mL for a 6-cm plate or 1 mL for a 35-mm tissue culture plate). Incubate cells at 37°C for 1 h or more, depending on the purpose of the experiment. Long incubation times, such as overnight labeling, may require supplementing the labeling medium with 10% DMEM or 10% DMEM/1% serum, because these longer labeling times may exhaust essential methionine/cysteine reservoirs and/or growth factors.

18. To optimize radioactive incorporation, check that the shelf in the tissue culture incubator is level and the radioactive medium is evenly distributed.

19. *S. aureus* Cowen A strain contains the protein A molecule in its membranes and is an excellent and economical reagent to adsorb nonspecific-binding proteins from cell lysates and can reduce the volume of the cell debris pellet in the next step.

20. Some experiments may require that individual samples be normalized to the same amount of total cell protein or to an amount of radioactivity incorporated into cellular proteins, such as when testing for an associated kinase activity or when quantitating apparent protein steady-state levels.

21. Slow-speed spins help to minimize background caused by high-molecular-weight protein complexes that may be pelleted at high speeds.

22. To reduce contaminating protein bands that nonspecifically bind to the inside walls of the microfuge tubes, it is recommended that following the final wash the immunocomplex-containing beads be resuspended and transferred to a new microfuge tube.

23. Protein complexes may contain proteins with widely varying molecular weights. It may be advantageous to run your samples on gradient SDS-polyacrylamide gels to best resolve both low- and high-molecular-weight proteins.

References

1. Moran, E. (1994) Cell growth control mechanisms reflected through protein interactions with the adenovirus E1A gene products. *Semin. Virol.* **5,** 327–340.
2. Harlow, E. and Lane, D. (1988) SDS polyacrylamide gel electrophoresis, in *Antibodies: A Laboratory Manual*, Cold Spring Harbor Laboratory, Cold Spring Harbor, NY, pp. 636–640.
3. Graham, F. L., Smiley, J., Russell, W. C., and Nairn, R. (1977) Characteristics of a human cell line transformed by DNA from human adenovirus type 5. *J. Gen. Virol.* **36,** 59–74.
4. Bonifacino, J. S. (1987) Biosynthetic labeling of proteins, in *Current Protocols in Molecular Biology,* Wiley, New York, Unit 10.18.
5. Neugebauer, J. M. (1990) Detergents: an overview, in *Guide to Protein Purification* (Deutscher, M. P., ed.) *Methods Enzymol.* **182,** 239–253.

9

Chromatin Immunoprecipitation to Study the Binding of Proteins to the Adenovirus Genome In Vivo

Jihong Yang and Patrick Hearing

Summary

The encapsidation of adenovirus DNA into virus particles depends on *cis*-acting sequences located at the left end of the viral genome. Repeated DNA sequences in the packaging domain contribute to viral DNA encapsidation, and several viral proteins bind to these repeats when analyzed using in vitro DNA–protein-binding assays. This chapter describes a chromatin immunoprecipitation approach to study the binding of viral proteins to packaging sequences in vivo. The technique is easily adaptable to study the interaction of any viral or cellular protein to Ad DNA or to cellular genomic DNA sequences. The assay permits accurate quantification over a wide range of DNA concentrations. The use of formaldehyde cross-linking to stabilize DNA–protein and protein–protein complexes formed in vivo allows the identification of macromolecular complexes found in living cells.

Key Words: Chromatin immunoprecipitation; ChIP; DNA–protein complex; adenovirus DNA.

1. Introduction

The encapsidation of adenovirus (Ad) DNA into virus particles depends on *cis*-acting sequences located at the left end of the genome (Ad5 nucleotides 230 to 380) *(1)*. Seven repeated sequences, termed A repeats because of their AT-rich content, are located within this domain and contribute to viral DNA packaging. A repeats A1, A2, A5, and A6 are the most important for packaging activity. A repeats contain a bipartite consensus motif (5'-TTTG N_8 CG-3'). Both the first and the second half-site of the consensus motif, as well as the eight-base-pair spacing between the half-sites, are critical for viral DNA packaging *(2,3)*. Two viral proteins, L1 52/55K and IVa2, have been found to play important roles in Ad packaging and virus assembly, although their exact roles in this process remain unclear *(4–7)*. The Ad L1 52/55K protein is found within

From: *Methods in Molecular Medicine, Vol. 131:*
Adenovirus Methods and Protocols, Second Edition, vol. 2:
Ad Proteins, RNA, Lifecycle, Host Interactions, and Phylogenetics
Edited by: W. S. M. Wold and A. E. Tollefson © Humana Press Inc., Totowa, NJ

immature virus particles, and this protein forms a physical complex with the Ad IVa2 protein *(8)*. In turn, the Ad IVa2 protein is essential for virus assembly and the formation of empty viral capsids *(7)*. Ad IVa2 is found in both empty and mature virus particles. The IVa2 protein in vitro binds to packaging A repeats 1 and 2 as well as A repeats 4 and 5 *(9)*. Collectively, these data demonstrate that the Ad L1 52/55K and IVa2 proteins play a key role(s) in the very early stages of the virus assembly process.

This chapter describes a chromatin immunoprecipitation (ChIP) approach that was used to study the binding of the Ad IVa2 and L1 52/55K proteins to wild-type and mutant packaging sequences in vivo using specific antisera directed against these products *(10)*. The method represents adaptations derived from protocols described in *(11,12)*. Viral chromatin was sheared by sonication to an average size of approx 500 base pairs, and the products of immunoprecipitation were quantified using real-time quantitative polymerase chain reaction (Q-PCR). This assay permits accurate quantification over a wide range of DNA concentrations. Discussion of a similar method for studying DNA–protein complexes on Ad DNA was recently described *(13)*. The use of formaldehyde cross-linking to stabilize DNA–protein and protein–protein complexes formed in vivo allows the identification of macromolecular complexes found in living cells.

2. Materials

1. N52.E6 cells (which express Ad E1A and E1B proteins) *(14)*.
2. Cells were grown in α-modification of Eagle medium (α-MEM) supplemented with 10% bovine calf serum, 2 mM glutamine, and penicillin/streptomycin.
3. Formaldehyde: 37% formaldehyde (Fisher, cat. no. F79-500).
4. Glycine.
5. Phosphate-buffered saline (PBS).
6. Sodium dodecyl sulfate (SDS) lysis buffer: 50 mM Tris-HCl, pH 8.0, 10 mM ethylene diamine tetraacetic acid (EDTA), 1% SDS.
7. Protease inhibitors (working concentrations in SDS lysis buffer, **item 6**): phenylmethylsulfonyl fluoride (PMSF; 1 mM), aprotinin (1 µg/mL), and pepstatin A (1 µg/mL).
8. Branson Sonifier 450 (VWR).
9. Proteinase K.
10. IP dilution buffer: 16.7 mM Tris-HCl, pH 8.0, 167 mM NaCl, 1.2 mM EDTA, 0.01% SDS, 1.1% Triton X-100.
11. MinElute™ PCR purification kit (Qiagen, cat. no. 28004).
12. Salmon sperm DNA.
13. Polyclonal or monoclonal antibodies.
14. RIPA wash buffer: 50 mM Tris-HCl, pH 8.0, 750 mM NaCl, 5 mM EDTA, 0.1% SDS, 1% Triton X-100, 0.1% sodium deoxycholic acid.

15. LiCl wash buffer: 10 mM Tris-HCl, pH 8.0, 0.25 M LiCl, 1 mM EDTA, 1% SDS, 0.5% Triton X-100, 1% sodium deoxycholate.
16. TE buffer: 10 mM Tris-HCl, pH 8.0, 1 mM EDTA.
17. Elution buffer: 50 mM NaHCO$_3$, 1% SDS.
18. Salmon sperm DNA/protein A agarose (Upstate Biotechnology, cat. no. 16-157C).
19. Qiagen PCR purification kit.
20. LightCycler-FastStart DNA Master SYBR Green I kit (Roche, cat. no. 3003230).
21. Barrier pipet tips.
22. Appropriate primer pairs.
23. 0.5-mL PCR tubes.

3. Methods

3.1. Cultured Cells and Virus Infections

Any cell line that is permissive for Ad infection may be used in these analyses. In our work, we utilized N52.E6 cells, which express Ad E1A and E1B proteins *(14)*. Approximately 10^7 cells are present per 100-mm dish. Include one extra dish to be used solely for estimation of cell number. Purified Ad particles were prepared by CsCl equilibrium gradient centrifugation.

1. Infect cells with 100 virus particles/cell for 1 h at 37°C.
2. Remove the virus inoculum, wash the cells twice, and add fresh medium.
3. Incubate infections for 18–24 h, although any suitable time point may be used depending on the nature of the analysis.

3.2. Formaldehyde Crosslinking

1. Aspirate medium from infected cell monolayers and add 10 mL of prewarmed serum-free α-MEM medium per 100-mm dish.
2. Add 37% formaldehyde directly to culture medium to a final concentration of 1%. Incubate for 10 min at 37°C.
3. Stop the crosslinking reaction by the addition of glycine to a final concentration of 125 mM and incubate 5 min at room temperature.

3.3. Cell Lysis

1. Aspirate medium and wash cells twice using ice-cold PBS solution. Add 600 µL of PBS to each plate, scrape the cells from the dish, and transfer into appropriately sized centrifuge tube.
2. Pellet the cells for 5 min at 1000g at 4°C.
3. Resuspend the cell pellet in 600 µL of SDS lysis buffer containing protease inhibitors (*see* **Note 1**). Incubate the resuspended cells 10 min on ice.

3.4. Sonication (see Note 2)

1. Shear the chromatin to 200–1000 base pairs in length with 3 sets of 20- to 30-s pulses, output control at 5, duty cycle at constant (Branson Sonifier). Keep sample tubes on ice between each sonication set for at least 2 min.

2. Centrifuge samples for 10 min at 25,000*g* at 4°C.
3. Collect supernatant, divide into 500-µL aliquots in microfuge tubes, and store samples at –80°C.
4. To check the lengths of DNA fragments following sonication and determine total DNA concentration:
   a. Take 50 µL of each sample to reverse the crosslinking by adding 500 µg/mL Proteinase K; incubate at 65°C overnight.
   b. Recover DNA by phenol/chloroform extraction and ethanol precipitation. Wash DNA with 70% ethanol, air dry.
   c. Suspend pellet in 50 µL of TE. Read absorbance at 260 nm for a 1:20 dilution of sample.
   d. Run 2 µg of DNA on a 0.8% agarose gel along with molecular-weight DNA standards for size comparisons.

3.5. Normalization of Input Chromatin by Real-Time PCR

1. Reverse crosslinks on 20 µL sample as described in **Subheading 3.4.**, **step 4**.
2. Purify the chromatin by using MinElute™ PCR purification kit.
3. Elute DNA using 50 µL of elution buffer.
4. Use 2 µL of each sample as DNA template to do real-time PCR (*see* **Subheading 3.8.**).

Q-PCR gives accurate quantification of target DNA, which allows one to adjust and standardize the input DNA concentration for each ChIP sample.

3.6. Immunoprecipitation

All manipulations are performed at 4°C, and all wash buffers contain protease inhibitors (*see* **Subheading 2.**, **item 7**, and **Note 1**).

1. Aliquot predetermined volume of chromatin (100 µg total DNA) to 1.5-mL microfuge tubes and add IP Dilution Buffer containing protease inhibitors (*see* **Subheading 3.3.**) to a final volume of 1 mL.
2. Add 80 µL of salmon sperm DNA/protein A agarose to preclear samples. Incubate for 1 h at 4°C with rotation.
3. Spin samples in a microfuge at full speed for 1 min at 4°C.
4. Transfer supernatants to new tubes with 15 µL of polyclonal antibody or 2 µg of monoclonal antibody (*see* **Note 3**). Incubate at 4°C overnight with rotation.
5. On the next day, add 60 µL of salmon sperm DNA/protein A agarose, and incubate at 4°C for 1 h with rotation.
6. Pellet the beads in a microfuge at 4000 rpm (1500*g*) for 1 min.
7. Carefully aspirate supernatant.
8. Wash five times with 1 mL of RIPA wash buffer. With each wash, rotate the samples at 4°C for 10 min., spin in a microcentrifuge at 4000 rpm (1500*g*) for 1 min, and remove supernatant.
9. Using this technique, perform the following additional washes of the samples: twice with 1 mL of LiCl wash buffer and twice with 1 mL of TE buffer.

10. Resuspend beads in the final pellet with 100 µL of TE buffer using a blunt 200-µL pipet tip and transfer beads into a new 1.5-mL microfuge tube.
11. Spin samples in a microfuge at 4000 rpm (1500g) for 1 min, discard supernatant, and add 150 µL of freshly prepared elution buffer.
12. Incubate at room temperature for 15 min with rotation.
13. Centrifuge samples in a microfuge at 4000 rpm (1500g) for 1 min and transfer supernatant to a new microfuge tube.
14. Repeat elution and combine the supernatants from the two elution steps into the same tube. Reverse formaldehyde crosslinking by adding 500 µg/mL Proteinase K and incubate at 65°C overnight.

3.7. Purify Immunoprecipitated Chromatin

Use a Qiagen PCR purification kit according to manufacturer's procedure and elute DNA at the last step with 50 µL elution buffer.

3.8. Real-Time PCR (Q-PCR)

Use LightCycler-FastStart DNA Master SYBR Green I kit (Roche) according to the manufacturer's instructions. Consult the kit instruction manual to optimize primer annealing temperature and $MgCl_2$ concentration experimentally. Use barrier pipet tips for all of the following steps. Always include control reactions: a negative control with primers but no input DNA and a series of positive controls for a standard curve. If using different primer pairs for the analyses, standard curves must be established for each primer pair used (*see* **Note 4**).
Prepare the PCR master mix in 0.5 mL PCR tube:

Component	Volume	Final concentration
H_2O	12.4 µL	
$MgCl_2$ (25 mM)	1.6 µL	3 mM*
Forward primer (10 mM)	1.0 µL	0.5 µM
Reverse primer (10 mM)	1.0 µL	0.5 µM
LightCycler-FastStart enzyme, SYBRGreen I and reaction mix	2.0 µL	1X
Total volume 18.0 µL		

*The LightCycler-FastStart DNA Master SYBR Green I contains $MgCl_2$ at 1 mM.

1. Mix gently, spin briefly, and transfer 18 µL into a PCR tube for each reaction.
2. Add 2 µL of DNA template. Mix and spin again.
3. Use forceps to insert glass capillary into centrifuge adaptors in the cooling block box.
4. Transfer the total 20-µL reaction mixture into the plastic container at the top of the capillary. Seal each capillary with a stopper. Always wear new gloves when handling the stopper.

5. To transfer the reaction mix from the plastic container at the top of the capillary into the glass tube, centrifuge adaptors with capillary in a microfuge at 3000 rpm (700*g*) for 5 s. Do not exceed a centrifugation force of 700*g*.
6. Put centrifuge adaptors with capillary back in the cooling block.
7. Load the capillaries from the adaptors into the sample carousel by dropping capillaries into the hole, and push down softly until you feel it snap into final position.
8. Place the carousel containing the samples in LightCycler and close the lid.
9. Click on LightCycler3 front screen icon, enter Run and Open Experiment. Programming:

Program 1: Preincubation and Denaturation of the Template DNA

Cycle program data	Value
Cycles	1
Analysis mode	None
Temperature targets	Segment 1
Target temperature (°C)	95
Incubation time (h:min:s)	10:00
Temperature transition rate (°C/s)	20
Secondary target temperature (°C)	0
Step size (°C)	0
Step delay (Cycles)	0
Acquisition mode	None

Program 2: Amplification of the Target DNA

Cycle program data	Value			
Cycles	38			
Analysis mode	Quantification			
Temperature targets	Segment 1	Segment 2	Segment 3	Segment 4
Target temperature (°C)	95	61	72	80
Incubation time (s)	5	5	10	1
Temperature transition rate (°C/s)	20	20	20	20
Secondary target temperature (°C)	0	0	0	0
Step size (°C)	0	0	0	0
Step delay (Cycles)	0	0	0	0
Acquisition mode	None	None	None	Single

Program 3: Melting Curve Analysis for Product Identification

Cycle program data	Value		
Cycles	1		
Analysis mode	Melting curves		
Temperature targets	Segment1	Segment 2	Segment 3
Target temperature (°C)	95	70	95
Incubation time (h:min:s)	0	15	0
Temperature transition rate (°C/s)	20	20	0.1
Secondary target temperature (°C)	0	0	0
Step size (°C)	0	0	0
Step delay (cycles)	0	0	0
Acquisition mode	None	None	Continuous

Program 4: Cooling the Rotor and Thermal Chamber

Cycle program data	Value
Cycles	1
Analysis mode	None
Temperature targets	Segment 1
Target temperature (°C)	40
Incubation time (h:min:s)	30
Temperature transition rate (°C/s)	20
Secondary target temperature (°C)	0
Step size (°C)	0
Step delay (cycles)	0
Acquisition mode	None

10. After programming, enter "Save Experiment File" to save a new file.
11. Enter "Edit Samples" and indicate sample's name, type, and the known concentration for standards in the spaces provided.
12. Enter "Done."
13. Enter "Run" and a window will appear to name this run and save the experiment.
14. Enter "Done" again. The computer will display real-time Run conditions screen; you can exit run, end program, add more cycles or edit samples any time.
15. When the run is finished, enter "Quantification" or "Melting Curve" to display the analysis and print the report.

Quantification of the recovery of specific DNA sequences for each immunoprecipitation reaction may be obtained from these results by integration of each result with the standard curve (e.g., DNA recovered for each IP in nanograms). The fold-enrichment of DNA recovered for each sample may be obtained using these values.

4. Notes

1. Add protease inhibitors to buffer just prior to use. PMSF has a half-life of approx 30 min in aqueous solution.
2. An important parameter is the quality of the crosslinked DNA, which needs to be effectively sonicated in order to assure the relevance of the results with specific antibodies and the target DNA segment of interest.
3. Another important parameter is the quality of the antibody used for the chromatin immunoprecipitation reactions. Antibodies of high specificity and affinity are required for optimal results. Preimmune serum should be used to verify specificity with polyclonal antibodies, and isotype-matched nonspecific monoclonal antibody should be used to verify specificity with monoclonal antibodies.
4. The third important parameter is to faithfully follow the extensive washing regimen during the immunoprecipitations to reduce background.
5. The fourth important parameter is the quality of the primer pairs used in Q-PCR. It is important that primer pairs with matching annealing temperatures (generally 55–60°C) and a high degree of sequence specificity be used (20 base pairs for Ad DNA, longer for the analysis of genomic DNA).
6. The final important parameter is to be extremely careful during the Q-PCR analysis in establishing standard curves to assure that the results with ChIP samples are within the linear range of the analysis.

References

1. Ostapchuk, P. and Hearing, P. (2003) Regulation of adenovirus packaging. *Curr. Top. Microbiol. Immunol.* **272,** 165–185.
2. Schmid, S. I. and Hearing, P. (1997) Bipartite structure and functional independence of adenovirus type 5 packaging elements. *J. Virol.* **71,** 3375–3384.
3. Schmid, S. I. and Hearing, P. (1998) Cellular components interact with adenovirus type 5 minimal DNA packaging domains. *J. Virol.* **72,** 6339–6347.
4. Gustin, K. E. and Imperiale, M. J. (1998) Encapsidation of viral DNA requires the adenovirus L1 52/55-kilodalton protein. *J. Virol.* **72,** 7860–7870.
5. Hasson, T. B., Soloway, P. D., Ornelles, D. A., Doerfler, W., and Shenk, T. (1989) Adenovirus L1 52- and 55-kilodalton proteins are required for assembly of virions. *J. Virol.* **63,** 3612–3621.
6. Zhang, W., Low, J. A., Christensen, J. B., and Imperiale, M. J. (2001) Role for the adenovirus IVa2 protein in packaging of viral DNA. *J. Virol.* **75,** 10,446–10,454.
7. Zhang, W. and Imperiale, M. J. (2003) Requirement of the adenovirus IVa2 protein for virus assembly. *J. Virol.* **77,** 3586–3594.
8. Gustin, K. E., Lutz, P., and Imperiale, M. J. (1996) Interaction of the adenovirus L1 52/55-kilodalton protein with the IVa2 gene product during infection. *J. Virol.* **70,** 6463–6467.
9. Zhang, W. and Imperiale, M. J. (2000) Interaction of the adenovirus IVa2 protein with viral packaging sequences. *J. Virol.* **74,** 2687–2693.

10. Ostapchuk, P., Yang, J., Auffarth, E., and Hearing, P. (2005) Functional interaction of the adenovirus IVa2 protein with adenovirus type 5 packaging sequences. *J. Virol.* **79,** 2831–2838.

11. Schepers, A., Ritzi, M., Bousset, K., et al. (2001) Human origin recognition complex binds to the region of the latent origin of DNA replication of Epstein-Barr virus. *EMBO J.* **20,** 4588–4602.

12. Wells, J., Graveel, C. R., Bartley, S. M., Madore, S. J., and Farnham, P. J. (2002) The identification of E2F1-specific target genes. *Proc. Natl. Acad. Sci. USA* **99,** 3890–3895.

13. Spector, D. J., Johnson, J. S., Baird, N. L., and Engel, D. A. (2003) Adenovirus type 5 DNA-protein complexes from formaldehyde cross-linked cells early after infection. *Virology* **312,** 204–212.

14. Schiedner, G., Hertel, S., and Kochanek, S. (2000) Efficient transformation of primary human amniocytes by E1 functions of Ad5: generation of new cell lines for adenoviral vector production. *Hum. Gene Ther.* **11,** 2105–2116.

10

Assaying Protein–DNA Interactions In Vivo and In Vitro Using Chromatin Immunoprecipitation and Electrophoretic Mobility Shift Assays

Pilar Perez-Romero and Michael J. Imperiale

Summary

Many events in the viral life cycle involve protein binding to defined sequences on the viral chromosome. Chromatin immunoprecipitation allows the detection of the in vivo interaction of specific proteins with specific genomic regions. In this technique, living cells are treated with formaldehyde to crosslink neighboring protein–protein and protein–DNA molecules. The crosslink with formaldehyde is reversible and covers a short distance (2 Å); the components that are crosslinked are therefore in close proximity. Nuclear fractions are isolated, and the genomic DNA is sheared to reduce the average DNA fragment size to around 500 bp. These nuclear lysates are used in immunoprecipitations with an antibody against the protein of interest. The DNA bound to the studied protein is enriched after the immunoprecipitation. After reversal of the crosslinking, the resulting DNA and proteins can be independently studied.

The electrophoretic mobility shift assay provides a rapid method to study DNA-binding protein interactions in vitro. This assay is based on the observation that complexes of protein and DNA migrate through a nondenaturing polyacrylamide gel more slowly than free DNA fragments. The assay is performed by incubating a purified protein, or a complex mixture of proteins, with a ^{32}P end-labeled DNA probe containing the protein-binding site. The reaction products are analyzed on a nondenaturing polyacrylamide gel. The specificity of the DNA-binding protein for the putative binding site is established by competition experiments using specific and nonspecific nonradiolabeled DNA probes. The components of the complexes can be identified with antibodies to the protein of interest.

Key Words: Chromatin immunoprecipitation; immunoprecipitation; protein–DNA interaction; adenovirus; virus assembly; DNA packaging; gel shift; radiolabeled DNA probe; electrophoretic mobility shift assay.

From: *Methods in Molecular Medicine, Vol. 131:*
Adenovirus Methods and Protocols, Second Edition, vol. 2:
Ad Proteins, RNA, Lifecycle, Host Interactions, and Phylogenetics
Edited by: W. S. M. Wold and A. E. Tollefson © Humana Press Inc., Totowa, NJ

1. Introduction

Association between proteins and DNA is crucial for DNA replication, transcription, recombination, and DNA packaging in adenovirus (Ad) assembly *(1–11)*. Over the years different methods have been developed to covalently bind protein to DNA *in situ* by treating living cells with crosslinking reagents. Formaldehyde crosslinking is reported to occur between exocyclic amino groups and endocyclic imino groups of DNA bases and the side-chain α-amino groups of amino acids *(12,13)*. Crosslinking performed with formaldehyde is fully reversible, allowing further analysis of both DNA and proteins *(14)*. The combination of crosslinking and immunoprecipitation has become a useful method for different studies related to chromatin *(15–17)*. The chromatin immunoprecipitation method (ChIP) allows the identification of the in vivo interaction of a known protein with its DNA-binding site. Recently, using ChIP assays, we and others have examined binding of proteins to the Ad packaging sequence *(3,4)*. The electrophoretic mobility shift assay (EMSA) is used to detect single- and double-stranded DNA-binding proteins from cell nuclear extracts or specific purified proteins in vitro *(18)*. It has been used extensively in the characterization of transcription factors. In Ad studies, EMSAs have been essential to identify proteins that specifically bind to genomic DNA involved in packaging and initiation of transcription of early and late gene *(6,19–23)*. These techniques are broadly applicable to the study of other protein–DNA interactions during the viral life cycle.

2. Materials

1. Hypotonic buffer: 5 mM PIPES, pH 8.0, 85 mM KCl, 0.5% NP-40, 0.5 mM phenylmethylsulfonyl fluoride (PMSF). Store at 4°C.
2. Lysis buffer: 1% sodium dodecyl sulfate (SDS), 10 mM ethylene diamine tetraacetic acid (EDTA), 50 mM Tris-HCl, pH 8.1. Store at 4°C.
3. Protease inhibitors: 1 μM PMSF, 5 μg/mL aprotinin, 5 μg/mL leupeptin. Added to solutions as required, not stable for more than a few hours.
4. ChIP dilution buffer: 1% Triton X-100, 2 mM EDTA, 150 mM NaCl, 20 mM Tris-HCl, pH 8.1. Store at 4°C.
5. Wash buffer I: 0.1% SDS, 1% Triton X-100, 2 mM EDTA, 20 mM Tris-HCl, pH 8.1, 150 mM NaCl. Store at 4°C.
6. Wash buffer II: 0.1% SDS, 1% Triton X-100, 2 mM EDTA, 20 mM Tris-HCl, pH 8.1, 500 mM NaCl. Store at 4°C.
7. Wash buffer III: 0.25 M LiCl, 1% NP-40, 1% sodium deoxycholate, 1 mM EDTA, 10 mM Tris-HCl, pH 8.1. Store at 4°C.
8. TE (Tris-EDTA) buffer: 10 mM Tris-HCl, pH 8.0, 1 mM EDTA.
9. Elution buffer: 1% SDS, 0.1 M NaHCO$_3$. Make fresh as required by adding the NaHCO$_3$ to the solution containing the SDS. Store at room temperature, as the salts precipitate at 4°C.
10. Specific and control antibodies.

11. Rabbit IgG (Sigma).
12. Protein A or G sepharose (Amersham Biosciences). Protein G and protein A have different IgG-binding specificities, dependent on the origin (species) of the IgG. The Sepharose more appropriate for the antibodies used should be determined.
13. Glycogen (Roche).
14. Salmon sperm DNA (Invitrogen).
15. 2.5 M Glycine. Store at room temperature.
16. Formaldehyde, 37%, reagent grade (Sigma).
17. Dounce homogenizer.
18. Fetal bovine serum (FBS).
19. Dulbecco's modified Eagle's medium (DMEM).
20. Phosphate-buffered saline (PBS).
21. Trichloroacetic acid.
22. 2X SDS sample buffer: 100 mM Tris-HCl, pH 6.8, 200 mM dithiothreitol (DTT), 4% SDS, 0.2% bromophenol blue, 20% glycerol. Store at –20°C.
23. EMSA nuclear extract buffer A: 10 mM N-2-hydroxyethylpiperazine-N'-2-ethanesulfonate (HEPES), pH 7.9, 10 mM KCl, 1.5 mM MgCl$_2$, 0.5 mM DTT, 0.5 mM PMSF.
24. EMSA nuclear extract buffer B: 20 mM HEPES, pH 7.9, 25% glycerol, 420 mM NaCl, 1.5 mM MgCl$_2$, 0.2 mM EDTA, 0.5 mM DTT, 0.5 mM PMSF.
25. Protein assay reagent (Bio-Rad).
26. T4 polynucleotide kinase (Invitrogen).
27. Poly(dI-dC)·poly(dI-dC) (Amersham Pharmacia Biotech).
28. γ-^{32}P ATP at 3000 Ci/mmol (Amersham Pharmacia Biotech).
29. 30% Acrylamide–0.8% *bis*-acrylamide solution.
30. X-ray film from Kodak (Rochester, NY).
31. EMSA binding buffer: 10 mM HEPES, pH 7.9, 20 mM KCl, 3 mM MgCl$_2$, 10 mM EDTA, 12% glycerol, and 1 mM DTT. Prepare in 1-mL aliquots and freeze at –20°C. Add bovine serum albumin (BSA) at 100 ng/µL at time of use.
32. 1X Annealing buffer: 10 mM Tris-HCl, pH 8.0, 1 mM EDTA, 50 mM NaCl.
33. 5X Forward buffer: 350 mM Tris-HCl, pH 7.6, 50 mM MgCl$_2$, 500 mM KCl, 5 mM BME.
34. 10X TBE buffer: 900 mM Tris base, 900 mM boric acid, 20 mM EDTA.
35. 10X TG buffer: 250 mM Tris base, 1.9 M glycine, 10 mM EDTA.
36. N,N,N',N'-Tetramethyl-ethylenediamine (TEMED).
37. 10% Ammonium persulfate. Dissolve in distilled water and filter-sterilize. Make 500-µL aliquots and keep at –20°C until used.
38. Secondary horseradish peroxidase-conjugated antibody.
39. ECL plus Western Blotting Detection System (Amersham-Pharmacia Biotech).

3. Methods

The methods described below outline (1) Ad infection of susceptible human cells and preparation of crosslinked nuclear extracts, (2) immunoprecipitation of the crosslinked molecules, (3) analysis of the immunoprecipitated DNA

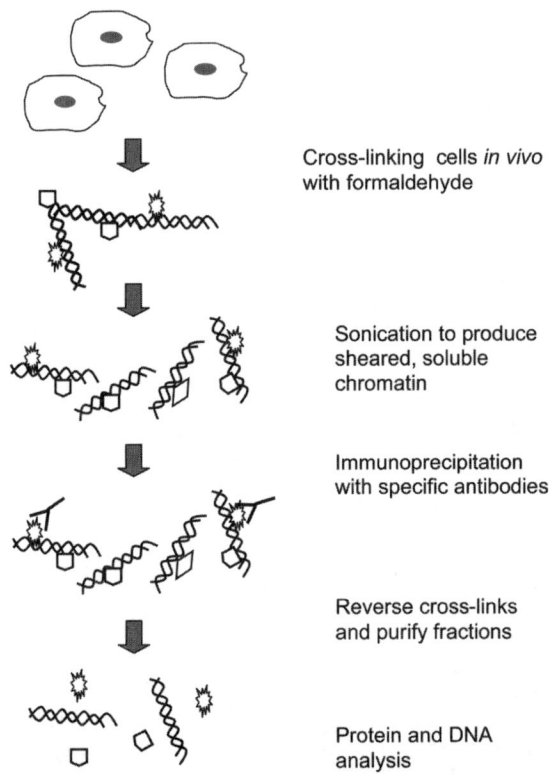

Cross-linking cells *in vivo*
with formaldehyde

Sonication to produce
sheared, soluble
chromatin

Immunoprecipitation
with specific antibodies

Reverse cross-links
and purify fractions

Protein and DNA
analysis

Fig. 1. Schematic of chromatin immunoprecipitation methods. Infected cells are crosslinked in vivo with formaldehyde. Nuclear extracts are isolated and chromatin is sheared by sonication. Antibodies against the proteins of interest are used for immunoprecipitations. The isolated protein–DNA complexes are de-crosslinked and fractions purified for further analysis. (Adapted from **ref. 15**.)

complexes (**Fig. 1**), (4) preparation of the proteins for EMSAs, (5) preparation of the DNA probes, and (6) gel shift assays.

Because of the nature of viral infections, a major concern can be cross-contamination between samples, so we recommend being extremely rigorous during the whole process, especially when ChIP assays, which rely on polymerase chain reaction (PCR), are performed (*see* **Note 1**).

3.1. Cell Infection and Extract Preparation

Any cell types susceptible to Ad can be used for this experiment. The conditions described below have been optimized for 293 cells, Ad-transformed human embryonic kidney cells (*24*).

3.1.1. Cell Infection

1. Grow 293 cells in two 15-cm plates to 85% confluency (~2 × 10^7 cells). One of the dishes will be used for Ad5 infection and the second as a mock-infected control.
2. Remove cell medium from the dishes and inoculate one dish with Ad serotype 5 (Ad5) at 10 PFU/cell in 4 mL of DMEM medium supplemented with 2% FBS for 2 h at 37°C. Add to the second dish Ad-free DMEM medium supplemented with 2% FBS. Cells are rocked every 10–15 min.
3. After this time, remove medium and add 20 mL of DMEM medium supplemented with 10% of FBS, and incubate at 37°C. The length of infection can be altered depending on the particular protein–DNA interaction of interest.

3.1.2. Cell Crosslinking

1. Remove the medium from the cells (*see* **Subheading 3.1.1.**, **step 3**) and add 5 mL of 1% formaldehyde in PBS. If many of the infected cells are detached from the dish as a result of the virus-induced cytopathic effect, scrape cells and centrifuge at 2000*g* for 5 min, then add 1% formaldehyde in PBS to the pellet and resuspend carefully. Incubate with the formaldehyde solution for 10 min at room temperature.
2. Stop crosslinking by adding 500 μL of 1.25 *M* glycine (to a final concentration of 125 m*M*) and incubate for 5 min at room temperature.
3. If the crosslinking was performed in the monolayer, scrape cells from the dishes and centrifuge cells at 2000*g* for 5 min.
4. Wash twice with PBS containing protease inhibitors. After each wash, centrifuge cells at 2000*g* for 5 min.

3.1.3. Preparation of Nuclear Extracts

1. After the last PBS wash (*see* **Subheading 3.1.2.**, **step 4**), resuspend the pellet in 1 mL of hypotonic buffer and incubate for 10 min on ice.
2. Transfer the cells to a glass Dounce homogenizer, and lyse them with 25 strokes of a tight-fitting pestle (*see* **Note 1**).
3. To isolate the nuclear fraction from the cytoplasm, after homogenization (*see* **Subheading 3.1.3.**, **step 2**) centrifuge samples at 10,000*g* for 1 min at 4°C and discard the supernatant containing the cytoplasmic fraction.
4. Wash the pellets containing the nuclei, to avoid any cytoplasmic residue, twice with 1 mL of hypotonic buffer, resuspending and centrifuging as in **Subheading 3.1.3.**, **step 3**.
5. After the last wash, resuspend the nuclear pellet in 400 μL of lysis buffer containing protease inhibitors and incubate on ice for at least 30 min but not more than 1 h.

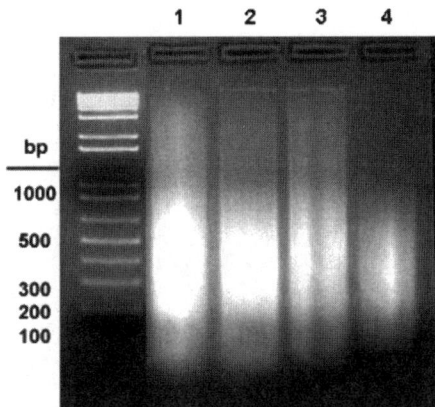

Fig. 2. Optimization of the DNA shearing conditions. De-crosslinked input DNA (from **Subheading 3.2.2.**, **step 4**) was subjected to sonication (550 Sonic Dismembrator, Fisher Scientific) on ice under different conditions. Lane 1: 40% output, 2 pulses of 10 s. Lane 2: 30% output, 2 pulses of 10 s. Lane 3: 30% output, 3 pulses of 10 s. Lane 4: 20% output, 4 pulses of 10 s. The optimal average chromatin fragment size is approx 500 bp and smaller, as shown in lane 4.

3.1.4. Sonication

The chromatin is sheared by sonication (*see* **Note 1**). The conditions should be determined empirically and the chromatin fragment size analyzed by running the sheared, de-crosslinked (*see* **Subheading 3.2.2.**) DNA on an agarose gel before performing the immunoprecipitation (**Fig. 2**); it is critical to break the DNA to fewer than 500-bp fragments to determine as exactly as possible the sequences with which the proteins associate. Factors affecting sonication include the volume of the sample, depth of the sonication probe, sonication strength, and the duration of sonication *(15,25)*. Sample volume should not exceed 1 mL, and the sonication depth can be increased if the samples are in microfuge tubes.

1. Set the sonicator (550 Sonic Dismembrator, Fisher Scientific) at 20% output. Give four pulses of 10 s on ice to the nuclear lysates (*see* **Subheading 3.1.3.**, **step 5**), leaving 1 min between pulses. The average Ad chromatin fragment size is about 500 bp and smaller using these conditions (*see* **Fig. 2**, lane 4).
2. Sonicate a solution of 1 µg/µL of salmon sperm DNA, which will be used later to prevent nonspecific binding.
3. Centrifuge sonicated samples (**Subheading 3.2.2.**, **steps 1** and **2**) for 10 min at 10,000*g* at 4°C to remove cell debris. Save the supernatants in clean microfuge tubes as crosslinked input samples. The input samples can be stored at –80°C for several months.

3.2. Immunoprecipitation

Immunoprecipitation is one of the most critical steps in the procedure. The ability of the antibodies used in the ChIP assay to immunoprecipitate the protein of interest should be determined previously in nonfixed cells. The amount of antibody used as well as the optimal time of immunoprecipitation should be determined empirically. Polyclonal antibodies are preferred over monoclonal antibodies to overcome epitope masking *(15–17)*.

3.2.1. Immunoprecipitations and Reverse Crosslinking

1. For immunoprecipitations, dilute 100 µL of crosslinked input samples (from **Subheading 3.1.4.**, **step 3**) 10 times with ChIP dilution buffer containing protease inhibitors.
2. To reduce nonspecific binding *(25)*, preclear lysates by adding 40 µL protein A or protein G Sepharose (50% slurry, Amersham Biosciences) and 2 µg of sonicated salmon sperm DNA (*see* **Subheading 3.1.4.**, **step 2**) and incubate 2 h at 4°C with rotation.
3. Centrifuge mixture at 1000*g* in a refrigerated microcentrifuge for 1 min. Discard the pellet and move the supernatant to a clean tube without taking any Sepharose beads.
4. Add antibody to the precleared samples (*see* **Subheading 3.2.1.**, **step 3**) and incubate 2 h to overnight with rotation at 4°C.
5. Collect immunocomplexes (*see* **Subheading 3.2.1.**, **step 4**) by adding 50 µL of protein G (or protein A) Sepharose (Amersham Biosciences) and incubating for at least 1 h at 4°C with rotation.
6. Centrifuge the beads at 1000*g* for 1 min in a refrigerated microcentrifuge and discard the supernatant.
7. Wash the beads for 10 min with rotation, once with each of the washing buffers (I, II, and III), and twice with TE. After each wash, centrifuge as in **Subheading 3.2.1.**, **step 6** and discard the supernatant.
8. Elute immunocomplexes from the beads by incubating with two 250-µL aliquots of elution buffer for 10 min at room temperature with rotation.
9. After each elution, centrifuge the beads at 10,000*g* for 1 min in a room temperature microcentrifuge and save the supernatant. Do not centrifuge at 4°C; salts in the elution buffer will precipitate. Combine the two sequential eluted supernatants in a single tube.
10. Reverse crosslinking by adding 50 µL of 2 *M* NaCl to the 500 µL eluted complexes (*see* **Subheading 3.2.1.**, **step 9**) to a final concentration of 200 m*M* and heating the samples for 4 h at 65°C.
11. In parallel prepare de-crosslinked input sample controls by mixing 100 µL of the crosslinked input samples (*see* **Subheading 3.1.4.**, **step 3**) with 400 µL of elution buffer. For de-crosslinking, treat the diluted input samples as in **Subheading 3.2.1.**, **step 10**.

3.2.2. Isolation of DNA

To isolate immunoprecipitated and input chromatin, de-crosslinked samples (*see* **Subheading 3.2.1, steps 10** and **11**) are treated with Proteinase K. Two hundred and fifty microliters of the de-crosslinked input samples will be used to isolate the input DNA (*see* **Note 2**).

1. Add 20 μg of Proteinase K and incubate for 1 h at 45°C.
2. Extract DNA from the protein fraction by adding 2 vol of phenol:chloroform: isoamyl alcohol (25:24:1), homogenize by vortexing for a few seconds and spinning down for 5 min at 10,000*g*. Collect the supernatant and repeat once.
3. Recover DNA by adding 20 μg of glycogen to the supernatant (*see* **Subheading 3.2.2., step 2**) followed by standard ethanol precipitation *(18)*, and resuspend the resulting DNA pellet in 30 μL of distilled water. We refer to the DNA obtained from the immunoprecipitations as ChIP DNA and that obtained from the input samples as input DNA (**Fig. 2**).

3.2.3. Isolation of the Proteins

Two hundred and fifty microliters of the de-crosslinked samples (*see* **Subheading 3.2.1., steps 10** and **11**) are used to isolate immunoprecipitated and input proteins (*see* **Note 3**).

1. Add an equal volume of 100% trichloroacetic acid to the samples and incubate for 20 min on ice.
2. Centrifuge at 4°C for 10 min at 10,000*g*, discard the supernatant, and wash the pellet with 200 μL of cold acetone.
3. Centrifuge at room temperature for 5 min at 10,000*g* and air-dry the pellet.
4. Dissolve in 30–50 μL 2X SDS sample buffer.

3.3. Analysis

3.3.1. Polymerase Chain Reaction

The ChIP DNAs (*see* **Subheading 3.2.2., step 3**) are used as a template for PCR amplification to detect if the DNA site in study was crosslinked to the immunoprecipitated proteins. To demonstrate the specificity of the immunoprecipitated DNA, the ChIP DNAs are used as templates with primers to amplify other regions of the Ad genome. Design primer pairs that yield products approx 200–300 bp in length, smaller than the average size of the fragmented chromatin. The input DNAs (*see* **Subheading 3.2.2., step 3**) are used as control templates to confirm the presence of all the sequence targets before immunoprecipitation.

1. Prepare a master mixture for each primer set. Templates used for PCR will include ChIP DNAs from the different immunoprecipitations and input DNA controls (*see* **Subheading 3.2.2., step 3**).

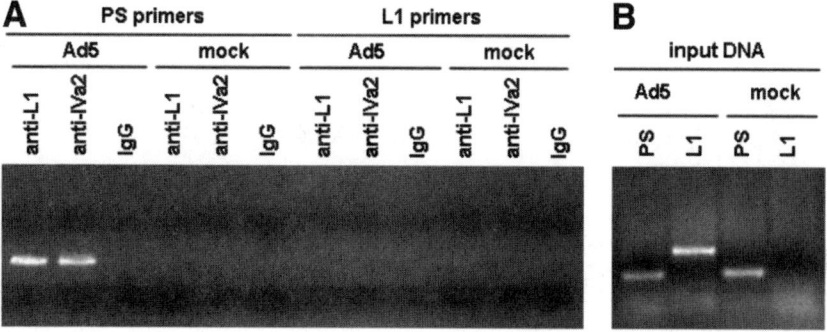

Fig. 3. The L1 52/55 kDa and IVa2 proteins interact in vivo with the packaging sequence. Chromatin immunoprecipitation (ChIP) assays performed in 293 cells infected with Ad5 at 10 PFU/cell and mock-infected cells. (**A**) Crosslinked nuclear lysates were used in immunoprecipitations with antibodies to L1 or IVa2 or control IgG. Polymerase chain reaction using ChIP DNAs as template confirmed the presence of the packaging sequence (PS) when immunoprecipitations were performed with either anti-L1 or anti-IVa2 antibodies. No PS is detected neither in the IgG control or mock-infected ChIP DNAs. No product was obtained in any reaction when the same ChIP DNA templates were used with primers to amplify the L1 open reading frame (ORF). (**B**) Control PCR to confirm presence of the PS and L1 ORF fragment in the adenovirus 5 input chromatin DNA. The PS in the mock-infected 293 cells is from the left end of the viral genome, which has integrated into the cellular chromosome *(24)*.

2. Each of the 50 µL PCR reactions must contain 0.2 m*M* dNTP, 2 m*M* MgSO$_4$, 0.2 µ*M* of each primer, 1–2.5 U Taq polymerase. Add 2 µL of the templates to each tube containing the PCR mixture.
3. Perform the PCR reaction using the following parameters: 1 cycle: 94°C for 3 min; 20–30 cycles: 94°C for 1 min, 50–55°C (depending on the annealing temperature of the primer used) for 30 s, 68–72°C (depending of the Taq polymerase used) for 30 s; 1 cycle: 68–72°C for 5 min *(18)* (*see* **Note 4**).
4. Analyze the PCR reaction products by 2% agarose gel electrophoresis of a 10-µL aliquot of the total reaction. The products should be visible by UV transillumination of the ethidium bromide-stained gel (**Fig. 3**; *see* **Note 5**).

3.3.2. Western Blot

1. Boil isolated proteins (input proteins and ChIP complexes; *see* **Subheading 3.2.3., step 4**) in 2X SDS sample buffer and separate them in a 10% SDS-polyacrylamide gel.
2. Transfer to nitrocellulose membrane using standard methods *(18)*.
3. Test the presence of the immunoprecipitated proteins by probing for the proteins of interest. To avoid antibody cross-reaction, use for Western detection an antibody from a different species to the one used for immunoprecipitations.

4. Run as well control input proteins (*see* **Subheading 3.2.3.**) to detect the protein before the immunoprecipitation and to compare with the immunoprecipitated proteins.
5. Visualize proteins using a secondary horseradish peroxidase-conjugated antibody and chemiluminescence detection as recommended by the manufacturer (Amersham-Pharmacia Biotech; *see* **Note 5**).

3.4. Preparation of the Proteins From Cell Nuclear Extracts for EMSA

The use of a complex mixture of proteins from Ad-infected cells for EMSA allows one to determine proteins that bind to a specific probe compared to a mock-infected cell control. However, the complexity of the mixture does not allow one to determine whether it is a direct interaction or other proteins are involved. In contrast, the use of purified proteins for EMSA allows one to determine whether a specific protein binds directly to a specific target DNA sequence (*see* **Note 6**).

3.4.1. Preparation of Cell Nuclear Extracts

Cells in 10-cm dishes infected with Ad5 at 10 PFU/cell or mock-infected cells are used to extract the nuclear proteins at the desired time postinfection (*see* **Subheading 3.1.1.**).

1. Scrape infected and mock-infected cells from the monolayer, move the cell suspension to a conical 15-mL tube, and centrifuge at 2000*g* for 5 min. Discard supernatant.
2. Wash cell pellets twice by resuspending carefully in 5 mL of PBS.
3. After last wash (*see* **Subheading 3.4.1.**, **step 2**), resuspend the cell pellet in 4 pellet volumes of nuclear extract buffer A and incubate on ice for 1 h.
4. Transfer the cells to a Dounce homogenizer and lyse them with 20 strokes of a tight-fitting pestle (*see* **Note 7**).
5. Move the homogenized samples (*see* **Subheading 3.4.1.**, **step 4**) to a microfuge tube and centrifuge them at 2000*g* for 5 min at 4°C.
6. Wash the pellet containing the nuclei with 1 mL of buffer A to eliminate cytoplasmic contaminants.
7. Resuspend nuclei pellet (*see* **Subheading 3.4.1.**, **step 6**) in 3 pellet volumes of nuclear extract buffer B and incubate on ice for 30 min.
8. Centrifuge mixture (*see* **Subheading 3.4.1.**, **step 7**) at 12,000*g* for 30 min and collect the supernatant in a new tube.
9. Determine protein concentration by standard Bradford assay (*18*).
10. Store at –80°C in aliquots to avoid repetitive freezing and thawing of the protein mixture.

3.5. Preparation of the DNA Probe

3.5.1. Design of the DNA Probe

The target DNAs used in EMSA as probes are linear fragments containing the binding sequence of interest.

1. For short oligonucleotides (20–50 bp), oligonucleotides can be purchased from a manufacturer (*see* **Note 8**).
2. For longer probes (100–500 bp), design specific primers and perform a standard PCR to amplify the DNA of interest from a known template (*see* **Note 9**).

3.5.2. Annealing Complementary Pairs of Oligonucleotides

The efficiency of oligonucleotide annealing is critical for the quality of the EMSA results. Single-stranded oligonucleotides are annealed with the complementary strand. The efficiency of the annealing depends on the salt concentration and the rate of temperature decrease, especially when the oligonucleotides are GC-rich or may form hairpin structures.

1. Prepare 100 ng/µL DNA solution as a concentrated stock. The oligonucleotides are more stable when stored at a high concentration. Dilute and prepare as needed.
2. Mix an equal amount of concentrated complementary oligonucleotides to a final concentration of 50 ng/µL in 1X annealing buffer.
3. Incubate mixture at 95°C for 5 min and slowly cool down the sample to room temperature to avoid formation of secondary structures. Incubations can be performed using a thermocycler with a simple protocol: 1 cycle: 95°C for 5 min; 70 cycles: 95°C (–1°C/cycle) for 1 min, 1 cycle: 4°C hold.

3.5.3. Radiolabeling of the Probe

Polynucleotide kinase, which catalyzes the transfer of the γ-phosphate of ATP to a 5'- hydroxyl group of DNA is used to radiolabel the probe *(26)*.

1. In a microfuge tube, mix the following components in a total 30-µL reaction: 100 ng of double-stranded DNA probe (*see* **Subheading 3.5.2.**, **step 3**), 6 µL of 5X forward buffer, 5 µL γ-^{32}P ATP (10 µCi/µL, 3000 Ci/mmol), 10 U of T4 polynucleotide kinase (Invitrogen), and distilled water to bring the volume to 30 µL.
2. Give a short pulse to the samples in a microcentrifuge to bring all the components to the bottom of the tubes.
3. Incubate reaction at 37°C for 1 h.
4. The reaction can be stopped (although this is not absolutely required) by adding EDTA to 5 m*M* final concentration or by heat inactivation for 10 min at 65°C.
5. Purify radiolabeled DNA (*see* **Subheading 3.5.3.**, **step 4**) from unincorporated nucleotide by using a G-25 Sephadex spin column (Roche), following the manufacturer's directions.

6. After purification of the probe, measure incorporated radioactivity by adding 1 µL of the radiolabeled probe (*see* **Subheading 3.5.3., step 5**) to liquid scintillation solution and measuring radioactivity in a scintillation counter.

3.6. Gel Shift Assay

The conditions described below have been optimized for detection of DNA binding of the Ad IVa2 protein, which is around 50 kDa in size, to the packaging sequence *(3,21)*.

3.6.1. Acrylamide Gel Preparation

Nondenaturing TBE–acrylamide gels or TG–acrylamide gels are used to resolve DNA–protein complexes from free DNA probe. The running buffer composition influences DNA–protein complexes detected by EMSA *(27)*. The percentage of acrylamide used in the gel varies depending on the size of the protein in study. We routinely use 4% acrylamide in 0.5X TBE buffer when using nuclear extracts and 5% acrylamide in 0.5X TG buffer for purified proteins.

1. Prepare 50 mL per gel of acrylamide solution containing 4–5% acrylamide, 2.5 mL of 10X TBE buffer or 10X TG buffer (to a final concentration of 0.5X), 250 µL 10% ammonium persulfate, 50 µL TEMED, and distilled water to complete final volume.
2. Set up a standard vertical unit (Hoefer SE600, 8 × 16 × 24 cm; Amersham Pharmacia Biotech) for gel electrophoresis. This system allows one to control the temperature with a built-in heat exchanger, which is key for reproducible electrophoresis results.
3. Clean the plates with distilled water to avoid ionic detergent residues. Cast a single or multiple gels, depending of the number of samples. Mark and number position of wells with indelible marker to assist in loading.
4. Pre-run the gel in 0.5X TBE buffer or 0.5X TG buffer for 1 h at 150 V before loading the samples.

3.6.2. DNA Binding Reaction

While the gel is pre-running **(Subheading 3.6.1., step 4)**, set up the binding reactions. Binding conditions used for EMSA must be optimized for the protein of interest. Conditions include salt concentration, temperature, pH, metal requirement, glycine concentration, and nonspecific DNA *(27,28)*.

Before performing the final experiment, the optimal amount of protein used to detect DNA–protein complexes must be determined by titrating either the nuclear extracts (from 500 ng to 4 µg) or purified proteins (from picomolar to nanomolar). Titrate as well the amount of salmon sperm DNA or poly(dI-dC)·poly(dI-dC) (200 ng–1 µg), which blocks nonspecific DNA binding, cold probe (50–200X excess) for competition experiments, and antibodies (1/10–1/100 dilution) to the specific protein of interest for supershift assays.

1. Prepare protein stock solutions to half of the final concentration so the same volume of protein is consistently added to the reaction. Use EMSA binding buffer to dilute proteins (*see* **Note 10**).
2. Prepare probe master mixture containing 2.5 ng of radiolabeled probe (*see* **Subheading 3.2.3.**, **step 6**), 400 ng of poly(dI-dC)·poly(dI-dC) for purified proteins, or 1 μg of poly(dI-dC)·poly(dI-dC) for nuclear extracts, and adjust final volume with EMSA binding buffer. Calculate final volume to be able to add 4 μL of the probe master mixture to each EMSA reaction.
3. For supershifts, prepare antibody stock solutions to add 2 μL to the EMSA reaction. Use specific antibody for the protein of interest (**Fig. 4**). As negative controls use an antibody against a nonrelated protein, preimmune serum, or an immunoglobulin of the same species subclass as the antibody of interest to demonstrate specificity. Antibodies are usually used at high-concentration: 1/10–1/20 dilution (*see* **Note 11**).
4. For competition experiments, prepare a stock probe mixture containing excess of cold DNA probe to be added to the EMSA reaction (**Fig. 4B**).
5. Mix all the components for each reaction. For each of the conditions set up reactions as follows:
   a. Lane 1: protein + ^{32}P probe.
   b. Lane 2: protein + ^{32}P probe + cold competitor.
   c. Lane 3: protein + ^{32}P probe + specific antibody.
   d. Lane 4: protein + ^{32}P probe + non-related antibody.
   e. Lane 5: protein + ^{32}P probe + control IgG.
   f. Lane 6: ^{32}P probe.
6. Adjust the final volume with EMSA binding buffer to 10 μL for purified protein, or 20 μL for nuclear extracts.
7. Incubate 10–20 min at room temperature (sometimes incubation on ice may work better) and then load onto the acrylamide gel. Do not add dye to the reactions, as the dye might interfere with the DNA–protein binding. Instead, add bromophenol blue dye in a blank lane to follow the rate of migration.
8. Run at 200 V for 2 h. Stop the electrophoresis before the dye reaches the end of the plate (approx 1.5 cm from the edge) to avoid radioactive contamination of the running buffer.

3.6.3. Gel Drying and Detection

1. Place the gel on a 3MM Whatman filter paper and dry in a gel dryer for 1 h at 80°C.
2. To visualize complexes, expose the dried gel to film (*see* **Note 12**).

4. Notes

1. Cross-contamination between samples can occur in some of the steps of the process such as douncing and sonication. In the case of virus-infected cells, any

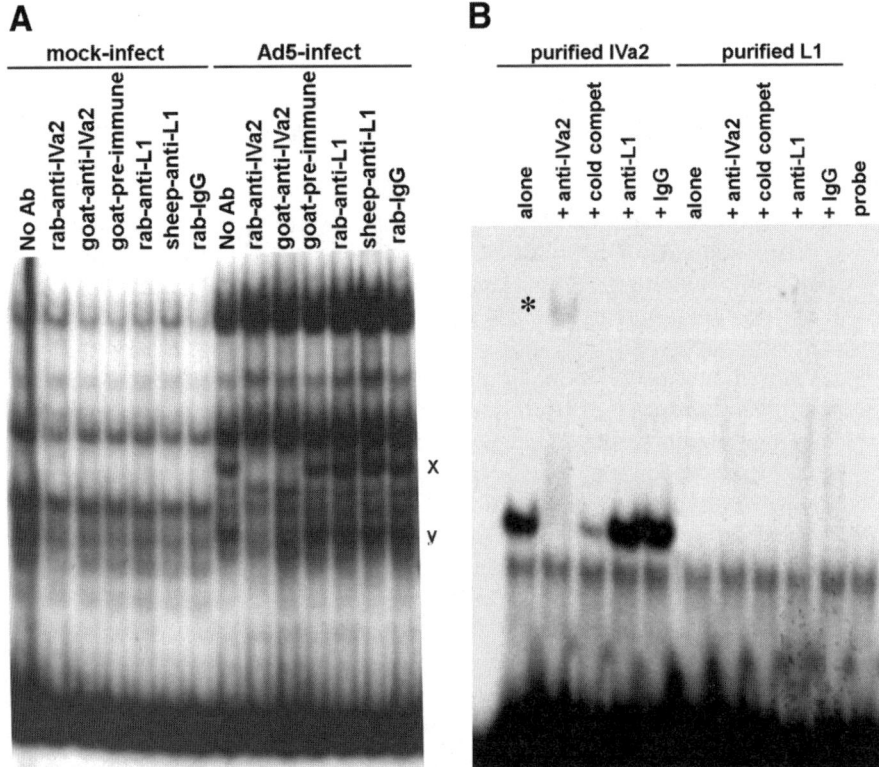

Fig. 4. Electrophoretic mobility shift assays were performed with a ^{32}P-labeled probe containing the binding site for the IVa2 protein. (**A**) Supershift analyses were performed using nuclear extracts prepared from mock-infected and adenovirus (Ad)5-infected 293 cells and rabbit (rab) and goat antibodies to IVa2, rabbit anti-L1 52/55 kDa, and as a control, goat pre-immune serum and rabbit IgG. Complexes x and y are specific in the Ad5-infected 293 nuclear extracts as compared with mock-infected nuclear extracts. Furthermore, IVa2 protein is part of both complexes x and y, as demonstrated by supershift analysis with antibody to IVa2 but not with antibody to L1 52/55 kDa protein, pre-immune serum, or rabbit IgG (*3,21*). (**B**) Electrophoretic mobile shift assay was performed using purified proteins. A supershifted band labeled with an asterisk is detected when anti-IVa2 antibody, but not anti-L1 or IgG, was used for supershifts. Specificity of the interaction is demonstrated as well by using 100X excess of cold probe.

small contamination will show up at the PCR analysis, but it will probably not be detectable at the protein level. We recommend using a different Dounce homogenizer for each sample, when possible, and rigorous cleaning of the sonication tip in between samples. Some authors have reported shearing the DNA by endonucleave treatment. However, fixed cells are highly resistant to restriction enzymes or DNase I treatment. As an alternative to sonication, DNA can be sheared by passing the samples through a series of hypodermic needles, 20 times for each diameter. The process is repeated with decreasing needle diameters (18G, 21G, and 25G).

2. DNA isolation: To isolate the immunoprecipitated DNA after de-crosslinking, one can use a PCR purification kit (Qiagen). Following the manufacturer's protocol, the DNA is separated from the proteins in the mixture. To elute the DNA from the column add 30 μL of distilled water.

3. Protein isolation: alternatively, proteins can be isolated from the first phenol–chloroform phase by the addition of BSA, 10 M H_2SO_4, and acetone overnight at –20°C, followed by centrifugation at 14,000g and washing with cold acetone *(16)*. To avoid excessive dilution of the immunoprecipitated complexes in the elution step, two consecutive elutions can be performed by adding 50 μL of an alternative elution buffer (containing 50 mM Tris-HCl, pH 8.0, 1% SDS, 10 mM EDTA). If the protein concentration is high enough after the elution from the beads, the 2X SDS sample buffer can be added directly to the de-cross-linked samples to be analyzed.

4. PCR conditions: Depending on the antibody used for the immunoprecipitation, the amount of DNA in the template sample might be different. Conditions for the primer pair used, the number of cycles, and the amount of template necessary in each case for the PCR must be empirically determined. We recommend performing the PCR with decreasing amount of template to determine the minimum required. Different Taq polymerases can be used; we recommend the use of High Fidelity Taq polymerase, which yields the least background in our hands.

5. ChIP assay limitations: Although the ChIP assay is a useful tool to study in vivo protein–DNA interactions, it has several limitations. It is a qualitative approach, and it does not determine if the interaction of a protein with a DNA target site is direct or is mediated by another protein(s). Furthermore, while the co-precipitated protein–DNA complex will contain the binding site of interest, the DNA fragment size of up to 500 bp expands the target site at both sides. To determine whether the interaction of the protein with the DNA might be within the expected binding site or in any of the surrounding sequences, various primer pairs can be used to map the exact DNA sequence.

6. Engineering a tagged version of the protein of interest makes the purification process faster and easier.

7. The homogenization of the cells to obtain the nuclei can be monitored by examining the samples at the microscope. Nuclei must be intact. The nuclei can be stained with Trypan blue; intact cells will not take up the dye while nuclei will.

8. For short oligonucleotides (20–50 bp), synthesis of the oligonucleotides can also be carried out by using short primers at each of the ends and extending using the Klenow fragment of DNA polymerase following a standard protocol *(18)*.

9. Large probes can also be obtained by preparing restrictions fragments containing the DNA binding site of interest.

10. Other EMSA binding buffers can be used, such as 80 mM KCl, 10% glycerol, 15 mM Tris-HCl, pH 7.9, 0.2 mM EDTA, 0.4 mM DTT, and 100 ng/µL BSA.
11. The use of DTT in the mixture for supershift reaction may affect the activity of the antibody.
12. Protein–DNA complexes can also be analyzed by exposing the dried gel to a PhosphorImager plate.

Acknowledgments

We thank the members of the Imperiale laboratory for help with this work and discussions. This work was supported by an award from the American Heart Association to P.P-R and by R01 AI52150 from the NIH to M.J.I.

References

1. Zhang, W., Low, J. A., Christensen, J. B., and Imperiale, M. J. (2001) Role for the adenovirus IVa2 protein in packaging of viral DNA. *J. Virol.* **75,** 10,446–10,454.
2. Zhang, W. and Imperiale, M. J. (2003) Requirement of the adenovirus IVa2 protein for virus assembly. *J. Virol.* **77,** 3586–3594.
3. Perez-Romero, P., Tyler, R. E., Abend, J. R., Dus, M., and Imperiale, M. J. (2005) Analysis of the interaction of the adenovirus L1 52/55-kilodalton and IVa2 proteins with the packaging sequence in vivo and in vitro. *J. Virol.* **79,** 2366–2374.
4. Ostapchuk, P., Yang, J., Auffarth, E., and Hearing, P. (2005) Functional interaction of the adenovirus IVa2 protein with adenovirus type 5 packaging sequences. *J. Virol.* **79,** 2831–2838.
5. Ostapchuk, P., Diffley, J. F., Bruder, J. T., Stillman, B., Levine, A. J., and Hearing, P. (1986) Interaction of a nuclear factor with the polyomavirus enhancer region. *Proc. Natl. Acad. Sci. USA* **83,** 8550–8554.
6. Lutz, P., Puvion-Dutilleul, F., Lutz, Y., and Kedinger, C. (1996) Nucleoplasmic and nucleolar distribution of the adenovirus IVa2 gene product. *J. Virol.* **70,** 3449–3460.
7. Hearing, P. and Shenk, T. (1986) The adenovirus type 5 E1A enhancer contains two functionally distinct domains: one is specific for E1A and the other modulates all early units in cis. *Cell* **45,** 229–236.
8. Hasson, T. B., Soloway, P. D., Ornelles, D. A., Doerfler, W., and Shenk, T. (1989) Adenovirus L1 52- and 55-kilodalton proteins are required for assembly of virions. *J. Virol.* **63,** 3612–3621.
9. Hasson, T. B., Ornelles, D. A., and Shenk, T. (1992) Adenovirus L1 52- and 55-kilodalton proteins are present within assembling virions and colocalize with nuclear structures distinct from replication centers. *J. Virol.* **66,** 6133–6142.
10. Gustin, K. E. and Imperiale, M. J. (1998) Encapsidation of viral DNA requires the adenovirus L1 52/55-kilodalton protein. *J. Virol.* **72,** 7860–7870.
11. Evans, J. D. and Hearing, P. (2003) Distinct roles of the adenovirus E4 ORF3 protein in viral DNA replication and inhibition of genome concatenation. *J. Virol.* **77,** 5295–5304.

12. McGhee, J. D. and von Hippel, P. H. (1975) Formaldehyde as a probe of DNA structure. II. Reaction with endocyclic imino groups of DNA bases. *Biochemistry* **14**, 1297–1303.
13. Chaw, Y. F., Crane, L. E., Lange, P., and Shapiro, R. (1980) Isolation and identification of cross-links from formaldehyde-treated nucleic acids. *Biochemistry* **19**, 5525–5531.
14. Jackson, V. (1999) Formaldehyde cross-linking for studying nucleosomal dynamics. *Methods* **17**, 125–139.
15. Orlando, V., Strutt, H., and Paro, R. (1997) Analysis of chromatin structure by in vivo formaldehyde cross-linking. *Methods* **11**, 205–214.
16. Das, P. M., Ramachandran, K., vanWert, J., and Singal, R. (2004) Chromatin immunoprecipitation assay. *Biotechniques* **37**, 961–969.
17. Kuo, M. H. and Allis, C. D. (1999) In vivo cross-linking and immunoprecipitation for studying dynamic protein:DNA associations in a chromatin environment. *Methods* **19**, 425–433.
18. Ausubel, F. M., Brent, R., Kingston, R. E., et al. (2003) *Current Protocols in Molecular Biology* (Chanda, V. B., ed), Wiley-Liss, Hoboken, NJ.
19. Lutz, P. and Kedinger, C. (1996) Properties of the adenovirus IVa2 gene product, an effector of late-phase-dependent activation of the major late promoter. *J. Virol.* **70**, 1396–1405.
20. Tribouley, C., Lutz, P., Staub, A., and Kedinger, C. (1994) The product of the adenovirus intermediate gene IVa2 is a transcriptional activator of the major late promoter. *J. Virol.* **68**, 4450–4457.
21. Zhang, W. and Imperiale, M. J. (2000) Interaction of the adenovirus IVa2 protein with viral packaging sequences. *J. Virol.* **74**, 2687–2693.
22. Kovesdi, I., Reichel, R., and Nevins, J. R. (1986) E1A transcription induction: enhanced binding of a factor to upstream promoter sequences. *Science* **231**, 719–722.
23. Kovesdi, I., Reichel, R., and Nevins, J. R. (1986) Identification of a cellular transcription factor involved in E1A trans-activation. *Cell* **45**, 219–228.
24. Graham, F. L., Smiley, J., Russell, W. C., and Nairn, R. (1977) Characteristics of a human cell line transformed by DNA from human adenovirus type 5. *J. Gen. Virol.* **36**, 59–74.
25. Spencer, V. A., Sun, J. M., Li, L., and Davie, J. R. (2003) Chromatin immunoprecipitation: a tool for studying histone acetylation and transcription factor binding. *Methods* **31**, 67–75.
26. Maxam, A. M. and Gilbert, W. (1977) A new method for sequencing DNA. *Proc. Natl. Acad. Sci. USA* **74**, 560–564.
27. Roder, K. and Schweizer, M. (2001) Running-buffer composition influences DNA-protein and protein-protein complexes detected by electrophoretic mobility-shift assay (EMSA). *Biotechnol. Appl. Biochem.* **33**, 209–214.
28. Andersen, R. D., Taplitz, S. J., Oberbauer, A. M., Calame, K. L., and Herschman, H. R. (1990) Metal-dependent binding of a nuclear factor to the rat metallothionein-I promoter. *Nucleic Acids Res.* **18**, 6049–6055.

11

Identifying Functional Adenovirus–Host Interactions Using Tandem Mass Spectrometry

Anuj Gaggar, Dmitry Shayakhmetov, and André Lieber

Summary

We describe a systematic, high-throughput approach to identify proteins involved in functional adenovirus (Ad)–host interactions in vitro and in vivo. We were particularly interested in identifying cellular proteins that interact with fiber knob, which is the moiety within the Ad capsid responsible for high-affinity attachment of virus to cellular receptors. We used recombinant fiber knob domains from members of group C and B Ads to purify virus interacting proteins from cell membrane lysates and from human and mouse plasma. Using tandem mass spectrometry, we identified a number of candidate Ad-interacting proteins, including functional cellular receptors and previously unknown interacting partners such as complement component C4-binding protein and other blood proteins that presumably are involved in Ad infection after intravenous virus application. The ability of these proteins to bind to Ad was further confirmed using in vitro protein binding assays as well as infection competition assays. The approach of using a structural protein can be universally applied for a variety of viral and nonviral pathogens and can reveal host cell factors critical in viral infection, immune evasion, and tissue specificity. This information is also a prerequisite to assess in vivo safety and efficacy of Ad-based gene transfer vectors.

Key Words: Serotype 5; serotype 35; fiber; tandem mass spectrometry; C4BP; Coxsackie and adenovirus receptor; CD46.

1. Introduction

Pathogens, viral or otherwise, remain a major public health concern, accounting for high levels of morbidity and mortality throughout the world (*1*). Despite knowing the modes of transmission, clinical manifestations, and, in some cases, the entire genome of many of these pathogens, we still struggle to understand structures and mechanisms involved in pathogen–host interaction. For many emerging pathogens, such as the SARS virus, an important first goal

From: *Methods in Molecular Medicine, Vol. 131:*
Adenovirus Methods and Protocols, Second Edition, vol. 2:
Ad Proteins, RNA, Lifecycle, Host Interactions, and Phylogenetics
Edited by: W. S. M. Wold and A. E. Tollefson © Humana Press Inc., Totowa, NJ

is to identify how the virus chooses which cells to infect and what molecules facilitate this recognition (2–4). This information can provide insight into how the virus affects the host (through defined cellular signaling, for example) and can aid in the development of novel therapeutics against the virus.

The binding of a pathogen to its cellular receptor, however, is only one aspect of how a pathogen interacts with a host. For example, many pathogens travel via the bloodstream to their target tissues and while in transit can complex to one of many molecules in the blood that can serve to deliver the pathogen to a host tissue or even protect the pathogen from immune surveillance and subsequent destruction (5,6). Elucidating these pathways of infection is crucial for understanding the tropism of pathogens and developing therapies for infectious diseases.

Adenovirus (Ad) is a double-stranded DNA virus that is known to cause a variety of benign infections in immunocompetent individuals and more severe diseases in the immunocompromised (7,8). Ad has also recently been used extensively as a gene-delivery vector in part because of its ability to transduce a wide range of target cells. There are 51 known serotypes of human Ad, and these are grouped into groups A to F (9). Viruses from different groups have different cell tropism, in part because of the usage of different attachment receptors. For many Ads, including serotypes from all groups except B, the Coxsackie and adenovirus receptor (CAR) can serve as a primary cellular attachment protein (10–12). Group B Ads, however, do not use CAR, and until recently their cellular receptor remained unknown (13–15).

It is becoming increasingly clear that knowledge of cellular receptors for Ad does not fully explain their tropism when applied systemically into hosts (16,17). After intravenous injection, CAR-interacting Ads can accumulate in liver macrophages despite the lack of functional receptors on these cells. Recent studies show that this process of liver infection is mediated by blood factors, and the identification of these proteins will be important in predicting and controlling Ad infection in vivo (18).

While it is clear that these host–pathogen interactions are important, there has not yet been a systematic, high-throughput way of identifying these interactions. Until now, investigators have relied on methods such as the yeast-two hybrid screening to identify interacting proteins. Although this method is powerful and can give insight into protein–protein interactions, it has limitations stemming from the expression of mammalian proteins in a yeast system. For many proteins, mammalian-specific modifications are central to their biological function. Further, proteins often function in complexes with other proteins or in multimers, and this type of analysis is difficult in two-hybrid systems. Other researchers have relied on scientific intuition, painstaking biochemical analysis of a variety of proteins, or fortuitous discovery to identify candidate proteins *(3,4,14)*.

Here we employed tandem mass spectrometry (MS/MS) to help facilitate identification of possible interactions of host proteins with Ad. We used Ad from serogroup C (type 5) and serogroup B (type 35) to identify interacting proteins. We first produced recombinant protein of the viral polypeptide of these Ads known to be important for host interactions. We then conjugated these proteins to agarose beads and used this complex to purify proteins from cellular membrane proteins or freshly isolated mouse plasma. The purified proteins were then analyzed using MS/MS to identify candidate proteins. Once these proteins were identified, additional assays were used to verify that proteins identified can truly interact with the viral polypeptide. We describe here techniques used to study the interaction between Ad serotype 5 or 35 fiber knobs and cell or plasma proteins. Similar studies can be performed with other serotypes and with other capsid proteins.

2. Materials

1. pQE30 bacterial expression system (Qiagen).
2. Ad fiber knob cDNA (*see* **Note 1**).
3. *Escherichia coli* strain M15pREP4 (Qiagen).
4. Oligonucleotide DNA primers.
5. Restriction enzymes, T4 DNA ligase.
6. Agarose and DNA electrophoresis apparatus.
7. LB medium.
8. Ampicillin and Kanamycin.
9. Isopropyl-β-D-thio-galactopyranoside.
10. Ni^{2+}-NTA agarose beads (Qiagen).
11. Dulbecco's phosphate-buffered saline (PBS).
12. Dialysis bags, molecular-weight cutoff = 14,000 (Spectrum Labs).
13. Streptavidin–agarose.
14. Phenylmethylsulfonyl fluoride (PMSF).
15. BiotinTag Biotinylation Kit (Sigma).
16. Sephadex-G15 columns.
17. Resuspension solution: 250 mM sucrose, 0.1% Triton X-100.
18. Dounce homogenizer.
19. Protease inhibitor cocktail (Sigma).
20. Membrane protein buffer: 1 mM Tris-HCl, pH 7.5, + 2% protease inhibitor cocktail.
21. RIPA buffer: 150 mM NaCl, 1% NP-40, 0.5% sodium deoxycholate, 50 mM Tris-HCl, pH 8.0, 1 mM ethylene diamine tetraacetic acid (EDTA),pH 8.0, 1 mM PMSF.
22. Avertin (2-2-2 tribromoethanol) (Sigma).

23. Saline.
24. Tween-20.
25. Imidazole.
26. 2X Laemmli buffer.
27. Sodium dodecyl sulfate–polyacrylamide gel electrophoresis (SDS-PAGE) apparatus.
28. Extravidin, horseradish peroxidase-conjugated avidin (Sigma).
29. ECL Western blotting detection kit (Amersham).
30. BCA protein concentration assay kit (Pierce).
31. Microcon filters, YM-10 (Millipore).
32. Ammonium bicarbonate.
33. Ultrapure acetonitrile.
34. Destaining solution: 15 mM $K_3Fe(CN)_6$ and 50 mM $Na_2S_2O_3$.
35. Trypsin.
36. Cover solution: 50 mM NH_4HCO_3, 5 mM $CaCl_2$.
37. 50% Acetonitrile.
38. 1% Tetrafluoroacetic acid.
39. Zip-Tip reverse-phase C18 columns (Millipore).
40. Sequencing grade-modified porcine trypsin (Promega).
41. Glacial acetic acid.
42. Finnigan LCQ liquid chromatography/MS (Finnigan).
43. MS/MS data analysis program (Sequest, distributed by Thermo Finnigan, and Mascot, distributed by Matrix Science).
44. PBS-T: PBS containing 0.1% Tween-20.
45. PBS-T plus 5% milk.

3. Methods

For the successful identification of proteins that interact with a known viral ligand, the following methods can be used to:

1. Produce recombinant virus bait protein.
2. Conjugate bait to appropriate agarose beads.
3. Use bait-bead complexes to precipitate proteins from cell lysates or plasma/serum.
4. Identify interacting proteins with mass spectrometry.
5. Verify interactions by viral competition experiments (**Fig. 1**).

3.1. Expression of Recombinant Viral Proteins

The methods used to produce recombinant viral proteins thought to be important for protein interactions is described elsewhere *(19)*. We describe here the production of recombinant fiber knob. Briefly, the following steps are for producing adenoviral fiber knob domains with a 6-histidine tag at the N-terminus of the protein.

1. Use PCR primers directed to the fiber knob domain of the Ad fiber, including the last repeat of the shaft (*see* **Note 1**), to amplify the region and then digest using

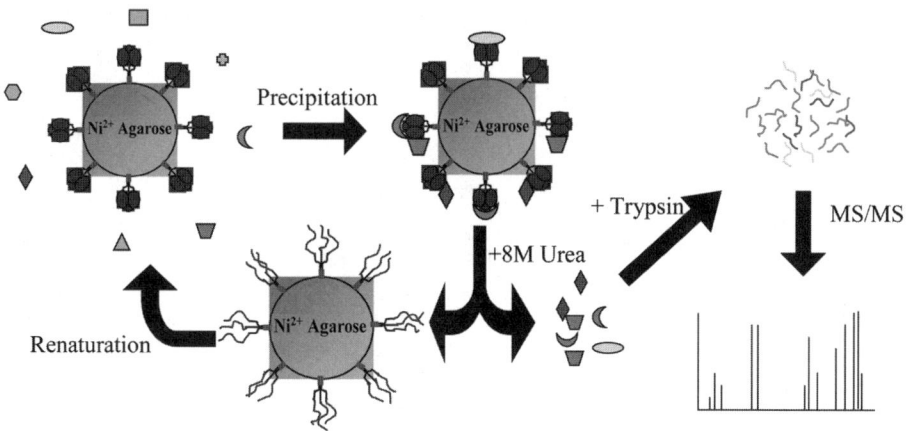

Fig. 1. Schematic showing conjugation of viral fiber knobs via six-His tag to Ni^{2+} agarose beads, binding of specific proteins and not unrelated proteins and subsequent elution of bound proteins with 8 M urea followed by trypsin digestion and tandem mass spectrometry analysis.

    appropriate restriction enzymes (in this case, *Hin*dIII and *Bam*HI) to facilitate in-frame-cloning into pQE30, which will place a 6-histidine tag at the N-terminus. (Primers to amplify Ad5 fiber knob were 5'-TTT AAG GAT CCG GTG CCA TTA CAG TAG GAA-3' and 5'-TAT ATA AGC TTA TTC TTG GGC AAT GTA TGA-3' and those for Ad35 were 5'-TTT AAG GAT CCG GTG ACA TTT GTA TAA AGG ATA G-3' and 5'-TAT ATA AGC TTA GTT GTC GTC TTC TGT AAT-3') *(19,20)*.

2. After cloning and verification by sequencing, transfect the resulting plasmid into M15(pRep4) bacteria.
3. After 1 h of growth at 37°C, induce protein expression by adding 1 mM isopropyl-β-D-thio-galactopyranoside and incubating for an additional 4–5 h.
4. After lysing bacteria and separating cellular debris, use Ni^{2+}-NTA agarose beads to purify expressed proteins under nondenaturing conditions as described by the manufacturer.
5. Dialyze the resulting proteins overnight in PBS + 17% glycerol and store at –20°C.

3.2. Conjugation of Viral Protein to Agarose

    Because there are several ways to conjugate the viral protein to agarose, we describe here two methods, each with advantages and disadvantages. One strategy involves the addition of a biotin tag onto the recombinant protein, and the other takes advantage of the 6-histidine tag present at the N-terminus of the molecule. Relative advantages and disadvantages of each method are discussed in **Note 1**.

3.2.1. Conjugation of Biotinylated Viral Protein to Streptavidin–Agarose

1. Mix 100 μL of dialyzed fiber knob domain in a 1.5-mL Eppendorf tube with 10 μL of biotinylation reagent (BiotinTag, prepared according to manufacturer's protocols) for 30 min at room temperature on a shaker.
2. Stop the reaction with the addition of 10 μL of 1 M Tris buffer, pH 7.4, and purify the tagged protein using Sephadex-G15 columns to remove excess biotinylation reagent.
3. Measure the protein concentration of the biotinylated fiber knob domain using the BCA protein concentration assay reagent.
4. Place 5–10 μg of the biotinylated Ad fiber knob domain in 200 μL of ice-cold PBS and mix with 20 μL streptavidin agarose beads in a shaker for 1–2 h at 4°C.
5. Pellet the mixture at 150g in a microcentrifuge and wash with 1 mL ice-cold PBS twice to remove unbound fiber knob protein. Store the resultant pellet at 4°C.

3.2.2. Conjugation of His-Tagged Viral Protein to Ni^{2+}-NTA Agarose Beads

1. Mix 10–20 μg dialyzed fiber knob domain with 50 μL of Ni^{2+}-NTA agarose beads slurry in 500 μL total volume, made up with ice-cold PBS, for 2 h at 4°C on a shaker.
2. Pellet the mixture at 150g in a microcentrifuge and wash with 1 mL ice-cold PBS twice to remove unbound fiber knob protein.

3.3. Isolation of Cellular Membrane Proteins

In order to identify membrane proteins capable of binding to a viral polypeptide, cell membrane proteins were prepared according to established protocols *(21)*. Briefly, the methods used were as follows:

1. Grow cells known to be susceptible to virus infection (e.g., HeLa cells, a cervical carcinoma-derived cell line) to confluence in standard 150-mm tissue culture dishes (*see* **Note 2**).
2. Scrape cells from each dish into 5 mL of ice-cold PBS + 1 mM PMSF. Repeat for at least six dishes and keep the cell suspension on ice.
3. Combine cell suspensions and pellet cells in a centrifuge at 300g to remove excess supernatant and resuspend the resultant pellet in 5 mL of ice-cold resuspension solution.
4. Lyse cells with 30 strokes of a Dounce homogenizer on ice, and pellet at 400g for 15 min to remove cellular debris.
5. Collect the supernatant and centrifuge at 100,000g for 1 h in Beckman ultraclear centrifuge tubes with total volume of 13.2 mL made up with ice-cold PBS.
6. Resuspend the pellet (containing membrane proteins) in 100 μL per 150-mm dish of membrane protein buffer.

3.4. Isolation of Serum Proteins

In order to identify proteins within mouse plasma that are capable of interacting with the Ad fiber knob domain, we isolated fresh mouse plasma according to the following protocol.

1. Anesthetize C57 BL/6 mice with intraperitoneal Avertin injections.
2. Collect 800 µL of blood from cardiac puncture from each mouse into a 1-mL syringe containing 200 µL of saline + 20 mM EDTA.
3. Immediately centrifuge the sample for 5 min at 2000g.
4. Transfer the supernatant (plasma) to a new tube and add Tween-20 and imidazole to final concentrations of 0.1% (v/v) and 5 mM, respectively.

3.5. Precipitation of Interacting Proteins to a Viral Polypeptide

Once the viral protein has been conjugated to the agarose beads (*see* **Subheading 3.3.**), precipitation of proteins is done.

1. Preclear freshly isolated plasma proteins or cell membrane proteins by incubating with either 100 µL of unconjugated Ni^{2+}-NTA agarose beads or 50 µL of unconjugated strepavidin–agarose beads (depending on labeling technique of the viral polypeptide) in a total volume of 1 mL made up with ice-cold PBS at 4°C for 2 h (*see* **Note 3**).
2. Centrifuge the sample at 150g for 5 min. Add the supernatant, containing precleared proteins, to bead-conjugated fiber knob domains (*see* **Subheading 3.2.**) and place on a shaker at 4°C for 2 h.
3. Pellet agarose beads at 300g and wash with ice-cold PBS (*see* **Note 4**) at least four times to reduce contamination with unbound proteins (*see* **Note 5**).

3.6. Isolation of Proteins From Complexes

Proteins bound to the Ad fiber knob domain are eluted from the complexes by one of two different techniques. For precipitations using the biotinylated fiber knob, protein complexes are disrupted by boiling. For those complexes formed using a 6-histidine tag, elution was done in the presence of 8 M urea to disrupt protein–protein interactions. Each application is suited for specific situations (as discussed in **Note 1**).

3.6.1. Isolation of Precipitated Proteins From Biotin–Streptavidin Complexes

1. Following the last wash step from **Subheading 3.5.**, **step 3**, boil the complexes with 20 µL 2X Laemmli Buffer for 5 min.
2. Load the entire sample for separation using electrophoresis in 7% polyacrylamide SDS-PAGE.
3. Silver stain the protein gel according to established protocols (*see* **Note 6**) (*22*).

4. Cut bands corresponding to proteins of interest from the gel, and destain with 400 μL of destaining solution for 8 min with vigorous shaking.

5. After four 8-min washes with ultrapure water, use pure acetonitrile to dry gel pieces with a Speed-vac for 30 min.

6. Add 0.5–1.0 μg of trypsin at 100 ng/μL in 1 m*M* HCl and add cover solution as needed to keep gel bits moist on ice for 1 h.

7. Discard excess solution and add 30 μL of cover solution and incubate for 12–15 h at 37°C.

8. After incubation, collect the supernatant and sonicate sample with 30 μL of 20 m*M* NH_4HCO_3 for 15 min and collect supernatant.

9. Extract residual peptides with three successive sonications in the presence of 30 μL of 50% acetonitrile + 1% tetrafluoroacetic acid, collecting the supernatant after each sonication.

10. Dry the samples in a Speed-vac to about 5 μL and use exchange chromatography resins (Zip-Tip) to purify peptides from salts that can interfere with subsequent analysis.

11. Analyze peptides using high-performance liquid chromatography separation followed by MS/MS analysis.

3.6.2. Isolation of Precipitated Proteins From 6-His-Ni^{2+}-NTA Agarose Complexes

1. Following the last wash step (**Subheading 3.5., step 3**), incubate complexes with 200 μL of 8 *M* urea, which disrupts protein–protein interactions but leaves 6-His-Ni^{2+}-NTA interactions intact.

2. In order to allow for optimal trypsin activity, urea is removed by increasing the volume to 500 μL using PBS and concentrating the solution using Microcon filters with a molecular weight cutoff of 3000 Daltons (*see* **Note 7**).

3. Digest concentrated protein samples in solution overnight using 0.5–1.0 μg trypsin in a total volume of 50 μL with final concentrations of 100 m*M* NH_4HCO_3 and urea concentration less than 2 *M* in an incubator.

4. Quench the reaction after 12–14 h with the addition of 2 μL glacial acetic acid and store frozen until ready for analysis.

5. Recover peptides by using exchange chromatography resins (Zip-Tip) to purify peptides from salts that can interfere with subsequent analysis.

6. Analyze peptides using high-performance liquid chromatography separation followed by MS/MS analysis.

Figure 2 shows a silver-stained SDS-PAGE separation of proteins from **Subheading 3.6.2., step 1**, along with identified proteins from MS/MS analysis.

3.7. Verification of Interactions

To further determine if proteins found through the above process can in fact interact with the Ad fiber knob, several addition verification experiments were done. Using cell membrane proteins, CD46 was determined to be a candidate

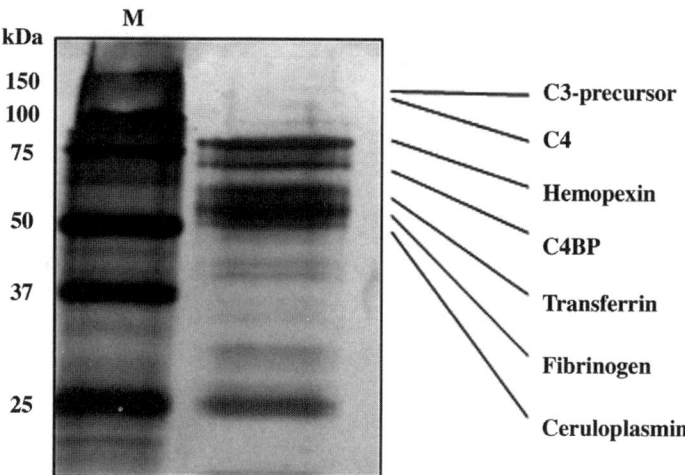

Fig. 2. Silver-stained gel showing molecular-weight markers (M) and proteins isolated from viral knob precipitations along with protein identity revealed by tandem mass spectrometry.

cellular receptor for group B Ads. Detailed experiments, including infection competition studies with soluble CD46 or reduction of infection with siRNA knockdown of receptor expression, verifying this finding can be found elsewhere *(15)*. Further verification of interacting proteins from mouse serum was performed *(18)*. In the first experiment, we used virus overlay protein blot assays to assess binding of purified proteins to Ad fiber knob domains. We then used one of the identified proteins, C4-binding protein (C4BP), in further analyses of binding to determine if it could compete for virus infection in vitro.

3.7.1. Verifying Protein–Protein Binding Using Virus Overlay Protein Blot Assays

1. Immobilize 5–10 µg of recombinant proteins identified by MS/MS analysis on nitrocellulose membrane.
2. Incubate membrane overnight with 5% milk in PBS + 0.1% Tween (PBS-T) at 4°C.
3. Wash membrane once with PBS-T.
4. Dilute biotinylated fiber knob domain (**Subheading 3.2.1., step 3**) in PBS-T + 0.5% milk to final concentration of 1–2 µg/mL and add to washed membrane for 1 h with gentle shaking at room temperature
5. After washing several times with PBS-T, add avidin diluted 1:1000 in PBS-T + 0.5% milk and incubate for 30 min at room temperature.
6. Wash several times with PBS-T and develop using ECL reagent (as per manufacturer's directions).

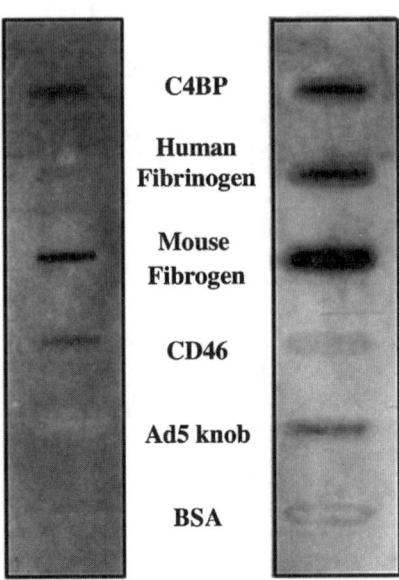

Fig. 3. Slot blot assay showing a panel of proteins probed with adenovirus (Ad)35 (left panel) or Ad5 knobs (right panel) to assess binding. BSA, bovine serum albumin.

Results in **Fig. 3** represent analysis using the Ad5 or Ad35 fiber knobs as probes.

3.7.2. Infection Competition Experiments With C4BP and Ad5/35

Chinese hamster ovary (CHO) cells are refractory to Ad infection; however, when stably transfected to express Ad attachment receptors, they become permissive for infection with the virus. While infection of CHO-CAR with Ad5 in the presence of physiological concentrations of C4BP and other plasma proteins did not demonstrate change in virus infectivity (data not shown), C4BP could efficiently compete infection of CHO-C2 cells (CHO cells expressing the C2 isoform of CD46) with an Ad35 fiber-containing vector (Ad5/35) (**Fig. 4A**) *(23)*.

1. Plate 2×10^5 CHO-C2 cells in 12-well plates and allow them to grow overnight.
2. Use purified Ad5/35 expressing green fluorescence protein (GFP) under the control of the cytomegalovirus promoter at a multiplicity of infection of 10 PFU/cell and add to cells in the presence of increasing amounts of C4BP in a total volume of 500 µL growth medium and incubate at 37°C for 3 h (*see* **Note 9**) *(24)*.
3. Wash cells several times with PBS and incubate in growth medium overnight.
4. Assess GFP fluorescence of cells using fluorescence microscope and/or flow cytometry analysis.

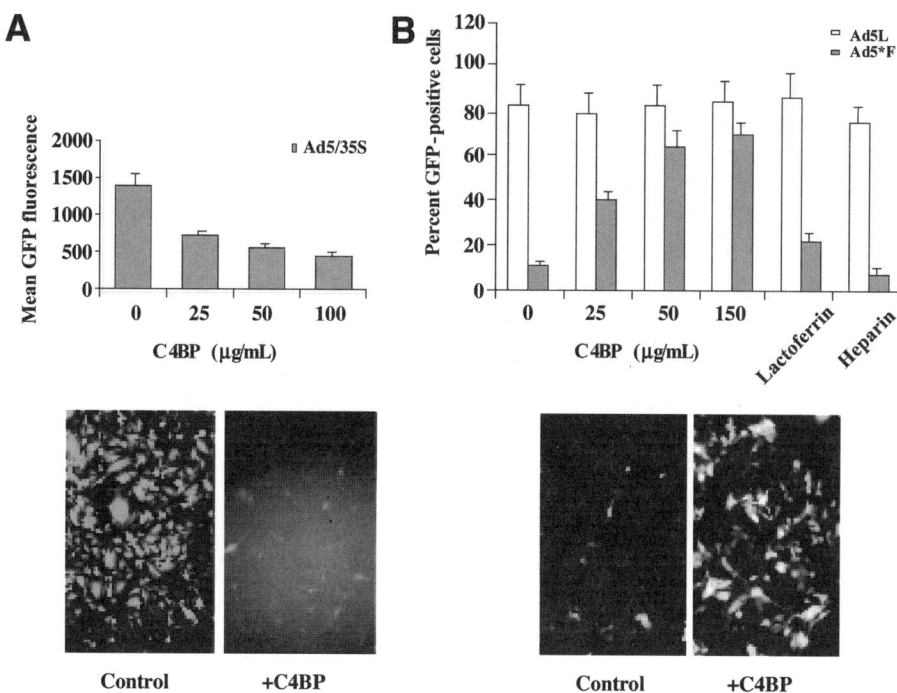

Fig. 4. (**A**) Infection competition of Chinese hamster ovary (CHO)-C2 cells in the presence of C4 binding protein (C4BP). Quantification of CHO-C2 cell infection by Ad5/35 by fluorescence-activated cell sorting in the presence of various concentrations of C4BP is shown in the upper panel. Representative cells infected with Ad5/35 with or without 100 μg/mL of C4BP are shown in the lower panel. (**B**) Infection enhancement of HepG2 cells in the presence of C4BP. Quantification of HepG2 infection by Ad5*F by FACS at various concentrations of C4BP and at highest C4BP levels in the presence of lactoferrin or heparin (**upper panel**) with representative cells infected with Ad5*F with or without 100 μg/mL of C4BP (**lower panel**).

3.7.3. Infection Competition Experiments With C4BP and Ad5*F

To analyze whether the infectivity of Ad can be modulated by interactions with plasma proteins identified through our approach, we infected human hepatocarcinoma cells, HepG2 (*see* **Note 10**), with an Ad5-based vector, Ad5*F (possessing a single point mutation abolishing its binding to its cellular receptor CAR *[16,25]*), in the presence of purified plasma proteins. This analysis revealed that when cells were infected with Ad5*F, few cells expressed the virus-encoded transgene GFP (**Fig. 4B**, 0 μg/mL C4BP). However, when Ad5*F infection was conducted in the presence of C4BP, a marked increase in vector infectivity toward these cells was observed (**Fig. 4B**, C4BP).

1. Plate 2×10^5 HepG2 cells in 12-well plates and allow to grow overnight.
2. Use purified Ad5*F at a multiplicity of infection of 50 PFU/cell and add to cells in the presence of increasing amounts of C4BP (up to 250 µg/mL, which is the physiological level in human plasma) in a total volume of 300 µL growth medium and incubate at 37°C for 3 h.
3. Wash cells several times with PBS and incubate in growth medium overnight.
4. Assess GFP fluorescence of cells using fluorescence microscope or flow cytometry analysis.

Here we demonstrate that using a straightforward high-throughput approach, novel functional interactions of a virus with host factors can be identified. Using Ad receptor-interacting fiber knob domain as bait, we recovered and identified a protein, C4BP, from whole mouse plasma that modulates virus infectivity toward both susceptible (CHO-C2 for Ad5/35) and resistant (HepG2 for Ad5*F) cell types. It should be noted that the other proteins identified through MS/MS analysis were confirmed to bind to Ad fiber knob domain in slot-blot assays, and future studies will determine the functional role of these interactions. Despite the complexity of mouse plasma protein composition, we were able to rapidly identify new Ad-interacting partners, which will have implications for Ad vector development. The approach of using a structural protein to identify host factors critical in viral infection, immune evasion, and tissue specificity can be universally applied for a variety of viral and nonviral pathogens and can reveal new targets for prospective therapies.

4. Notes

1. Ad DNA for amplification of the fiber knob domain was extracted using standard DNA extraction techniques from purified viral particles of wild-type Ad of various serotypes. Fiber knob domains were amplified using primers encompassing the last repeat of the shaft domain and the entire fiber knob domain. The last repeat of the shaft is necessary because it is required for proper trimerization of the recombinant protein.
2. Biotinylation of proteins using the described methods adds a biotin molecule to the lysine residues throughout the peptide. This can theoretically aid in recovery because a single protein can have multiple biotinylation sites that can be used for precipitation, allowing for increased yield. Disadvantages of this technique include an added manipulation of the purified protein. Specifically, biotinylation of lysine residues that are important in protein–protein interactions might affect the outcome of this method. Testing if the biotinylated protein can still bind to a protein of interest (by either competitive experiments or virus overlay protein blot assays) can help to confirm if a protein retains functionality after biotinylation. The use of the 6-histidine tag also has advantages and disadvantages. The advantages include less manipulation of the protein because it has an incorporated tag as well as the fact that it is likely not involved in interactions because it is (in this example) far from putative interaction sites. Another advan-

tage is that elution of precipitated proteins can be done without contamination with bait proteins (*see* **Fig. 1**). The disadvantages include a potentially weak signal, with only one bait protein per polypeptide, that may not be readily accessible in nondenatured conditions. Further, the Ni^{2+}-agarose beads are not compatible with certain conditions that may be required for protein purification (such as increased EDTA solutions).

3. Using epithelial cell lines such as HeLa cells can often produce large amounts of keratins that tend to contaminate MS/MS samples. The choice of cell type can be determined by identifying those cells likely to produce abundant receptor molecules (e.g., by infectivity assays).

4. Depending on the type of beads used, a second preclearing round can be done, especially if the beads appear to bind to proteins abundant in plasma.

5. To reduce nonspecific binding to Ni^{2+}-NTA beads, add 10 mM imidazole to the wash solution. Also, if nonspecific background binding is high, consider washing samples with PBS in the presence of a nonionic detergent, such as 0.1% NP-40.

6. It is especially important to wash many times when using serum proteins because the abundant proteins like albumin can often contaminate samples even after four washes if less abundant proteins are being analyzed.

7. For better likelihood of successful identification by MS/MS analysis, do not fix the gel with formaldehyde during silver staining.

8. Urea can disrupt trypsin activity and, if this contaminant is too high, will result in suboptimal peptide recovery. A second concentration step can be performed in the same microcon container by again increasing the volume to 500 μL using PBS and concentrating to 50 μL. Also, the molecular-weight cutoff of the filter can be adjusted to enrich in particular weights of target proteins.

9. These experiments utilized the convenience of pseudotyped vectors with an easily assayable transgene. These experiments can also be performed with wild-type Ad using cytopathic effect as a primary outcome instead of GFP transgene expression.

10. The choice of HepG2 as a cell line was made because of previous observations that liver infection in vivo by Ad5-based vectors is not CAR dependent and likely is mediated by other serum factors. Using a vector abolished for CAR-binding, we are better able to assay the contribution of our newly discovered blood factors to infection of these human liver-derived cells in the absence of CAR binding.

References

1. World Health Organization. (2003) World Health Report, Geneva, Switzerland.
2. Li, W., Moore, M. J., Vasilieva, N. et al. (2003) Angiotensin-converting enzyme 2 is a functional receptor for the SARS coronavirus. *Nature* **426,** 450–454.
3. Tatsuo, H., Ono, N., Tanaka, K., and Yanagi, Y. (2000) SLAM (CDw150) is a cellular receptor for measles virus. *Nature* **406,** 893–897.
4. Wang, X., Huong, S. M., Chiu, M. L., Raab-Traub, N., and Huang, E. S. (2003) Epidermal growth factor receptor is a cellular receptor for human cytomegalovirus. *Nature* **424,** 456–461.

5. Agnello, V., Abel, G., Elfahal, M., Knight, G. B., and Zhang, Q. X. (1999) Hepatitis C virus and other flaviviridae viruses enter cells via low density lipoprotein receptor. *Proc. Natl. Acad. Sci. USA* **96,** 12,766–12,771.

6. Hilgard, P. and Stockert, R. (2000) Heparan sulfate proteoglycans initiate dengue virus infection of hepatocytes. *Hepatology* **32,** 1069–1077.

7. Ruuskanen, O., Meurman, O., and Akusjarvi, G. (2002) Adenoviruses, in *Clinical Virology* (Hayden, F. G., ed.), ASM Press, Washington, DC, pp. 515–535.

8. Hierholzer, J. C. (1992) Adenoviruses in the immunocompromised host. *Clin. Microbiol. Rev.* **5,** 262–274.

9. Shenk, T. (2001) Adenoviridae, in *Fields Virology*, 4th ed. (Knipe, D. M. and Howley, P.M., eds.), Vol. 2, Lippincott Williams and Wilkins, Philadelphia, PA, pp. 2265–2301.

10. Bergelson, J. M., Cunningham, J. A., Droguett, G., et al. (1997) Isolation of a common receptor for Coxsackie B viruses and adenoviruses 2 and 5. *Science* **275,** 1320–1323.

11. Roelvink, P. W., Lizonova, A., Lee, J. G., et al. (1998) The coxsackievirus-adenovirus receptor protein can function as a cellular attachment protein for adenovirus serotypes from subgroups A, C, D, E, and F. *J. Virol.* **72,** 7909–7915.

12. Tomko, R. P., Xu, R., and Philipson, L. (1997) HCAR and MCAR: the human and mouse cellular receptors for subgroup C adenoviruses and group B coxsackieviruses. *Proc. Natl. Acad. Sci. USA* **94,** 3352–3356.

13. Sirena, D., Lilienfeld, B., Eisenhut, M., et al. (2004) The human membrane cofactor CD46 is a receptor for species B adenovirus serotype 3. *J. Virol.* **78,** 4454–4462.

14. Segerman, A., Atkinson, J. P., Marttila, M., Dennerquist, V., Wadell, G., and Arnberg, N. (2003) Adenovirus type 11 uses CD46 as a cellular receptor. *J. Virol.* **77,** 9183–9191.

15. Gaggar, A., Shayakhmetov, D. M., and Lieber, A. (2003) CD46 is a cellular receptor for group B adenoviruses. *Nat. Med.* **9,** 1408–1412.

16. Alemany, R. and Curiel, D. T. (2001) CAR-binding ablation does not change biodistribution and toxicity of adenoviral vectors. *Gene Ther.* **8,** 1347–1353.

17. Martin, K., Brie, A., Saulnier, P., Perricaudet, M., Yeh, P., and Vigne, E. (2003) Simultaneous CAR- and alpha V integrin-binding ablation fails to reduce Ad5 liver tropism. *Mol. Ther.* **8,** 485–494.

18. Shayakhmetov, D. M., Gaggar, A., Ni, S., Li, Z. Y., and Lieber, A. (2005) Adenovirus binding to blood factors results in liver cell infection and hepatoxicity. *J. Virol.* **79,** 7478–7491.

19. Zinn, K. R., Douglas, J. T., Smyth, C. A., et al.. (1998) Imaging and tissue biodistribution of 99mTc-labeled adenovirus knob (serotype 5). *Gene Ther.* **5,** 798–808.

20. Shayakhmetov, D. M., Li, Z. Y., Ternovoi, V., Gaggar, A., Gharwan, H., and Lieber, A. (2003) The interaction between the fiber knob domain and the cellular attachment receptor determines the intracellular trafficking route of adenoviruses. *J. Virol.* **77,** 3712–3723.

21. Wu, E., Fernandez, J., Fleck, S. K., Von Seggern, D. J., Huang, S., and Nemerow, G. R. (2001) A 50-kDa membrane protein mediates sialic acid-independent binding and infection of conjunctival cells by adenovirus type 37. *Virology* **279,** 78–89.

22. Gharahdaghi, F., Weinberg, C. R., Meagher, D. A., Imai, B. S., and Mische, S. M. (1999) Mass spectrometric identification of proteins from silver-stained polyacrylamide gel: a method for the removal of silver ions to enhance sensitivity. *Electrophoresis* **20,** 601–605.

23. Oglesby, T. J., White, D., Tedja, I., et al. (1991) Protection of mammalian cells from complement-mediated lysis by transfection of human membrane cofactor protein and decay-accelerating factor. *Trans. Assoc. Am. Physicians* **104,** 164–172.

24. Shayakhmetov, D. M., Papayannopoulou, T., Stamatoyannopoulos, G., and Lieber, A. (2000) Efficient gene transfer into human CD34(+) cells by a retargeted adenovirus vector. *J. Virol.* **74,** 2567–2583.

25. Roelvink, P. W., Mi Lee, G., Einfeld, D. A., Kovesdi, I., and Wickham, T. J. (1999) Identification of a conserved receptor-binding site on the fiber proteins of CAR-recognizing adenoviridae. *Science* **286,** 1568–1571.

12

The Use of Cell Microinjection for the In Vivo Analysis of Viral Transcriptional Regulatory Protein Domains

Maurice Green, Andrew Thorburn, Robert Kern, and Paul M. Loewenstein

Summary

Microinjection of mammalian cells provides a powerful method for analyzing in vivo functions of viral genes and viral gene products. By microinjection, a controlled amount (ranging from several to many thousands of copies) of a viral or cellular gene, a protein product of a gene, a polypeptide fragment encoding a specific protein domain, or an RNA molecule can be delivered into a target cell and the functional consequences analyzed. Microinjection can be used to deliver antibody targeted to a specific protein domain in order to analyze the requirement of the protein for specific cell functions such as cell cycle progression, transcription of specific genes, or intracellular transport.

This chapter describes examples of the successful use of microinjection to probe adenovirus E1A regulatory mechanisms. Detailed methods are provided for manual and semiautomatic microinjection of mammalian cells as well as bioassay protocols for microinjected cells including immunofluorescence, colorimetic, *in situ* hybridization, and autoradiography.

Key Words: Cell microinjection; adenovirus protein domains; gene regulation; transcription activation; transcription repression; early gene 1A; *in situ* hybridization; immunofluorescence; manual microinjection; semiautomatic microinjection.

1. Introduction

Microinjection of mammalian cells provides a powerful method for analyzing in vivo functions of viral genes and viral gene products. By microinjection, a controlled amount (ranging from several to many thousands of copies) of a viral or cellular gene, a protein product of a gene, a polypeptide fragment encoding a specific protein domain, or an RNA molecule can be delivered into a target cell and the functional consequences analyzed. Injection of DNA into

From: *Methods in Molecular Medicine, Vol. 131:*
Adenovirus Methods and Protocols, Second Edition, vol. 2:
Ad Proteins, RNA, Lifecycle, Host Interactions, and Phylogenetics
Edited by: W. S. M. Wold and A. E. Tollefson © Humana Press Inc., Totowa, NJ

the nuclei of cultured mammalian cells provides a sensitive bioassay for gene expression. The product of a single injected gene copy can be detected using standard immunofluorescence, immunochemical, and autoradiographic techniques *(1)*. The direct analysis of protein function by microinjection has been facilitated by the ability to produce biologically active recombinant proteins and protein fragments encoded by many interesting genes.

Another valuable use of microinjection is to deliver antibody targeted to a specific protein domain in order to analyze the requirement of the protein for specific cell functions such as cell cycle progression, transcription of specific genes, or intracellular transport. Antibodies introduced by microinjection into mammalian cells are highly specific. Several other strategies have used microinjection successfully, including the delivery and functional analysis of antisense RNA and the determination of morphological alterations in response to an injected molecule, including cell transformation. Cell microinjection has been uniquely used to address mechanistic questions, such as whether the function of a microinjected protein requires the activation of cellular genes or whether it interacts directly with cellular factors.

There have been increasing numbers of publications describing the use of cell microinjection to provide insights into important biological questions. This has been facilitated by the availability of semi-automated injection equipment and computerized microcapillary pullers, which have made cell microinjection a reliable and routine laboratory procedure.

1.1. Cell Microinjection Strategies to Analyze Adenovirus Gene Functions

The response of a cell to the expression of a microinjected gene, the introduction of a microinjected protein or protein fragment, or the microinjection of an antibody can be measured by a variety of methods. Below are some of the cell microinjection strategies used to probe adenovirus (Ad) gene function.

1.1.1. Transcriptional Activation by an Ad E1A Protein Fragment Encoding Conserved Domain 3: Immunofluorescence Analysis of the Protein Product of an Activated Gene

The Ad 2 (5) early gene 1A (*E1A*) encodes two overlapping oncoproteins of 289 and 243 amino acid residues (referred to as 289R and 243R, respectively). Analysis of the expression of Ad E1A mutants showed that the CR3 domain (conserved amino acid residues 140–188) unique to the Ad 289R protein is required for transactivation of early viral genes (for review, *see* **ref. 2**). To determine directly whether the amino acid residues within CR3 are sufficient for transactivation, a chemically synthesized E1A peptide containing CR3 was microinjected into HeLa cells together with a reporter plasmid, pE2, which

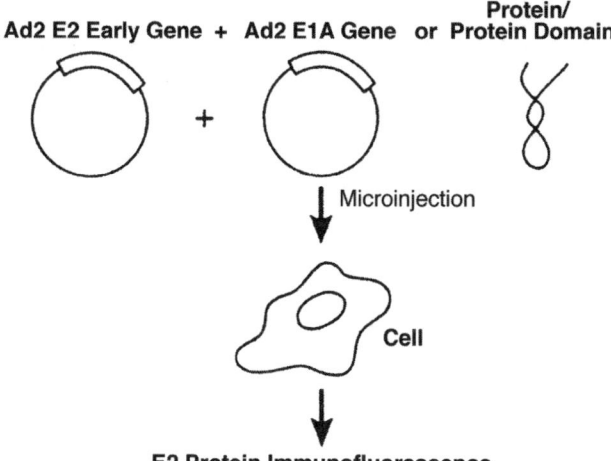

Fig. 1. Cell microinjection–transcriptional activation assay. The schematic shows coinjection of the Ad2 *E1A* gene or an E1A protein domain with the Ad2 E2 early gene reporter. Expression of E2 is measured 18 h after microinjection by immunofluorescence with E2-specific antibody.

encodes the Ad2 *E2* early gene (*see* **Fig. 1** for schematic). Immunofluorescence analysis for the pE2 product (the 73K DNA-binding protein) showed that CR3 efficiently activated the *E2* gene *(3)*. Thus by microinjection, the E1A CR3 peptide was shown to be the smallest known protein domain functioning as a transcriptional activator.

1.1.2. Microinjection of the Ad E1A Transactivation Peptide (CR3) to Determine Whether Transcriptional Activation Requires De Novo Protein Synthesis: In Situ Hybridization Analysis of the RNA Product of an Activated Gene

There are two possible mechanisms by which the E1A CR3 domain might transactivate early genes. First, CR3 might induce the expression of cellular gene(s), resulting in the increased synthesis of a specific transcription factor required for transactivation. Second, CR3 might directly activate a preexisting transcription factor. The first mechanism but not the second requires cellular protein synthesis. A microinjection strategy to directly test these alternative models was designed. A plasmid expressing the *E2* gene (pE2) or pE2-CAT (the CAT gene driven by the promoter of Ad E2) was coinjected with the CR3 peptide in the presence or absence of the protein synthesis inhibitor cycloheximide, and the formation of the RNA product measured by *in situ* hybridization with an ^{35}S-labeled CAT DNA or E2 RNA probe (*see* **Fig. 2** for schematic).

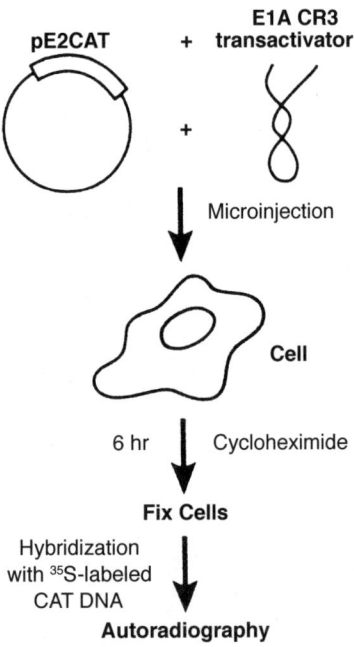

Fig. 2. Cell microinjection *in situ* hybridization assay for gene activation in the absence of protein synthesis. The schematic shows the assay for activation of the E2 promotor fused to the CAT reporter (pE2CAT) by the E1A PD3 (protein domain 3) transcriptional activator in the presence of the protein synthesis inhibitor cycloheximide or anisomycin. Expression of the CAT gene is measured by *in situ* hybridization followed by autoradiography.

Under conditions where protein synthesis was effectively blocked, CR3 efficiently activated the E2 promotor, resulting in the accumulation of E2 or CAT RNA (*see* **Fig. 3**). Thus, cell microinjection uniquely demonstrated that the E1A CR3 domain transactivates early viral genes by a direct mechanism involving interaction with a preexisting cellular factor(s) *(4)*.

1.1.3. Microinjection of an Ad E1A Protein Fragment to Analyze E1A Transcriptional Repression: Immunofluorescence Analysis of the Protein Product of a Repressed Gene

The E1A 243R protein, which is identical to the 289R protein except that it lacks CR3, encodes a potent transcription repression function, which targets a set of cellular genes that regulate cellular growth and differentiation (for review, *see* **ref. 2**). To determine whether the Ad CR1 domain (amino acids 40–80) and the nonconserved N terminus (amino acids 1–39) together are sufficient for repression of a target gene, a recombinant protein containing only

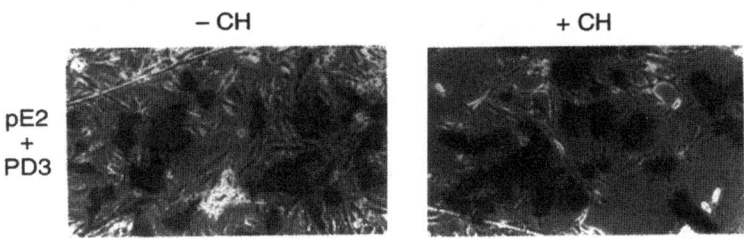

Fig. 3. Effect of the protein synthesis inhibitor, cycloheximide, on transactivation of the E2 promoter by the adenovirus E1A peptide, PD3. As shown, E2 is activated both in the absence (-CH) and presence (+CH) of cycloheximide as measured by *in situ* hybridization with an ^{35}S-labeled E2 DNA probe. The amount of CH added (+CH) is sufficient to inhibit protein synthesis by 98% *(4)*.

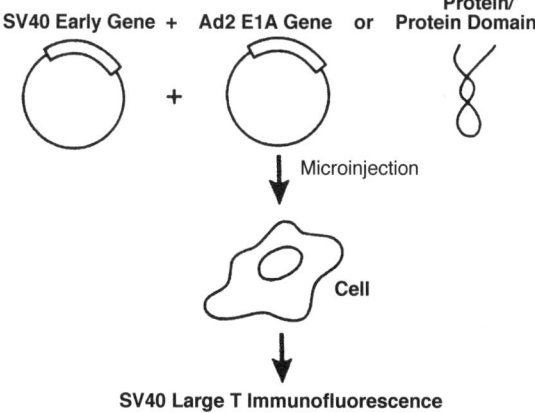

Fig. 4. Cell microinjection–transcription repression assay. The schematic shows comicroinjection of the SV40 early gene reporter with the Ad2 *E1A* gene or an E1A protein domain. Expression of the reporter is measured by immunofluorescence with antibody specific for the SV40 T antigen.

the first 80 N-terminal residues of E1A (E1A 1-80) was coinjected with the E1A repressible SV40 promoter plasmid, p1-11, and the formation of SV40 T antigen was measured by immunofluorescence (*see* **Fig. 4** for schematic). Included in each microinjection mixture was 1% tetramethylrhodamine dextran, which served as marker for successfully injected cells. As shown in **Fig. 5**, most cells injected with the SV40 plasmid synthesized T antigen (**Fig. 5A**), whereas most cells co-injected with p1-11 plus the E1A 1-80 protein were repressed in T antigen synthesis (**Fig. 5C**). Co-injection of p1-11 with a mutant E1A 1-80 protein containing a deletion in amino acids 4–25 did not repress

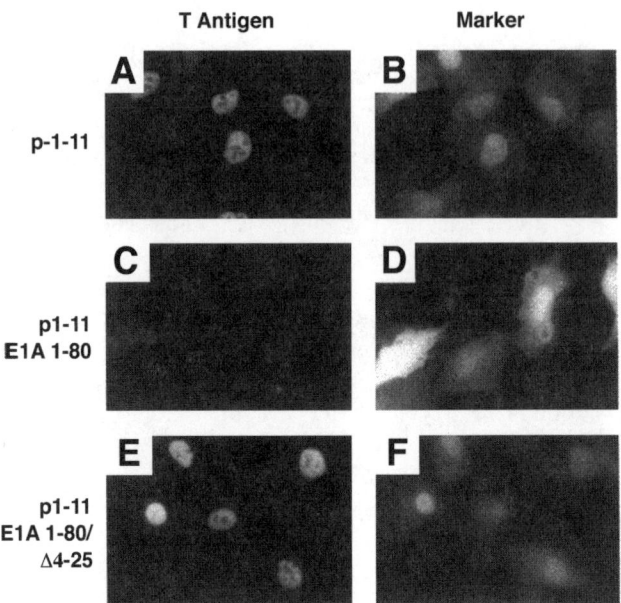

T Antigen   **Marker**

Fig. 5. Transcription repression of the SV40 T antigen by the cell microinjection assay. Cells were injected with SV40 T antigen expressing plasmid p1-11 alone (**A** and **B**), p1-11 and E1A 1-80 polypeptide (**C** and **D**), or p1-11 and E1A 1-80 Δ4-25 polypeptide (**E** and **F**). Included in each microinjection mixture was 1% tetramethylrhodamine dextran, which served as a marker for microinjected cells. The left-hand panels show cells staining positively for T antigen by FITC-conjugated second antibody. The right-hand panels show the same field viewed through filters allowing observation of rhodamine-containing cells. As shown, microinjection of the recombinant E1A 1-80 protein represses expression of T antigen, whereas microinjection of the deletion mutant, E1A 1-80 Δ4-25, does not *(5)*.

transcription (**Fig. 5E**). These microinjection experiments show that the N-terminal 80 amino acids are sufficient for E1A repression and that a sequence within the E1A N-terminal 4–25 residues is required for the repression function *(5)*.

1.1.4. Microinjection of Viral Oncogenes and Oncoproteins to Study Their Role in Cell Cycle Progression: Measurement of Cellular DNA Synthesis by Incorporation of ^{3}H-Thymidine and Autoradiography or by Incorporation of BudR and Immunofluorescence

Expression of the Ad E1A 243R protein can induce cell cycle progression from the G_0-phase to the S-phase, as demonstrated by cell microinjection. Microinjection of Ad E1A DNA *(6)* or recombinant E1A proteins *(7)* into serum-starved quiescent cells induced the synthesis of cellular DNA, as mea-

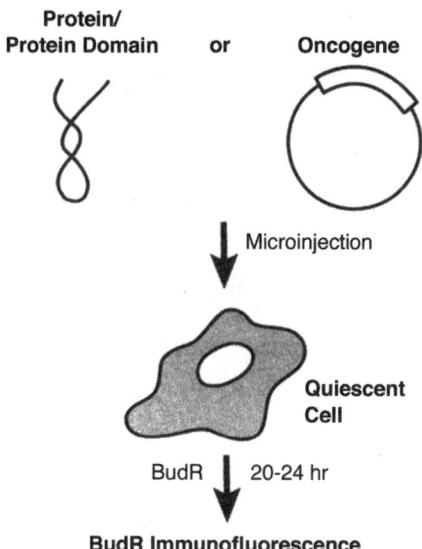

Fig. 6. Cell microinjection–cellular DNA induction assay. The schematic shows the microinjection of an oncogene or oncoprotein domain into serum-deprived quiescent cells. DNA induction is measured by the incorporation of BudR into cellular DNA followed by immunofluorescence detection of BudR-containing DNA with BudR-specific antibody.

sured by uptake of ^{3}H-thymidine followed by autoradiography. Thus, cell microinjection provides a direct method to examine the ability of viral genes and viral gene products to influence the cell cycle. Induction of DNA synthesis in microinjected cells can also be conveniently assayed by incorporation into cell DNA of the thymidine analog, BudR, followed by immunofluorescence analysis for BudR-containing DNA (*see* **Fig. 6** for schematic).

1.1.5. Microinjection of Specific Antibodies to Analyze Gene Requirements During Cell Cycle Progression: Measurement of Cellular DNA Synthesis by Incorporation of ^{3}H-Thymidine and Autoradiography or by Incorporation of BudR and Immunofluorescence

Ad E1A can transform cells in cooperation with other oncogenes such as *ras* *(8)*. Microinjection of antibodies specific to the Ras protein was shown to block progression from serum-starved G_0-phase cells to S-phase cellular DNA synthesis *(9)*. Of interest, microinjection with recombinant E1A protein was able to overcome the blockage caused by Ras antibody, indicating that E1A can substitute for *ras* and may act downstream of Ras in the cell cycle *(10)*. The temporal requirements for several cell proliferation-related genes in addition to *ras*, including *fos*, *fra-1*, and *fra-2* *(11,12)*, have been delineated by the microinjection of specific antibodies at various times during the cell cycle. Thus, microinjection of specific antibodies facilitates analysis of complex cellular regulatory functions.

1.1.6. Microinjection of Specific Antibody to Analyze the Role of a Putative Coactivator in the Transcription of Coinjected Promoter/ LacZ Constructs: Colorimetric Assay for LacZ-Expressing Cells

E1A 243R encodes functions required for the induction of cellular division and the inhibition of cell differentiation. The N-terminal region of E1A can interact with the p300/CBP (CREB-binding protein) family of proteins, which function as coactivators for the transcription of genes in pathways regulating cell division and cell differentiation. The transcription factor CREB, which is a component of the cyclic adenosine monophosphate (cAMP) signal transduction pathway, was thought to be a target of CBP. To prove this hypothesis, an assay for functional CREB was established by microinjection of a lacZ reporter construct driven by CRE (cAMP response element) into fibroblasts followed by staining for β-gal expression *(13)*. CREB activity was found to be blocked by coinjection of antibody against CBP, demonstrating that CBP is essential for functioning of the cAMP pathway. By a similar strategy using lacZ reporter constructs with upstream serum response element or TPA response element, it was shown that CBP was also essential for expression of two additional signal pathways *(13)*. Thus, microinjection of specific antibody provided convincing evidence that CBP functions in multiple signal transduction pathways, which might explain how E1A may regulate divergent pathways.

2. Materials
2.1. Cell Culture for Microinjection

For microinjection experiments, standard cell culture techniques are used to maintain stock cultures of cells, generally in Dulbecco's modified Eagle's medium (DMEM) supplemented with the appropriate serum, either fetal bovine serum (FBS) or calf serum. Cells are trypsinized and plated into 35-mm culture dishes containing cover slips marked to facilitate the location where cells are to be injected. Standard 22-mm^2 cover slips (Corning No.1) can be scribed with a diamond point pen. We scribe a cross over a 2- to 4-mm circle, yielding four separate quadrants, which can be used for different injection mixtures. The orientation of the four quadrants is established by scribing one of the lines of the cross with an extension to the left (similar in shape to the number "7"). The average quadrant will contain from 20 to 100 cells, depending on cell type and density. After scribing, cover slips are soaked for 1 h in ethanol and dry-sterilized at 180°C. Alternatively, the scribed slips can be removed from ethanol, carefully flamed, and placed in a sterile culture dish prior to plating cells. Cells plated on cover slips are grown in a humidified 5% CO_2 incubator until ready for microinjection. Cells should not be removed from the incubator for microinjection for more than 15 min at a time in order to minimize the rise in pH that occurs in carbonate-based DMEM. Alternatively, during microinjection cells can be maintained in carbonate-free DMEM buffered with 10 mM N-2-hydroxyethylpiperazine-N'-2-ethanesulfonic acid (HEPES), pH 7.4.

2.2. Preparation of Quiescent G_o-Phase Cells

For cell cycle studies, it is often desirable to synchronize cells in the G_o phase (quiescence) by serum starvation. Many cell lines can be made quiescent by allowing growth until cells are 50–70% confluent, washing in serum-free medium, and transferring to medium containing 0.5–1.0% serum for 1–4 d. These include, for example: murine cell lines C127 (1% FBS for 2–3 d), NIH 3T3 (0.5% calf serum for 1–2 d), Swiss 3T3 (2.5% FBS for 3–4 d); rat cell lines, rat1 (0.5% FBS for 2–3 d) and REF52 (0.5% FBS for 2–3 d); and human cell line Hs68 (0.5% FBS for 2–3 d). For each cell line and lot of serum, it is important to test the quiescence of the cells and their response to serum induction. We operationally define a cell population as "quiescent" when no more than 5–8% of the cells synthesize DNA, as determined by BudR incorporation during a 24-h period. To be useful for experimentation, more than 90% of the quiescent cells should enter the cell cycle and synthesize DNA after supplementation with 20% serum.

2.3. Microcapillaries for Microinjection

The "needles" used for microinjection are borosilicate glass capillaries drawn to a fine tip, 0.2–0.5 μm in diameter. The limiting factor in microinjection is often the quality of the drawn microcapillary. Excellent commercial microcapillaries in two styles (long or short shank) are available from Eppendorf (Femtotips) and are designed to operate with the Eppendorf capillary holder. For two reasons, every laboratory should consider having available a microprocessor-controlled capillary puller. If the laboratory does a large amount of microinjection, the cost will be significantly reduced. Second, by altering the parameters of the capillary puller, tips can be designed for each microinjection sample and cell type. In our laboratory, we use a Kopf (Tujunga, CA) model 750 vertical puller. This device allows alteration of the filament temperature and duration and strength of the pull, all of which affect the shape and size of the capillary tip. We use 10-cm borosilicate standard wall capillaries with an outer diameter of 1.2 mm (cat. no. GC120F10, Warner Instrument Corp). These capillaries contain a solid filament fused to the inner wall of the capillary, which facilitates top loading of samples by capillary action.

2.4. Preparation of Nucleic Acid and Protein Samples for Microinjection

Samples for microinjection should be free of particulates and aggregates to avoid clogging the microcapillary tip. Samples are centrifuged at 17,000g for 10 min prior to loading the microcapillary. It is generally not necessary to sterilize samples for microinjection experiments since minor bacterial contamination will not be a significant factor in the usual 1- or 2-d course of an experiment.

For DNA and RNA samples, researchers have successfully used a variety of buffers ranging from water to physiological phosphate-buffered saline (PBS). For protein samples, injection should be in the minimal buffer necessary to retain the biological activity of the injected molecule. A typical buffer is 50 mM HEPES, pH 7.4, containing 40 mM NaCl. The microinjection buffer alone should be tested for its effect on the injected cell. Typically, plasmid DNA is injected at concentrations from 10 to 500 ng/µL, which introduces about 20–1000 molecules per cell (assuming a M_r of 3×10^6 and an injection volume of 10 fL). RNA preparations are injected at 1–10 µg/µL. Biologically active proteins are often injected at 50–1000 ng/µL, which introduces 3000–60,000 protein molecules per cell (assuming a M_r of 50,000). IgG is generally injected at concentrations from 2 to 10 µg/µL.

2.5. Microinjection Markers

Single cell assays by microinjection have become more reliable because of the ability to easily identify cells that have been successfully injected. This is done by adding to the injection sample one of several classes of markers. Purified nonrelevant IgG (2–5 µg/µL) can be added to an injection sample and injected cells identified by immunofluorescence using fluorochrome-conjugated antibody against the species of injected IgG. One alternative is to add a fluorescent molecule to the injection mixture. We have used fluorochrome-conjugated dextrans (lysine-fixable) at 50–100 ng/µL—either dextran-fluorescein (molecular weight 70,000) or dextran-tetramethylrhodamine (molecular weight 70,000) from Molecular Probes (Invitrogen). The use of these fluorochromes requires separate centrifugation prior to assembly of the injection mixture when plasmid DNAs are used, because the formation of DNA/dextran complexes depletes the injection mixture of plasmid. Co-injection with dextran-conjugated fluorochromes permits facile identification by direct fluorescence of viable injected cells. Another excellent alternative is co-injection with a plasmid such as pEGFRN-1 (CLONTECH) (2–10 ng/µL), which expresses efficiently the green fluorescence protein (GFP). Fluorescence can be observed within several hours after cell microinjection by use of standard fluorescein isothiocyanate filters, thus identifying injected cells that are physiologically competent to transcribe and translate microinjected genes.

3. Methods

3.1. Microinjection Methods and Instrumentation

Microinjection of mammalian cells is best performed under phase contrast microscopy. The tip of the microcapillary loaded with the microinjection sample is positioned above the target cell by the use of a micromanipulator, and an appropriate volume is transferred into the cytoplasm or nucleus by air pressure supplied by a syringe or by an automatic injection device. Three types of micromanipulator setups can be used for cell microinjection. The first is the oil-type hydraulic micromanipulator that we use for manual microinjection; an excellent hydraulic unit is the Nikon/Narashige. The second type is a mechanical micromanipulator such as that manufactured by Leitz. The third type is the electromechanical micromanipulator, which we use for semi-automatic micro-injection. For the third type, the Eppendorf model 5171 micromanipulator coupled with the model 5246 transjector is highly recommended. The micromanipulator is microprocessor controlled, which allows highly reproducible microcapillary positioning. The transjector permits accurate delivery of a controlled volume of injected materials into each cell. Eppendorf's newer semi-automatic system uses the InjectMan NI2 micromanipulator and the FemtoJet 5247 microinjector. This newer system works in a fashion similar to that of the previous model 5171/5246 Eppendorf system. The use of an inverted microscope is desirable because of its ample stage and the large working distance between the stage and the condenser. It is advantageous to use a microscope equipped with UV fluorescence to monitor cells injected with fluorescent marker molecules. This permits identification of microinjected cells to assess cell survival and for experiments that involve multiple injections at various times. For teaching purposes, it is desirable to fit the microscope with a video camera linked to a high-resolution monitor.

We describe below in some detail protocols for manual microinjection as well as for the semi-automatic microinjection using the older Eppendorf system, which is still in use in many laboratories (5217/5246), and the newer Eppendorf system, which is currently available (NI2). These details provide insights into the microinjection methodology. Manuals accompanying microinjection equipment describe the equipment but do not (and are not intended to) teach the scientist how to microinject cells.

3.1.1. Manual Microinjection

3.1.1.1. POSITIONING THE MICROCAPILLARY ABOVE THE CELL FIELD

For manual microinjection, we use a Narishige MO-204 joystick micromanipulator connected to a hydraulic microdrive, onto which the microcapillary holder is attached. The hydraulic microdrive is connected to a mechanical Coarse movement manipulator, both of which are mounted on a Nikon Diaphot inverted phase contrast microscope. The Coarse manipulator moves the microdrive and attached microcapillary holder through coarse movements. The microscope and the micromanipulator are placed on a vibration isolation table (Technical Manufacturing Corp.).

To begin microinjection, connect the microcapillary loaded with the sample to be injected via tygon tubing to a Gilmont micrometer 2.0-mL threaded plunger syringe. Apply a positive pressure of approx 0.2 mL on the syringe, which has been filled with water. This will provide a holding pressure that prevents clogging of the microcapillary tip during subsequent manipulation. Carefully attach the microcapillary to the Narashige capillary holder (clamped to the microdrive at approx 45° angle to the plane of the microscope stage). Center the 35-mm dish containing cells to be injected on the stage of the microscope and focus the objective on the plane of the cells. The Coarse controls consist of three knobs, which direct movement in the *x*-axis (left and right), *y*-axis (front and back), and *z*-axis (up and down). Use the knobs on the Coarse movement manipulator to bring the microcapillary to the center of the culture dish and to lower its tip to just below the surface of the culture medium.

Looking through the microscope at ×100 magnification, focus on the microcapillary tip by moving the microcapillary back and forth with the joystick in the Y direction to locate the microcapillary as a moving shadow and raising the objective with the focusing knob until the tip of the microcapillary comes into view. Change the objective lens to give a final magnification of ×200 or ×400. Focus on the tip of the microcapillary and check that it is intact. In three- to four-step increments, first lower the focal plane with the focusing knob and then lower the microcapillary tip with the Coarse control until the tip is in focus. *Finally, when the target cells are in focus, bring the microcapillary tip down, leaving it just above the focal plane of the cells, i.e., the tip is kept out of focus to avoid penetrating the cells.*

3.1.1.2. MICROINJECTION OF CELLS

Use the joystick assembly controls for fine movement of the microcapillary and for microinjection. The joystick assembly of the Narashige micromanipulator contains two knobs at the base, which control fine movement separately in the *x*- or *y*-axis. Deflection of the joystick causes the microcapillary to move in a combined xy-axis. Rotation of the joystick raises or lowers the microcapillary in the z-axis. Use the two knob controls on the joystick assembly to center the microcapillary tip over the area of cells to be microinjected. Rotate the joystick to lower the tip to just above the surface of the cell to be injected, the "search plane" (the tip should appear slightly blurry in the search plane). Microinject cells by carefully lowering the tip to penetrate the nucleus or the cytoplasm of the cell as desired. The tip of the capillary, which is positioned at a 45° angle, will penetrate the cell vertically, i.e., at a 90° angle. Successful injections are indicated by gentle swelling of the nucleus or the cytoplasm. Move the microcapillary tip from cell to cell by appropriate deflection of the joystick. When all of the desired cells in a field are injected, move to the new area by use of the joystick knobs.

During cell microinjection it will be necessary to periodically regulate the pressure applied to the microcapillary by adjustment of the Gilmont syringe. Injections with the manual system are performed under constant pressure resulting in a constant flow of fluid. Therefore, injections into the nucleus will also deliver some sample into the cytoplasm. By constant pressure and by operator control of the degree of nuclear or cytoplasmic swelling, a moderate degree of control is obtained, delivering volumes of about 10–20 fL *(1)*. Although easy to learn, successful cell microinjection using the manual system requires patience and experience.

3.1.2. Eppendorf Semi-Automated Cell Microinjection System: Micromanipulator 5171 Combined With the Transjector 5246

The computerized Eppendorf system (Eppendorf North America, Westbury, NY) enables precise, reliable, and rapid positioning of microcapillaries for microinjection of mammalian cells. The system consists of two basic units: the micromanipulator, which electronically controls movement of the microcapillary, and the transjector, which regulates injection parameters. The micromanipulator contains three precision DC stepper motors that are controlled by the operator through a joystick and are capable of smooth movement, rapid or slow, in the x-, y-, and z-axes. Cells may be automatically injected either in the *Axial mode* (the cell is injected by movement of the microcapillary in the 45° angle) or the *non-Axial mode* (the cell is injected by the capillary being held at a 45° angle while it is lowered in the z-axis, as is done in manual microinjection). The transjector is programmable through a convenient keypad that communicates with the micromanipulator to execute automatic injection routines. The injection parameters, including injection pressure, injection time, and compensation pressure (a constant holding pressure to prevent backflow of medium into the microcapillary), are programmed to permit controlled injection of reproducible volumes of injection mixture into cells.

3.1.2.1. Filling and Mounting the Microcapillary

Swing the tool holder assembly containing the capillary holder towards the operator and screw in an injection-sample-loaded microcapillary as described below. For Eppendorf femtotips, insert the tip of an Eppendorf microloader containing 0.5–1.0 µL of the injection sample as far as possible into the femtotip and slowly empty the contents into the tip. Remove the protective cap and screw the loaded femtotip into the capillary holder. To load glass microcapillaries pulled in the laboratory, pipet the injection mixture onto the top of the capillary. After capillary action has drawn the liquid to the tip, insert the blunt end of the capillary into the 1.2-mm chuck assembly, tighten it, and insert the chuck assembly into the capillary holder. Return the capillary holder assembly to the default injection position.

3.1.2.2. Setting Parameters for Microinjection

Power on the transjector and the micromanipulator when the sample is ready to be microinjected. Using the transjector control keypad (T pad), set the valve to the valve position 1 (the default position) by cycling through the I1/I2 key on the T keypad, which selects the active injection port (select valve position X when there is no capillary mounted in order to prevent the bleeding of air through the open valve). The T pad will now display AUTO (automatic mode) and the settings in permanent memory for Pi (injection pressure), Pc (compensation pressure), I1, Ti (time of injection in fractions of a second), and n (number of injections). The most recently entered parameters in the transjector may have to be modified for different tips and injection samples in order to obtain satisfactory injections. Use the SET key on the transjector T pad to scroll through the Pi, Ti, and Pc options. Adjust the values for Pi, Ti, and Pc with the UP or DOWN key (followed by the SAVE key to place the new values in permanent memory if desired). After setting these parameters, press the INJECT MODE key to switch from the programming mode to the inject mode. For most microcapillaries and mammalian cells, use settings for Pi of 50–200 hPA, for Pc of 20 hPA, and for Ti of 0.2–0.8 s (settings of 80 for Pi, 20 for Pc, and 0.2 s for Ti are good starting points).

3.1.2.3. Positioning the Microcapillary Tip Above the Cells

Using the micromanipulator keypad (M pad), switch to the Fast mode with the SPEED toggle key and to Axial movement with the AXIAL key. The M pad display should now read Dynamic, Fast, AX (axial movement), In (inject mode), X, Y, and Z coordinates in µm (0 when the instrument is first turned on), and JB = INJ/IMP (joystick button in the inject/impale mode). If the above parameters are not displayed, enter the correct ones using the M pad keys.

Center a 35-mm culture dish containing the cover slip with cells to be injected above the objective on the microscope stage and focus on the target cells using ×100 magnification. Bring the tip of the microcapillary slightly past the center of the microscope objective (which is still focused on the target cells) by rotating the joystick knob clockwise to lower the microcapillary, in the simultaneous X/Z direction (i.e., a 45° angle because the Axial movement key is toggled on). At this point the tip of the microcapillary will be positioned slightly beyond the target cells and well above the surface of the culture medium.

After centering the microcapillary tip, toggle off the Axial movement key so that the microcapillary is now lowered in the Z rather than the X/Z direction. Carefully lower the microcapillary by clockwise rotation of the joystick until the tip is just below the surface of the culture medium. Switch to Slow mode using the SPEED toggle key on the M pad to permit a controlled approach to the plane of the cells. While viewing through the microscope, raise the objective with the focusing knob to focus on the tip of the microcapillary by moving the joystick back and forth in the Y direction to locate the microcapillary as a moving shadow and raising the objective until the tip comes into view. Change the objective to provide a final magnification of ×200 or ×400. Focus on the microcapillary tip and confirm that it is intact. Press the CLEAN key on the T pad and confirm that the injection mixture flows from the tip, as indicated by the schlieren pattern (the clean function flushes the microcapillary at maximum pressure of approx 9000 hPa). In three to four steps, alternately lower the focal plane with the focus knob and then lower the microcapillary with the joystick until its tip is in focus. *At the last step, when the target cells are in focus, carefully lower the microcapillary tip until it is slightly above the focal plane of the cells, i.e., keep the tip slightly out of focus to avoid penetrating the cells.*

3.1.2.4. PREPARATION FOR MICROINJECTION BY SETTING THE Z-LIMIT AND SEARCH PLANE

Check the injection parameters on the M pad by pressing the SET key and then the INJECT key. Use the SET key to toggle between the choices and use the UP or DOWN key as a toggle to change the choices. Ensure that INJ is on, IMP is off, and, for most injections, the Axial-injection mode is on (*see* **Note 1**). The speed of injection can be modified (300 µm/s is a good choice). Finally, press the ENTER key to exit the program mode. On the T pad display, N indicates the total number of injections attempted; set N to zero prior to each new series of injections.

The Z-LIMIT is the final depth of the microcapillary tip upon injection of adherent cells and is experimentally established for each series of injections. The Z-LIMIT must be set before microinjection can be performed. Carefully rotate the joystick clockwise to lower the microcapillary tip until it just touches the surface of the target cell. Set this temporary Z-LIMIT into memory by pressing SET, LIMIT, and ENTER on the M pad. Raise the microcapillary tip to the SEARCH PLANE with the joystick so that it is slightly above the surface of the cells. The image of the tip in the SEARCH PLANE will appear slightly out of focus. Attempt to inject a cell by positioning the tip slightly beyond the site of the intended injection and pressing the center button on the joystick to automatically inject the cell. The microcapillary tip will be withdrawn in the X direction, will pierce the cell in the Axial direction down to the Z-LIMIT, and will then be withdrawn to its original position in the SEARCH PLANE. If injection is successful, as indicated by a gentle swelling of the targeted nucleus or cytoplasm, the current Z-LIMIT is satisfactory. Most often, setting the Z-LIMIT will require several adjustments. If the tip passes through the cell entirely, as indicated by a white spot that persists, adjust the Z-LIMIT upward. If the tip does not puncture the cell, adjust the Z-LIMIT downward. To adjust the Z-LIMIT, press the LIMIT key, the UP or DOWN key (each press of the key adjusts the Z-LIMIT 160 nm), and then ENTER.

3.1.2.5. SEMIAUTOMATIC INJECTION OF ADHERENT CELLS ON COVER SLIPS

Cells are injected by positioning the microcapillary tip, while in the SEARCH PLANE, over the target site and pressing the button on the joystick. To inject the nucleus, position the tip of the capillary slightly beyond the center of the nucleus. To inject the cytoplasm, position the tip over the cytoplasm near the outer edge of the nucleus. Once an appropriate Z-LIMIT is set, cells in the same focal plane are readily injected without modification of the Z-LIMIT. However, uneven surfaces on the cover slip or culture dish or differences in cell morphology may require periodic readjustment of the Z-LIMIT.

Depending on the sample injected, the characteristics of the microcapillary tip, and the condition and type of target cells, it may be necessary to adjust the injection pressure (Pi) and injection time (Ti) to obtain gentle swelling of the target cell. Adjustment is accomplished by directly accessing the appropriate key on the T pad and adjusting to the desired value with the UP and DOWN keys followed by the INJECT MODE key. If debris adheres to the tip, it can often be removed by briefly pressing the CLEAN/HOME key on the M pad, which will withdraw the tip from the medium. Pressing of the CLEAN/HOME key again will return the microcapillary to its original position. If the tip becomes plugged, raise it slightly above the SEARCH PLANE by rotating the joystick counterclockwise and focus on the tip. Then press the CLEAN key on the T pad to expel liquid at 7000 hPa, which will unplug the tip, as visualized by the schlieren pattern under the microscope. After unplugging the tip, return to the SEARCH PLANE and continue cell injections. If the plug cannot be expelled, load and position a new microcapillary and continue microinjection.

When changing the culture dish and/or the microcapillary, press and hold down the CLEAN/HOME key on the M pad. This action will withdraw the microcapillary to the limits of the X and Y motors. When changing the microcapillary, close the transjector valve by selecting valve position X with the T pad. After changing a femtotip or a manually drawn capillary (no longer than the previously used capillary), press the CLEAN/HOME key again to return to within 700 μm of the original SEARCH PLANE. Then reestablish both the Z-LIMIT and the SEARCH PLANE. For a manually drawn microcapillary that is longer than the previous microcapillary, cancel the home function by pressing the CANCEL key. Then position the microcapillary above the cells and establish the Z-LIMIT and SEARCH PLANE as described above.

3.1.3. Eppendorf Semiautomated Cell Microinjection System: Micromanipulator InjectMan NI2 Combined With the FemtoJet 5247

The newer computerized Eppendorf NI2 system (Eppendorf North America, Westbury, NY) has recently become available and provides a more convenient system for microinjection of mammalian cells. The system consists of two basic units: the InjectMan NI2 micromanipulator, which electronically controls movement of the microcapillary and the FemtoJet microinjector, which regulates injection parameters like the 5171 system. The InjectMan NI2 micromanipulator also contains three precision DC stepper motors that are controlled by the operator through a joystick and are capable of smooth movement, rapid or slow, in the x-, y-, and z-axes. Cells can be automatically injected either in the Axial mode, the non-Axial mode, or the Step-Inject mode. In the Axial mode, the cell is injected by an axial movement of the microcapillary. The angles of the microinjection capillary and the axial movement are typically set at 45°, but can be set to an angle of inclination to the culture dish of between 15° and 75°. In the non-Axial mode, the microinjection capillary is held at an angle between 15° and 75°, whereas the cell is injected with a z-axis movement, which is at a 90° angle to the culture dish, similar to that used for manual microinjection. The Step-Inject mode allows the microinjection capillary to be set at an angle of between 15° and 75°, and the micromanipulator is programmed to move a set distance forward without using a Z-LIMIT. Because the Axial mode is the most commonly used injection mode for mammalian cells, it is the mode that we describe here. The FemtoJet is programmable and communicates with the micromanipulator to execute automatic injection routines. Like the 5171, the injection parameters are injection pressure (Pi), injection time (Ti), and compensation pressure (Pc). These parameters are set using variable speed control knobs on the front of the FemtoJet to permit the controlled injection of reproducible volumes of injection sample into cells.

3.1.3.1. Filling and Mounting the Microcapillary

The capillary holder is attached to the NI2 motor module. Swing the motor module toward the operator and install an injection sample loaded microcapillary as described below. Load Eppendorf Femtotips as described in **Subheading 3.1.2.1.** Screw the loaded tip into the Femtotip adapter and screw the Femtotip adapter to the front of the capillary holder. To load glass microcapillaries pulled in the laboratory, remove the Femtotip Adapter from the capillary holder and loosely attach the 1.0- to 1.1-mm Capillary Grip Head. Pipet the injection sample into the top of the microcapillary. After capillary action has drawn the liquid to the tip, insert the blunt end of the microcapillary into the Capillary Grip Head so that it passes completely through the O-rings. Tighten the Capillary Grip Head onto the capillary holder. Swing the NI2 motor module back into the default injection position.

3.1.3.2. Setting Parameters for Microinjection

Before powering on the FemtoJet, disconnect the Injection Line from the front of the console and leave it disconnected until the start-up program is complete and the normal operating display appears on the LCD panel (approx 3 min). Power on the InjectMan micromanipulator and the FemtoJet. Press the MENU key on the FemtoJet to close the injection port. Reconnect the Injection Line to the front of the console, attach the microinjection capillary as described in **Subheading 3.1.3.1.**, and press the MENU key to re-open the injection port. AUTO (automatic mode) should appear in the upper left of the display, and the last values used for the three injection parameters will appear along the bottom of the display. These parameters are: Pi, measured in hPa; Ti, measured in tenths of a s; and Pc, measured in hPa. Depending upon injection sample and capillary, it may be necessary to modify injection parameters from their previous settings in order to obtain satisfactory injections. These values can be changed when the injection valve is open and the normal operating display appears on the LCD panel. To modify these values, turn the Pi knob, Ti knob, and/or the Pc knob so that the desired values are displayed. The optimal injection parameters for each project need to be determined experimentally, but for most types of mammalian cells the Pi will be between 50 and 200 hPa, the Ti will be between 0.2 and 0.8 s, and the Pc between 20 and 50 hPa. A good starting point for use with Femtotips is a Pi of 80 hPa, a Ti of 0.2 s, and a Pc of 30 hPa.

3.1.3.3. Positioning the Microcapillary Tip Above the Cells

Using the micromanipulator keypad (M pad), switch from the FINE mode to the COARSE mode using the speed selector toggle key labeled Coarse/Fine. Press the AXIAL key to put the manipulator into axial movement. The LCD display on the M pad should now read COARSE and will display the X, Y, and Z coordinates of the microcapillary in μm (0,0,0 when the instrument is turned on). AXIAL will appear beneath the coordinates. The speed of movement of the microcapillary in COARSE mode can be adjusted using the thumb wheel located on the lower right side of the M pad. Center a 35-mm culture dish containing the cover slip with cells to be injected above the objective on the microscope stage and focus on the target cells using ×100 magnification (10X objective). With AXIAL on, twisting the joystick clockwise will lower the microcapillary axially in the simultaneous X/Z direction (i.e., a 45° angle). Using the joystick, visually bring the tip of the microcapillary slightly past the center of the microscope objective by rotating the joystick clockwise to lower the microcapillary. Adjust the position of the microcapillary as needed by moving the joystick left or right and toward the rear or front of the microscope.

After centering the microcapillary tip, toggle off the AXIAL movement key so the microcapillary will move in the true *z*-axis when the joystick is twisted. Carefully lower the microcapillary by clockwise rotation of the joystick until the tip is just below the surface of the culture medium. Switch to FINE mode using the speed selector toggle key to permit a controlled approach to the plane of the cells. The speed of movement of the microcapillary in FINE mode may be adjusted using the thumb wheel located on the lower right side of the M pad. In a stepwise manner, lower the microcapillary to near the plane of the cells as described in **Subheading 3.1.2.3.** *At the last step, when the target cells are in focus, carefully lower the microcapillary tip until it is slightly above the focal plane of the cells, i.e., keep the tip slightly out of focus to avoid penetrating the cells.* Focus on the microcapillary tip and ensure that it is intact. Press the CLEAN key on the front of the FemtoJet and confirm that the injection sample flows from the tip, as indicated by the schlieren pattern (the CLEAN function flushes the microcapillary at maximum pressure of ~7000 hPa).

3.1.3.4. Preparation for Microinjection by Setting the Z-LIMIT and Search Plane

Check the injection parameters on the M pad. The first time the InjectMan NI2 is used, it may need to be programmed with the correct parameters to inject mammalian cells. Press the MENU key to display the programming menu on the M pad display. Use the thumbwheel to move the cursor down to the INJECT heading and press the UP arrow key to display the INJECT submenu. If necessary, edit the values in the INJECT submenu to read as follows: InjSpeed, 480 um/s; Synchron, Immed; Search+L, OFF; LimAxis, Z-AXIS; and InjAxial, ON. To change the default injection speed, use the thumbwheel to highlight the InjSpeed header on the submenu. Press the UP arrow key and the cursor will highlight the speed value. Edit the InjSpeed value by turning the thumbwheel until the value 480 μm/s is displayed in the field, and press the

LIMIT key to save that value. The cursor will automatically move back to the header column. Use the thumbwheel to highlight the next header, and edit it in the same manner. When all the settings are correct, press the MENU key to exit the programming mode. InjectMan NI2 will maintain these new settings even if the power is interrupted. On the FemtoJet LCD panel, the value "n = #" where # indicates the total number of injections attempted; press the COUNT key to set n to zero prior to each new series of injections.

The Z-LIMIT is the final depth of the microcapillary tip upon injection of adherent cells and is experimentally established for each series of injections. The Z-LIMIT must be set before microinjection can be performed. First, focus the microscope precisely on the cells to be injected. Select one cell as a target and carefully rotate the joystick clockwise to lower the microcapillary tip until it just touches the surface of the target cell. Set this temporary Z-LIMIT into memory by pressing LIMIT on the M pad. Z-LIMIT will appear on the M pad display to indicate that a Z-LIMIT is in place. Raise the microcapillary tip to the SEARCH PLANE with the joystick so that it is slightly above the surface of the cells (10–20 µm). The tip of the microcapillary in the SEARCH PLANE will appear slightly out of focus. Once you have saved a Z-LIMIT, avoid refocusing the microscope, because the focal plane of the microscope will be the same as your Z-LIMIT. Without raising or lowering the tip from the SEARCH PLANE, attempt to inject a cell by positioning the tip above and slightly beyond the site of the intended injection and pressing the center button on the top of the joystick to automatically inject the cell. The microcapillary tip will be withdrawn in the X direction, will pierce the cell in the axial direction down to the Z-LIMIT delivering an injection, and will then be withdrawn to its original position in the SEARCH PLANE. If injection is successful, as indicated by a gentle swelling of the targeted nucleus or cytoplasm, the current Z-LIMIT is satisfactory. Most often, setting the Z-LIMIT will require several adjustments. If the tip passes through the cell entirely, as indicated by a white spot that persists, adjust the Z-LIMIT upward. If the tip does not puncture the cell, adjust the Z-LIMIT downward. To adjust the Z-LIMIT, press the UP or DOWN key on the M pad (each press of the key adjusts the Z-LIMIT 160 nm). To clear a Z-LIMIT from memory, press the LIMIT key again.

3.1.3.5. Semiautomatic Injection of Adherent Cells on Cover Slips

Semiautomatic injection of cells is done as described in **Subheading 3.1.2.5.** If it becomes necessary to alter the injection parameters, adjust by turning the appropriate knob on the FemtoJet. If debris adheres to the tip, it can often be removed by briefly pressing the HOME key on the M pad to withdraw the tip from the medium. Pressing the HOME key again will return the microcapillary to its original position. If the tip becomes plugged, raise it slightly above the

SEARCH PLANE by rotating the joystick counterclockwise and focus the microscope on the tip. Then press the CLEAN key on the FemtoJet to expel the plug at 7000 hPa as described in **Subheading 3.1.2.5.**

When changing the culture dish and/or the microcapillary, press and hold down the HOME key on the M pad. This action will withdraw the microcapillary to the outward limits of the X and Y motors. The motor module can then be swung towards the user for easy access. Before removing a microcapillary, close the FemtoJet injection valve by pressing the MENU key on the FemtoJet. When the new microcapillary is in place, re-open the injection valve by pressing the MENU key again. After changing a Femtotip or a manually drawn capillary (no longer than the previously used capillary), the capillary can be returned to the culture dish by pressing the HOME key again, as long as an OFFSET of at least 700 µm has previously been entered by editing the values under the Home heading in the programming menu. To edit the OFFSET value, press MENU on the M pad, select HOME, press UP, highlight OFFSET and press UP, enter a value of 700 µm with the thumbwheel, press LIMIT, then Menu. If there is no offset, or if the new microcapillary is longer than the one previously used, simply move the joystick to one side to clear the HOME function, and return the microcapillary into the culture dish as before. In either case, once a needle has been changed, reestablish both the Z-LIMIT and the SEARCH PLANE.

3.2. Analysis of Microinjected Cells

3.2.1. Staining Microinjected Cells With Fluorescent Antibodies Directed Against Specific Proteins

Most microinjection experiments ultimately involve staining of injected cells with fluorescent antibodies and analysis by fluorescence microscopy. Immunofluorescence is used to detect the protein product of a reporter gene or that of an endogenous gene whose expression has been modulated by the microinjected effector molecule. Microinjected cells on cover slips are fixed, permeabilized, and incubated with the primary antibody, which recognizes the protein of interest, and then stained with a second fluorochrome-conjugated antibody, which recognizes the primary antibody. Optimal fixation conditions for each antigen–antibody interaction must meet two main criteria: the cell structure should be preserved, and the antigenic structure should be maintained in a form that allows recognition by the antibody. The optimal fixative is determined empirically. There are two main classes of fixatives: organic solvents such as alcohols or acetone and crosslinkers such as formaldehyde or glutaraldehyde. Some commonly used organic fixatives are 100% methanol, 50% methanol/50% acetone, and 5% acetic acid/95% ethanol. Commonly used crosslinkers are 3.7% formaldehyde in PBS and 4.0% paraformaldehyde in PBS.

It is often necessary to block nonspecific interactions to prevent high fluorescent backgrounds. This is often achieved by incubating permeabilized cells with PBS containing 5% bovine serum albumin (BSA) for 30 min. Using PBS/ 0.1% Tween-20 in the blocking, washing, and incubation solutions will also help to maintain a high signal-to-noise ratio. Below is a general protocol that has been found useful for many antigens.

1. At the completion of the microinjection experiment, transfer cover slips (cell-side up) to new 35-mm dishes, rinse twice with PBS, and fix cells by incubation with 2 mL of 3.7% paraformaldehyde/PBS for 10 min. Aspirate the solution and permeabilize cells by incubation with 0.3% Triton X-100/PBS for 3 min. Rinse cover slips twice with PBS and store in PBS at 4°C until ready for staining.
2. Rinse cells on cover slips twice with PBS/0.1% Tween-20 and block cells by incubation in PBS/5% BSA for 30 min. Carefully pipet 50 µL of an appropriate dilution of primary antibody in PBS/0.1% Tween-20 onto the central area of injected cells (prior to use, centrifuge diluted antibody for 10 min at 14,000g to remove aggregates). Incubate cover slips in a humidified chamber for 60 min at 37°C.
3. Wash cells three times for 5 min with 3 mL of PBS/0.1% Tween-20 on a laboratory rocker.
4. Block cells with PBS/5% goat serum for 30 min.
5. Add to the cover slip 50 µL of an appropriate dilution of the secondary antibody in PBS/0.1% Tween-20, e.g., 1:100 dilution of goat antirabbit IgG conjugated with either fluorescein isothiocyanate or Texas red. Incubate in a humidified chamber for 60 min at 37°C.
6. Wash cover slips three times with PBS/0.1% Tween-20.
7. Using a fine forcep, dip each cover slip in water for 10 s, drain briefly against a Kimwipe, and place cell-side down on a drop (~25 µL) of mounting medium (*see* **Note 2**) placed on a microscope slide. Gently press the cover slip down (a pencil eraser works well) and remove excess fluid with a Kimwipe. Seal edges of the cover slip with clear nail polish for permanent storage. The cells are now ready for fluorescence microscopy and photography (*see* **Note 3**).

3.2.2. Colorimetric Assay to Detect Expression From Microinjected Promoter/LacZ Reporter Constructs

A facile method to analyze promoter function in vivo is to introduce into the cell by microinjection a chimeric reporter gene such as lacZ driven by the promoter of interest. Coinjection of this promoter/reporter with a candidate transcription factor, or with an antibody directed against an endogenous protein thought to modulate the expression of the promoter, provides a powerful strategy for analyzing transcriptional regulation in vivo. Expression of lacZ from the chimeric promoter construct is readily measured at the single-cell level by the colorimetric assay for β-gal expression described below.

1. Wash the cover slips containing the microinjected cells in 35-mm dishes twice with PBS.
2. Fix cells for 5 min with 3.7% formaldehyde/PBS. Wash twice with PBS.
3. Add 1 mL of freshly made X-Gal solution and incubate at room temperature for 15 min to overnight (*see* **Note 4**). To prepare 10 mL of X-Gal solution, add the following to 9.6 mL of PBS: 100 µL of 200 mM MgCl$_2$, 100 µL of 500 mM potassium ferrocyanide (210 mg/mL in PBS), 100 µL of 500 mM potassium ferricyanide (160 mg/mL in PBS), and 100 µL of X-Gal (5-bromo-4-chloro-3 indoyl-β-D-galactoside at 100 mg/mL in dimethyl formamide and store in the dark at –20°C). Mix well by inversion. If a precipitate forms, pass the solution through a syringe filter.
4. Wash cover slips with PBS, rinse with water, and mount cover slips cell-side down on a microscope slide with an appropriate mounting medium (*see* **Note 2**).

3.2.3. Analysis of Cellular DNA Synthesis by Immunofluorescence Detection of BudR Incorporation Into Cellular DNA

Studies directed at gaining insights into the molecules that regulate the cell cycle often depend upon the detection of S-phase DNA synthesis at various times after the microinjection of effector molecules into cells. Cellular DNA synthesis has been successfully monitored by the incorporation of ^{3}H-thymidine into DNA followed by autoradiography. A more facile and rapid procedure in current use is the incorporation of BudR into cellular DNA followed by immunofluorescence detection as described below.

1. Add 5-bromodeoxyuridine to cells on cover slips in 35-mm dishes with 2 mL of medium to a final concentration of 100 µM (4 µL of a 50 mM solution in PBS) and incubate for 20–24 h (*see* **Note 5**).
2. Wash cells twice with PBS and fix cells in –20°C methanol for 10 min. Rehydrate by incubating cells in PBS for at least 10 min. The cover slips can be stored in PBS at 4°C for several days.
3. Incubate cover slips for 30 min in 1.5 N HCl to denature cellular DNA. Wash cover slips three times for 3 min with PBS.
4. Incubate cover slips for 30 min at 37°C in a humidified incubator with 50 µL of primary antibody solution consisting of a 1:50 dilution in PBS/0.5% BSA of anti-BrdU monoclonal antibody (Becton Dickinson, cat. no. 7580).
5. Wash cover slips three times for 5 min with PBS.
6. Incubate cover slips for 30 min at 37°C with a secondary antibody solution consisting of 50 µL of a 1:50 dilution in PBS/0.5% BSA of Texas red-conjugated sheep anti-mouse immunoglobulin polyclonal antibody (Amersham, cat. no. N 2031).
7. Wash cover slips three times with PBS, rinse briefly with water, and mount each cover slip cell-side down on a microscope slide (*see* **Note 2**).

3.2.4. In Situ *Hybridization and Autoradiography to Measure Specific RNA Molecules in Microinjected Cells*

Studies on the transcription of microinjected genes, of reporter genes driven by microinjected promoter/reporter constructs, or of endogenous cellular genes require the use of *in situ* hybridization to detect specific RNA molecules. RNA formation can often be detected within 20 min after injection of a suitable expression plasmid. Hybridization is performed using radiolabeled DNA or antisense RNA probes followed by autoradiography with a low-melting-temperature photographic emulsion. Below we provide detailed protocols used in our laboratory based on the methods developed in the laboratory of R. Singer *(4,14)*. ^{35}S is preferable for radiolabeling probes used for *in situ* hybridization. DNA probes are prepared by nick translation or by random priming using ^{35}S-dCTP as labeled precursor and a commercial kit. Antisense RNA probes are prepared from an appropriate vector containing the gene of interest upstream of a T7, T3, or Sp6 promoter. ^{35}S-CTP is used as precursor with the appropriate RNA polymerase and a commercial kit (e.g., Promega, Madison, WI).

The use of nonradioactive detection of mRNA by *in situ* hybridization has become common. One of the most sensitive procedures involves labeling the DNA or RNA probe or an appropriate oligonucleotide with digoxigenin-tagged deoxyuridine triphosphate (DIG-11-dUTP) followed by immunological detection using antidigoxigenin antibody conjugated with alkaline phosphatase and colorimetric analysis. These procedures are reported to be as sensitive as radioactive detection. The methodology used for *in situ* hybridization with probes labeled by the digoxigenin is very similar to those described below with appropriate modifications (for example, *see* **ref. *15***). Roche Applied Biosciences offers reagents for nonradioactive *in situ* hybridization and provides an excellent application manual *(16)*, which provides detailed protocols.

3.2.4.1. FIXATION OF MICROINJECTED CELLS

1. Wash microinjected cells on 22-mm^2 cover slips twice with 3 mL of PBS.
2. Rinse cover slips with 3 mL of 4% paraformaldehyde/5 m*M* MgCl$_2$/PBS and incubate for 15 min with 3 mL of the same solution; 4% paraformaldehyde is prepared by heating paraformaldehyde (Sigma-Aldrich) in PBS at approx 60°C with stirring for about 3 h and then adding MgCl$_2$ to 5 m*M*, followed by filtration to remove particulates.
3. Rinse cover slips with 3 mL of 70% ethanol and incubate with 3 mL of 70% ethanol at 4°C for at least 1 h.

3.2.4.2. PREHYBRIDIZATION

1. Remove each cover slip from 70% ethanol with fine-tipped forceps, blot the edge against a Kimwipe, place carefully in a staining dish containing 10 mL of 5 m*M*

$MgCl_2$/PBS, and incubate for 10 min. Be sure to monitor which side of the cover slip contains the cells (mark one end of the staining dish and place cover slips with cells facing the marked end) (a VWR Scientific staining dish, cat. no. 25452-002, contains four pairs of grooves and can accommodate seven cover slips positioned in a zig/zag orientation).

2. For antisense RNA probes, acetylate cells by incubation for 10 min with 10 mL of 0.1 M triethanolamine-HCl, pH 8.0, containing 0.25% acetic anhydride (stock 0.1 M triethanolamine is brought to pH 8.0 with about 2 mL of concentrated HCl and stored at room temperature). Add 25 µL of anhydrous acetic anhydride to 10 mL of triethanolamine just before use. Rinse cover slips twice with 2X standard sodium citrate (SSC: 150 mM NaCl/15 mM Na_3 citrate, pH 7.0).

3. Aspirate the solution and incubate cover slips for 10 min in the staining dish with 10 mL of 0.1 M glycine/0.2 M Tris-HCl, pH 7.4.

4. Aspirate the solution completely and incubate cover slips for exactly 10 min in 10 mL of 50% molecular biology grade formamide/2X SSC/10 mM dithiothreitol (DTT) at 65°C (do this at the same time as the probe is being denatured, just prior to hybridization).

3.2.4.3. Hybridization

1. Dry the following mixture in a separate Eppendorf tube for each cover slip (Savant Speed Vac Concentrator or a lyophilizer). In order add to each tube: 2 µL of 10 mg/mL (10 µg) *Escherichia coli* tRNA (Boehringer); 2 µL of 10 mg/mL (10 µg) salmon sperm DNA (Sigma, sheared to about 300 bp by passing through a 26-gage needle); and 6–20 ng of either ^{35}S-labeled DNA or ^{35}S-labeled RNA probe (1–5 × 10^6 cpm).

2. Add 10 µL of formamide to the dried probe mixture in each Eppendorf tube.

3. Cap each Eppendorf tube and denature the probe by heating for 10 min at 90°C using a heating block.

4. Prepare 2X hybridization buffer by mixing 1/5 vol of 20X sodium saline citrate (SSC), 1/5 vol of BSA (20 mg/mL, Boehringer, Mol. Biol grade), 2/5 vol of 50% dextran sulfate (Pharmacia), and 1/5 vol of 1 M DTT. One at a time, remove an Eppendorf tube from the heating block, add 10 µL of 2X hybridization buffer, and gently pipet up and down five times with a 20-µL pipetman (the final concentration in the hybridization mixture is 2X SSC or 0.36 M Na+). Take every precaution to avoid air bubbles. If air bubbles are formed, centrifuge the tube for 10 s. Transfer the hybridization mixture in two installments (to avoid air bubbles) to a sheet of parafilm taped tightly on top of a glass plate. With a pair of fine forceps, remove the cover slip from the prehybridization mixture, drain briefly, and place it cell-side down on top of the 20-µL drop of hybridization mixture. Remove any air bubbles over the region of injected cells by gentle manipulation of the cover slip with the forceps. When all cover slips have been applied to the probe mixture, place a second sheet of parafilm over the cover slips to prevent drying, and incubate at 37°C for DNA probes or 45°C for RNA probes in a humidified incubator for 4 h to overnight.

3.2.4.4. Post-Hybridization Washing of Cover Slips After Hybridization With DNA Probes

1. Gently lift each cover slip from the parafilm with a pair of forceps as follows. To avoid breaking fragile cover slips, add a drop of 50% formamide/2X SSC to the edge of the cover slip, pierce the parafilm near the edge with sharp forceps, and gently break the surface tension by slowly raising the cover slip to allow the solution to enter the space between the cover slip and the parafilm. Incubate the cover slips in 10 mL of 50% formamide/2X SSC for 30 min at 37°C in a staining dish (seven slips per dish as described above).
2. Aspirate the solution and incubate the cover slips for 30 min in 10 mL of 50% formamide/1X SSC at 37°C.
3. Transfer each cover slip to individual 60-mm culture dishes containing 10 mL of 1X SSC and wash by rocking for 30 min. Repeat washes two or three times.
4. Dehydrate the cells on cover slips in a staining dish by sequential incubation for 5 min with 70%, 95%, and 100% ethanol. Air-dry for at least 30 min.
5. Mount each cover slip cell-side up at one end of a microscope slide using a drop of Pro-Texx mounting medium (American Scientific Products). Mounted cover slips are ready for autoradiography after about 1 h.

3.2.4.5. Post-Hybridization Washing of Cover Slips After Hybridization With RNA Probes

1. Gently lift each cover slip from the parafilm as described above and wash in individual 60-mm culture dishes by rocking for 60 min in 10 mL of 4X SSC/10 mM DTT.
2. Transfer cover slips to a staining dish. Dehydrate by 2-min sequential incubations in 70% ethanol and 95% ethanol containing 300 mM ammonium acetate, followed by 100% ethanol.
3. Incubate the cover slips for 10 min in 10 mL of 50% formamide/2X SSC/10 mM DTT at 50°C.
4. Rinse cover slips with 2X SSC.
5. Incubate cover slips for 30 min at 37°C with 10 mL of 20 µg/mL RNase A, prepared by adding 20 µL of 10 mg/mL stock RNase in water to 10 mL of RNase buffer (RNase buffer is 500 mM NaCl/10 mM Tris-HCl, pH 8.0/1 mM ethylene diamine tetraacetic acid).
6. Incubate cover slips for 10 min in 10 mL of RNase buffer at 37°C.
7. Incubate cover slips for 10 min in 10 mL of 2X SSC. Rinse twice with 2X SSC.
8. Incubate cover slips for 15 min in 10 mL of 0.1X SSC at 40°C.
9. Transfer each cover slip to a 60-mm culture dish containing 10 mL of 0.1X SSC and rock for 10 min.
10. Dehydrate cells in a staining dish for 2-min periods in 70% and 95% ethanol containing 300 mM ammonium acetate, followed by 100% ethanol. Air-dry for at least 30 min.
11. Mount cover slips cell-side up at one end of microscope slide using Pro-Texx mounting medium. The slides are ready for autoradiography after 1 h.

3.2.4.6. Autoradiography

1. In a darkroom under a safelight (Kodak Wratten no. 2 safelight containing a 25-W bulb), melt a 20-mL aliquot of Kodak NTB-2 emulsion in a 50-mL centrifuge tube in a 42°C water bath for at least 1 h (Kodak NTB-2 emulsion upon receipt from the vender is melted at 42°C in the dark for at least 2 h and 20-mL aliquots placed in plastic tubes, wrapped in aluminum foil, and stored at 4°C until use).
2. Carefully and slowly dip the cover slip end of the slide into the emulsion. Allow the mounted cover slip to drain vertically in a test tube rack for about 30 min in the absolute dark (as the emulsion dries, it becomes very light-sensitive, even to a safelight).
3. Place slides in a light-tight slide box containing a dessicant, wrap the box in aluminum foil, and expose the emulsion covered cover slips at 4°C for 2–4 d, depending on the strength of the radioactive signal.
4. After exposure, allow slides to warm to room temperature.
5. In a darkroom (no safelight), develop the emulsion covered cover slips mounted on slides in a slide staining tray for 5 min at approx 18°C in Kodak D-19 developer diluted 1:1 with water.
6. Rinse slides for 30 s in water at approx 18°C.
7. Fix slides for 5 min in Kodak fixer at approx 18°C.
8. Wash slides for 30 min in cold running water. Air-dry slides.
9. Score slides by microscopic observation under phase at a magnification of ×200–400. Heavily hybridized cells will have many exposed silver grains visible over the cell. Lightly hybridized cells will show individual grains, often best viewed out of phase, and can be quantitated by grain counting. *See* **Note 6** for combined immunofluorescence and autoradiographic analysis.

4. Notes

1. The Eppendorf micromanipulator 5171 provides the ability to automatically inject in two different modes: the Axial injection mode "on" and the Axial injection mode "off." In both modes, the capillary is held at a 45° angle. With the Axial mode "on," injection is automatically performed with the cell being pierced in the Axial direction, i.e., at a 45° angle, and then returned to its initial position at the Search Plane. With the Axial mode "off," the capillary (which is normally held at a 45° angle) is lowered vertically in the X direction, thus injecting the cell, and then is returned to its initial position. The Axial injection mode "on" is the preferred method for injecting most mammalian cells. However, the Axial mode "off" has been found to improve survival of certain cell types, including human cardiac myocytes and prostate cells.
2. Several commercially available mounting media are available for permanently attaching coverslips cell-side down to slides. For fluorescence microscopy, we recommend a SlowFadeTM-light Antifade kit from Molecular Probes. A permanent mounting medium useful for both fluorescence and X-Gal staining is Gelvatol, which can be used for fluorescent microscopy by the addition of DABCO as described *(17)*.

3. It is important to photograph all important data for permanent records.
4. High levels of β-galactosidase expression produce detectable blue cells within 60 min after addition of the X-Gal solution. For cells with less β-galactosidase activity, overnight staining may be required, and therefore stained cells should be monitored by phase microscopy to determine when to terminate X-Gal staining. Note that when coinjecting a fluorescent marker, X-Gal staining (caused by a blue precipitate) will quench the immunofluorescent signal from the marker. Because β-galactosidase activity is derived from an injected plasmid, all blue-stained cells have been successfully injected whether a fluorescent marker is seen or not. Thus, the total number of injected cells is the sum of blue-stained cells plus fluorescent cells (not exhibiting a blue stain).
5. BudR can be added to cells for different time periods, depending on the purpose of the experiment. For example, to measure quiescence or stimulation of quiescent cells by serum or microinjection of various oncogenes or mitogens, it is convenient to add BudR 4–8 h after microinjection, followed by incubation for 12–24 h after addition of BudR, followed by fixation and staining for BudR incorporation. In this manner, one can determine the cumulative number of cells that enter the S phase during the incorporation period. Pulse incorporation of BudR for shorter periods at various times after microinjection may be used to define more precisely the time periods when cellular DNA synthesis is occurring.
6. It is possible to detect both protein by immunofluorescence and RNA by *in situ* hybridization. First, do standard *in situ* hybridization analysis, but do not add photographic emulsion. Second, perform the antibody staining reaction. Third, coat the slide with photographic emulsion, expose, and develop. Under the microscope, one can now observe the same field by fluorescence microscopy for antibody-reacting cells and by phase microscopy for radioactive grains over the hybridization-positive cells.

Acknowledgments

We thank C. E. Mulhall for editorial assistance. The experiments on cell microinjection reported here were supported by Public Health Service grants CA29561 and Research Career Award AI04739 to M. G. from the National Institutes of Health.

References

1. Capecci, M. (1980) High efficiency transformation by direct microinjection of DNA into cultured mammalian cells. *Cell* **22,** 479–488.
2. Shenk, T. (2001) *Adenoviridae*: the viruses and their replication, in *Fundamental Virology*, Lippincott, Williams and Wilkins, New York, pp. 1053–1088.
3. Lillie, J. W., Loewenstein, P. M., Green, M. R., and Green, M. (1987) An adenovirus E1A protein region required for transformation and transcriptional repression. *Cell* **46,** 1043–1051.

4. Green, M., Loewenstein, P. M., Pusztai, R., and Symington, J. S. (1988) An adenovirus E1A protein domain activates transcription in vivo and in vitro in the absence of protein synthesis. *Cell* **53,** 921–926.

5. Song, C.-Z., Tierney, C. J., Loewenstein, P. M., et al. (1995) Transcriptional repression by human adenovirus E1A N-terminus/conserved domain 1 polypeptides in vivo and in vitro in the absence of protein synthesis. *J. Biol. Chem.* **270,** 23,263–23,267.

6. Stabel, S., Argos, P., and Philipson, L. (1985) The release of growth arrest by microinjection of adenovirus. *EMBO J.* **4,** 2329–2336.

7. Kaczmarek, L., Ferguson, B., Rosenberg, M., and Baserga, R. (1986) Induction of cellular DNA synthesis by purified adenovirus E1A proteins. *Virology* **152,** 1–10.

8. Ruley, H. E. (1983) Adenovirus early region 1A enables viral and cellular transforming genes to transform primary cells in culture. *Nature (London)* **304,** 602–606.

9. Mulcahy, L. S., Smith, M. R., and Stacey, D. W. (1985) Requirement for ras proto-oncogene function during serum-stimulated growth of NIH 3T3 cells. *Nature (London)* **313,** 241–243.

10. Dobrowolski, S., Harter, M., and Stacey, D. W. (1994) Cellular ras activity is required for passage through multiple points of the G_0/G_1 phase in BALB/c 3T3 cells. *Mol. Cell. Biol.* **14,** 5441–5449.

11. Riabowal, K. T., Vosatka, R. J., Ziff, E. B., Lamb, N. J., and Feramisco, J. R. (1988) Microinjection of fos-specific antibodies blocks DNA synthesis in fibroblast cells. *Mol. Cell. Biol.* **8,** 1670–1676.

12. Kovary, K., and Bravo, R. (1992) The jun and fos protein families are both required for cell cycle progression in fibroblasts. *Mol. Cell. Biol.* **11,** 4466–4472.

13. Arias., J., Alberts, A. S., Brindle, P., et al. (1994) Activation of cAMP and mitogen responsive genes relies on a common nuclear factor. *Nature (London)* **370,** 226–229.

14. Lawrence, J. B., and Singer, R. H. (1986) Intracellular localization of messenger RNAs for cytoskeletal proteins. *Cell* **45,** 407–415.

15. Kislauskis, E. H., Li, Z., Singer, R. H., and Taneja, K. L. (1993) Isoform-specific 3'-untranslated sequences sort α-cardiac and β-cytoplasmic actin messenger RNAs to different cytoplasmic compartments. *J. Cell Biol.* **123,** 165–172.

16. *Nonradioactive In Situ Hybridization Application Manual.* Roche Applied Sciences, Indianapolis, IN.

17. Harlow, E. and Lane, D. (1999) *Using Antibodies. A Laboratory Manual.* Cold Spring Harbor Laboratory, Cold Spring Harbor, New York.

13

Determination of the Transforming Activities of Adenovirus Oncogenes

Michael Nevels and Thomas Dobner

Summary

The last 50 yr of molecular biological investigations into human adenoviruses (Ads) have contributed enormously to our understanding of the basic principles of normal and malignant cell growth. Much of this knowledge stems from analyses of the Ad productive infection cycle in permissive host cells. Also, initial observations concerning the transforming potential of human Ads subsequently revealed decisive insights into the molecular mechanisms of the origins of cancer and established Ads as a model system for explaining virus-mediated transformation processes. Today it is well established that cell transformation by human Ads is a multistep process involving several gene products encoded in early transcription units 1A (*E1A*) and 1B (*E1B*). Moreover, a large body of evidence now indicates that alternative or additional mechanisms are engaged in Ad-mediated oncogenic transformation involving gene products encoded in early region 4 (*E4*) as well as epigenetic changes resulting from viral DNA integration. In particular, studies on the transforming potential of several *E4* gene products have now revealed new pathways that point to novel general mechanisms of virus-mediated oncogenesis. In this chapter we describe in vitro and in vivo assays to determine the transforming and oncogenic activities of the E1A, E1B, and E4 oncoproteins in primary baby rat kidney cells and athymic nude mice.

Key Words: Adenovirus; transformation; oncogenes; E1A; E1B; E4; tumorigenicity; baby rat kidney cells; nude mice.

1. Introduction

Oncogenic transformation of primary cells by human adenoviruses (Ads) is a multistep process that is initiated by the viral *E1A* gene. Autonomous ectopic expression of *E1A* gene products is sufficient to induce unscheduled cellular DNA synthesis and proliferation by virtue of their ability to interact with and modulate the function of several growth-regulatory cellular proteins that con-

From: *Methods in Molecular Medicine, Vol. 131:*
Adenovirus Methods and Protocols, Second Edition, vol. 2:
Ad Proteins, RNA, Lifecycle, Host Interactions, and Phylogenetics
Edited by: W. S. M. Wold and A. E. Tollefson © Humana Press Inc., Totowa, NJ

trol transcription and cell cycle progression, including several Rb family members and p300/CBP (reviewed in **ref. 1**). Consequently, upon *E1A* expression, primary cells with a limited life span can become immortal. The Ad type 5 (Ad5) *E1A* transcription unit encodes two major proteins of 289 and 243 amino acids derived from differentially spliced overlapping transcripts, referred to as 13S and 12S RNA, respectively. Whereas the 13S-derived protein (E1A-13S) is required for viral replication, the 12S product (E1A-12S) encodes all functions necessary for immortalization.

In general, *E1A*-immortalized cells are not completely transformed in that they grow slowly and not to high densities, are anchorage dependent, and usually are not tumorigenic. Instead, full manifestation of the transformed phenotype requires expression of a cooperating second oncogene (reviewed in **refs. 2** and **3**). Such cooperating oncogenes include cellular genes like activated *ras*. However, within the context of the adenoviral genome, the role of this second oncogenic determinant has been traditionally ascribed to *E1B*. This transcription unit encodes two major proteins, E1B-55kDa and E1B-19kDa, which are structurally unrelated but independently cooperate with E1A proteins in the complete oncogenic transformation of primary cells in culture. This function is, at least in part, a result of their ability to inhibit E1A-induced p53-dependent and p53-independent apoptotic cell death (reviewed in **ref. 4**). It appears that much of the oncogenic activity of E1B-55kDa is related to its interaction with the cellular tumor suppressor protein p53, although p53-independent mechanisms are likely to exist. The E1B-19kDa shares structural and functional similarities with the cellular anti-apoptotic Bcl-2 protein.

Over the past 10 yr we and others have shown that, besides *E1A* and *E1B*, human Ads contain an additional third oncogenic region, *E4* (reviewed in **ref. 5**). Specifically, the protein products of three open reading frames within this transcription unit—*E4orf1*, *E4orf3*, and *E4orf6*—have been implicated in oncogenic transformation. Recent work suggests that transformation by *E4orf1* differs considerably from the E1-encoded functions and is mediated through a novel mechanism that involves interactions with multiple PDZ-domain-containing proteins *(6,7)*. In contrast, the oncogenic activities of *E4orf3* and *E4orf6* resemble those of the E1B oncoproteins. They can individually cooperate with *E1A* to completely transform primary rodent cells in vitro. Moreover, they enhance transformation mediated by the *E1A* and *E1B* oncogenes. The transforming properties of *E4orf6* involve two distinct mechanisms, one of which is linked to degradation of the p53 protein in combination with E1B-55kDa *(8)*. The molecular basis of transformation by *E4orf3* is not known, but it appears not to be linked to this protein's ability to interact with cellular promyelocytic leukemia bodies (**ref. 9** and unpublished results).

The transforming activities of adenoviral *E1*- and *E4*-encoded oncogenes can be determined by DNA transfection of primary rodent cells. For this purpose, we routinely use primary kidney epithelial cells prepared from neonatal rats. These cells are commonly referred to as baby rat kidney (BRK) cells. They provide at least two advantages over other cell culture systems, including primary fibroblasts and immortalized cell lines. First, the use of BRK cells results in negligible background transformation. Second, because human neoplasia are predominantly (approx 90%) epithelial in origin, the results obtained with BRK (epithelial) cells may be extrapolated to human cancers.

2. Materials

1. Baby rats: a litter (approx 6–12 pups) of 5- to 7-d-old neonatal outbred Sprague Dawley (SD) rats can be purchased from Harlan SD (Indianapolis, IN).
2. Nude mice: athymic NMRI (*nu/nu*) mice can be obtained from Harlan SD.
3. Plasmids: mammalian expression plasmids (e.g., pcDNA3 derivatives) encoding Ad E1A12S/13S, E1B-55kDa, E1B-19kDa, *E4orf3*, *E4orf6* and activated *ras* have been described *(8,9)* and can be obtained from the authors.
4. Carrier DNA: purified DNA from fish sperm can be obtained from Roche (cat. no. 10 223 646 001). Resuspend DNA in sterile double-distilled water at 10 mg/mL and shear to an average fragment length of $\leq$10,000 bp (evaluated by agarose gel/ethidium bromide electrophoresis) by sonication. Store in aliquots at –20°C. Dilute 1:10 to make working stocks (1 mg/mL).
5. Phosphate-buffered saline (PBS) without Ca^{2+} and Mg^{2+}: dissolve 8.0 g NaCl, 0.20 g KCl, 1.60 g Na_2HPO_4 (or 1.78 g $Na_2HPO_4 \cdot 2H_2O$), and 0.24 g KH_2PO_4 in 900 mL of double-distilled water and adjust the pH to 7.2–7.4 with HCl or NaOH, if necessary. Make up the volume to 1.0 L, sterilize by filtering or autoclaving (20 min, 15 psi, liquid cycle), and store at room temperature.
6. 100X Collagenase-dispase solution: dissolve 100 mg collagenase/dispase powder (Roche, Mannheim, Germany; cat. no. 10 269 638 001) in 1 mL PBS. Filter-sterilize and store in aliquots at –20°C.
7. 2X *N*-2-Hydroxyethylpiperazine-*N*'-2-ethanesulfonic acid (HEPES)-buffered saline (HeBS): dissolve 8.2 g NaCl, 5.95 g HEPES, and 105 mg Na_2HPO_4 in 400 mL double-distilled water and adjust pH to exactly 7.05. The exact pH is extremely important for efficient transfection (the optimal pH range is 7.05–7.12). Add water to 500 mL, filter-sterilize, and store in aliquots at –20°C. Avoid repeated freezing and thawing. Test for transfection efficiency with each new batch (*see* **Note 1**).
8. 2.5 *M* $CaCl_2$ solution: dissolve 36.74 g $CaCl_2 \cdot 2H_2O$ (Sigma-Aldrich, St. Louis, MO; cell culture tested; cat. no. C 7902) in 500 mL of double-distilled water. Filter-sterilize and store in aliquots at –20°C. This solution can be frozen and thawed repeatedly.
9. 100X G418 solution: prepare 50 mg/mL active ingredient of G418 (Calbiochem; cell culture tested; cat. no. 345810) in double-distilled water, filter-sterilize, and store at 4°C protected from light.

10. Crystal violet solution: dissolve 5 mg crystal violet (Sigma-Aldrich) in 125 mL methanol and 375 mL double-distilled water.
11. Tissue culture media (Dulbecco's modified Eagle's medium [DMEM], trypsin–ethylene diamine tetraacetic acid solution) and tissue culture dishes can be purchased from Invitrogen (Carlsbad, CA) and Sarstedt (Newton, NC), respectively.

3. Methods

3.1. Preparation of BRK Cells

1. Anesthetize 5- to 7-d-old neonatal rats, normally a litter of 6–12 pups, by exposure to isoflurane. Alternatively, kill the animals by cervical dislocation (*see* **Note 2**). Place the pups on a layer of sterile paper towels and remove a section of the skin on the dorsal side (*see* **Fig. 1**). Open the abdominal cavity by an incision starting from the iliosacral region up to the sternum. Flap skin and adhering abdominal wall. Completely remove the gastrointestinal tract and liver. Remove both kidneys with sterile forceps. Transfer kidneys to a sterile Petri dish or 50-mL tube containing PBS.
2. Wash the kidneys with PBS and transfer to a fresh Petri dish containing a small volume of PBS.
3. Remove capsules and attached blood vessels with pointed forceps (*see* **Note 3**).
4. Transfer kidneys to a new Petri dish containing a small volume of PBS and mince into small pieces (approx 0.1–0.5 mm in diameter) using a sterile scalpel and/or scissors.
5. Transfer minced kidney pieces (~10 kidneys/25 mL PBS) to a 50-mL sterile conical tube. Add collagenase–dispase solution to a final concentration of 1 mg/mL.
6. Vortex tissue suspension and incubate for approx 1 h at 37°C. Vortex approximately every 15 min for approx 30 s.
7. Wait for 5 min to allow the clumps to settle, and transfer the supernatant to a new 50-mL sterile centrifuge tube kept on ice.
8. Add new PBS (~25 mL/10 kidneys) containing 1 mg/mL collagenase–dispase to the clumps (in original 50-mL tube), vortex, and incubate for another 1–3 h until the clumps are dispersed (*see* **Note 4**).
9. Pool the supernatant (**Subheading 3.1., step 7**) and second digest and centrifuge for 5 min at 800*g* and 4°C.
10. Remove the supernatant and suspend the cell pellet in a small volume of prewarmed (37°C) DMEM containing 10% fetal bovine serum and antibiotics.
11. Disperse the cells with the help of a pipet and dispense the amount of cells corresponding to one-half kidney into one 10-cm tissue culture dish, and incubate at 37°C in a CO_2 incubator overnight. Approximately 24 h after plating, cells can be either transfected with DNA directly or passaged once at a 1:4 ratio before transfection (*see* **Note 5**).

3.2. Transfection of BRK Cells

To analyze the immortalizing and transforming functions of human Ad *E1* and *E4* oncogenes, plasmids expressing the individual (viral) gene products

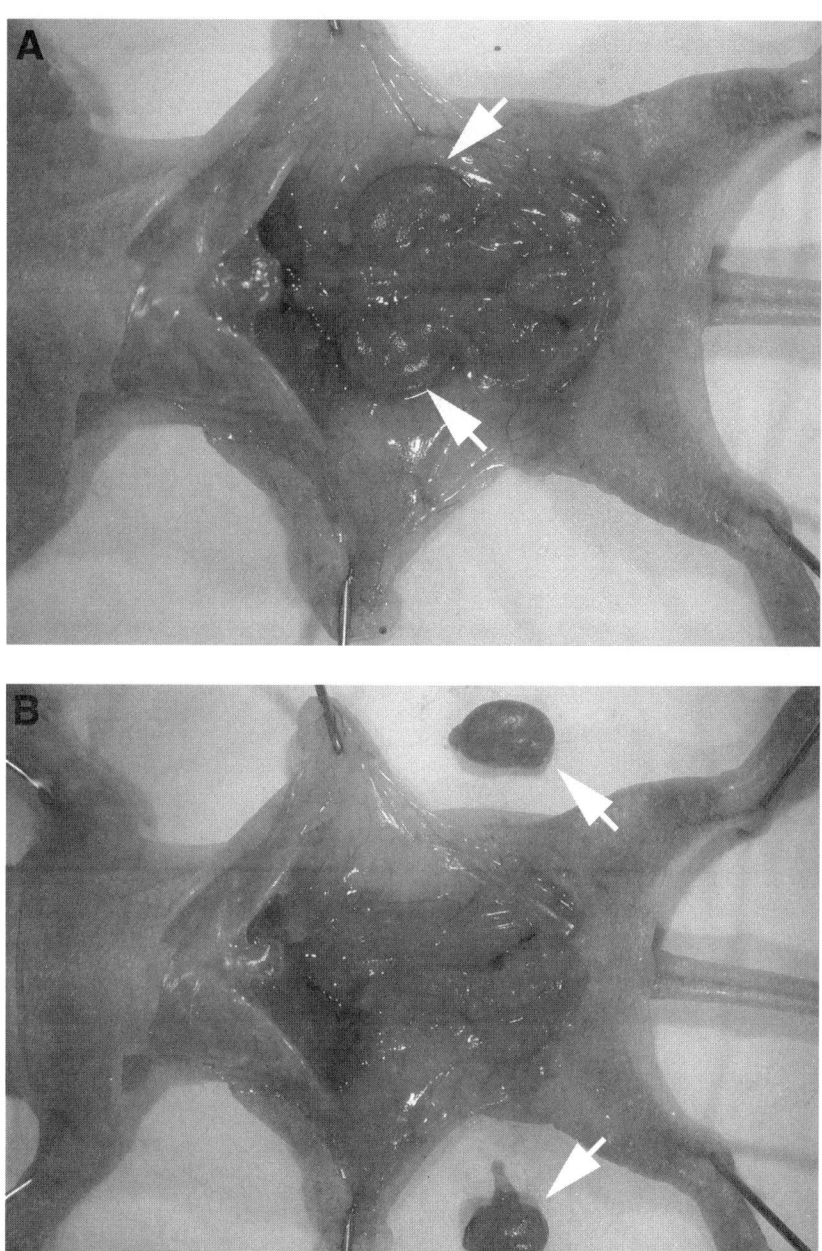

Fig. 1. (**A**) Illustration of the rat abdominal cavity after complete resection of the gastrointestinal tract and the liver. Both kidneys (**arrows**) are clearly exposed and can be removed with sterile forceps (**B**).

(e.g., pcDNA3 derivatives encoding E1A-12S, E1A-13S, E1B-55kDa, E1B-19kDa, *E4orf3*, or *E4orf6*) or combinations thereof (e.g., pAd5 XhoI-C 15 carrying a viral genomic fragment comprising the entire Ad5 *E1* region) are transfected into BRK cells. The use of plasmids carrying a selectable marker gene (e.g., the *neo* marker, which allows for selection with G418) facilitates clearing of nontransfected cells and abortive foci induced by *E1A*-mediated transient cell proliferation, thus markedly reducing background. However, selection for a marker gene eliminates not only transient but also stably transformed foci derived by "hit-and-run" transformation mechanisms that do not require permanent maintenance of oncogenes *(10)*. For cooperative transformation with strong transforming genes such as combinations of *E1A* with activated *ras*, the use of a dominant selection procedure is not generally needed. We routinely transfect BRK cells using a modified calcium phosphate protocol, but other transfection methods including liposome-mediated techniques may also be employed. BRK cells are also highly suited for gene transduction with adequate retrovirus vectors. The calcium phosphate protocol that we use is given below. Make sure to prepare duplicate or triplicate samples.

1. Mix 1–5 µg of plasmid DNA containing the transforming gene(s) with carrier DNA to give a total DNA amount of 20 µg. Add 50 µL of 2.5 M $CaCl_2$ solution and make up to 500 µL with sterile, double-distilled water in a 1.5-mL reaction tube. Place 500 µL of 2X HeBS buffer into a 5-mL (12 × 75 mm) polystyrene round-bottom snap cap tube.
2. Transfer the DNA–$CaCl_2$ mixture drop by drop to the 500 µL of 2X HeBS (with a 1-mL pipet and a plastic tip) while vortexing HeBS at medium speed.
3. Allow precipitate to sit for 20 min at room temperature.
4. Add the total volume (1 mL) of the precipitate to each 10-cm dish of cells containing 10 mL growth medium and mix gently.
5. Incubate at 37°C in a CO_2 incubator for at least 5 h or overnight.
6. Gently shake the dish to dislodge the residual precipitate and remove the medium. Wash once with 10 mL of PBS, feed with 10 mL of fresh medium, and incubate at 37°C.
7. Optional: add G418 (500 µg/mL) within the next few days.
8. Change medium every fourth day. In assays involving stronger transforming genes, such as the activated *ras* oncogene, the transfected cells can be stained 10–14 d after transfection. For weaker cooperating oncogenes, such as Ad E1B, cells can be stained 15–20 d after transfection.

3.3. Staining of Immortalized Colonies and Transformed Foci

The immortalized/transformed colonies become visible approx 1 wk after transfection. After periodical visual examination of the colonies, if the sizes of the colonies appear satisfactory, stain with crystal violet (or Giemsa) for quantitation and photography.

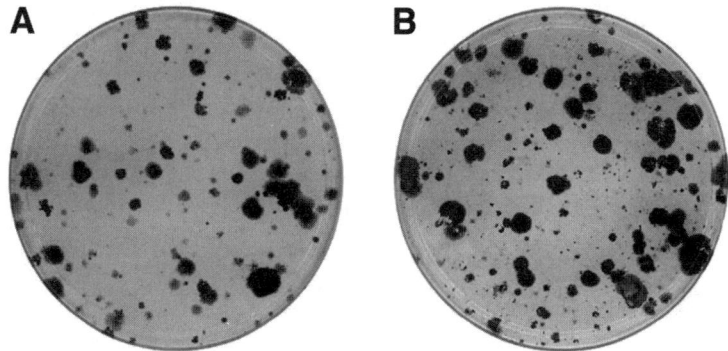

Fig. 2. Focus formation by (**A**) Ad5 *E1A* and *E1B*, or (**B**) *E1A*, *E1B*, and *E4orf6*. Baby rat kidney cells were transfected with 5 μg each of plasmids expressing *E1A/E1B* and *E4orf6* as indicated using the calcium phosphate precipitation technique. Transfected cells were stained with crystal violet 27 d after transfection and scanned.

1. One day prior to staining, feed cells with fresh medium.
2. Remove medium and wash once with 5 mL PBS.
3. Fix and stain the colonies with crystal violet solution for 5 min, and wash two times with tap water.
4. Dry the dishes, count the colonies, and photograph if necessary.

The immortalized and transformed colonies obtained with *E1A/E1B* and *E1A/E1B* plus *E4orf6* are shown in **Fig. 2**.

3.4. Characterization of Transformed Cells

To determine the extent of transformation, the transformed cells are generally assayed for their ability to form anchorage-independent colonies on semi-solid (soft-agar) medium in vitro and/or examined for their ability to form tumors in immunodeficient mice or immunocompetent syngeneic hosts. Here we describe simple tumorigenicity and malignancy assays in athymic nude mice.

3.4.1. Tumor Incidence and Tumor Growth in Athymic Mice

1. Grow cells on 10- or 15-cm tissue culture dishes. At the day of injection, dishes should contain 50–80% confluent monolayers. Each transformed cell line should be tested in at least five mice, and for each animal 1×10^6 cells are needed.
2. Wash monolayers two times with serum-free DMEM and scrape cells into serum-free medium using a sterile plastic cell lifter. Carefully transfer cell suspension to a conical tube and centrifuge for 5 min at 800*g*.
3. Remove supernatant and add serum-free medium to make up a suspension that contains 10^7 cells/mL (1×10^6 cells per injection).

4. Inject 100 μL of cell suspension into nude mice subcutaneously using a syringe equipped with a 26-gage needle.
5. Examine the site of injection weekly for development of tumor nodules to determine the latency period. After development of visible (palpable) tumors, measure the length and width of the tumors and calculate the approximate tumor area. Results are displayed as tumor incidence (no. of mice with tumors/total no. of animals) and tumor size (mean tumor area ± standard errors [mm^2]).

3.4.2. Tumor Histology and Metastasis Assay

For histological analysis, tumors as well as surrounding connective tissue and adhering skin regions are fixed in Bouin's solution and prepared for routine paraffin histology. Sections (5 μm) are stained with hematoxylin–eosin according to the Masson and Golder method modified by Jerusalem. Athymic mice injected with transformed cells that induce tumors can also be examined for formation of metastatic tumor nodules in the internal organs such as lungs or lymph nodes. If the burden of primary tumors is heavy for the animals (tumor area ≥100 mm^2), the primary tumors must be surgically removed, after which the animals are maintained for an additional 2–4 wk and autopsied.

4. Notes

1. The 2X HeBS solution can be rapidly tested by mixing 0.5 mL of 2X HeBS with 0.5 mL of 250 mM $CaCl_2$ and vortexing. A fine precipitate should develop that is readily visible in the microscope.
2. Neonatal Fisher rats are equally suitable. Kidneys may also be obtained from animals between 2 and 4 d after birth, which, however, will result in reduced tissue yields.
3. During preparation of BRK cells, as much of the capsule present in the kidney should be removed as possible. If it is not fully removed, the preparation will contain fibroblasts. Fibroblasts will grow faster than epithelial cells, causing increased background.
4. Try to minimize incubation periods because prolonged digestion of minced kidneys with collagenase–dispase will result in reduced viability and reduced plating efficiency of epithelial cells.
5. Continued passaging before transfection will result in decreased transformation efficiencies.

References

1. Endter, C. and Dobner, T. (2004) Cell transformation by human adenoviruses. *Curr. Top. Microbiol. Immunol.* **273**, 163–214.
2. Williams, J. F., Zhang, Y., Williams, M. A., Hou, S., Kushner, D., and Ricciardi, R. P. (2004) E1A-based determinants of oncogenicity in human adenovirus groups A and C. *Curr. Top. Microbiol. Immunol.* **273**, 245–288.

3. Chinnadurai, G. (2004) Modulation of oncogenic transformation by the human adenovirus E1A C-terminal region. *Curr. Top. Microbiol. Immunol.* **273,** 139–161.
4. White, E. (2001) Regulation of the cell cycle and apoptosis by the oncogenes of adenovirus. *Oncogene* **20,** 7836–7846.
5. Täuber, B. and Dobner, T. (2001) Adenovirus early E4 genes in viral oncogenesis. *Oncogene* **20,** 7847–7854.
6. Glaunsinger, B. A., Weiss, R. S., Lee, S. S., and Javier, R. T. (2001) Link of the unique oncogenic properties of adenovirus type 9 E4-ORF1 to a select interaction with the candidate tumor suppressor protein ZO-2. *EMBO J.* **20,** 5578–5586.
7. Thomas, D. L., Schaack, J., Vogel, H., and Javier, R. T. (2001) Several E4 region functions influence mammary tumorigenesis by human adenovirus type 9. *J. Virol.* **75,** 557–568.
8. Nevels, M., Rubenwolf, S., Spruss, T., Wolf, H., and Dobner, T. (2000) Two distinct activities contribute to the oncogenic potential of the adenovirus type 5 E4orf6 protein. *J. Virol.* **74,** 5168–5181.
9. Nevels, M., Täuber, B., Kremmer, E., Spruss, T., Wolf, H., and Dobner, T. (1999) Transforming potential of the adenovirus type 5 E4orf3 protein. *J. Virol.* **73,** 1591–1600.
10. Nevels, M., Täuber, B., Spruss, T., Wolf, H., and Dobner, T. (2001) "Hit-and-run" transformation by adenovirus oncogenes. *J. Virol.* **75,** 3089–3094.

14

Human Adenovirus Type 12

Crossing Species Barriers to Immortalize the Viral Genome

Walter Doerfler

Summary

When viruses cross species barriers, they often change their biological and pathogenetic properties. In the author's laboratory the nonproductive interaction of Syrian hamster cells with human adenovirus type 12 (Ad12) has been studied. Ad12 induces undifferentiated tumors in newborn hamsters (*Mesocricetus auratus*) at high frequency. Ad12 inefficiently enters hamster (BHK21) cells, and only small amounts of viral DNA reach the nucleus. Viral DNA replication and late transcription are blocked. In Ad12-induced tumor cells, multiple copies of viral DNA are chromosomally integrated. The integrated viral DNA becomes *de novo* methylated. Cellular DNA methylation and transcription patterns in Ad12-transformed cells and in Ad12-induced tumor cells are altered. These changes may be related to the oncogenic potential of Ad12 in hamsters. In this chapter, concepts and techniques for the study of the Ad12-hamster cell system are summarized.

Key Words: Abortive adenovirus type 12 infection; crossing species barriers; Ad12-induced hamster tumors; Ad12 DNA integration; *de novo* methylation of integrated foreign (Ad12) DNA; alterations of cellular DNA methylation patterns.

1. Introduction

Viruses have evolved as biological elements with a highly select coordination to specific hosts and/or host cells. There are numerous steps in the interaction between a virus and its host that have to be optimized for the virus to be capable of undergoing a fully productive replicative cycle in the most permissive cell system. These parameters range from the attachment of the viral particle on the cell surface to the mechanisms of release of the newly assembled virions from the infected cell or organism. This adaptive process involves a large number of different viral and cellular proteins, whose optimized activi-

From: *Methods in Molecular Medicine, Vol. 131:*
Adenovirus Methods and Protocols, Second Edition, vol. 2:
Ad Proteins, RNA, Lifecycle, Host Interactions, and Phylogenetics
Edited by: W. S. M. Wold and A. E. Tollefson © Humana Press Inc., Totowa, NJ

ties and interactions are required to proceed at a unique temporal schedule during the viral replication program. Any deviation from this—in evolutionary terms—presently normalized standard invariably will entail the loss of or decrease in viral reproduction.

There exists, however, a corollary in the philosophy of viral strategies. "Smart viruses" are those that abstain from killing the host cell, admittedly at the expense of maximal viral output, but at the gain of some other, usually long-term, advantage for this particular virus in an environment that has developed antiviral strategies. There are numerous ways for viruses to achieve the goal of long-term survival, if only for immortalizing their genomes and to scheme for better times or circumstances for the rescue of the informational essence of the virion.

A not infrequently chosen contingency resides in taking refuge in a different host, although the virus initially cannot anticipate whether such untested territory will be amenable to and prove useful for viral survival. In its endeavor to survive, however, the viral parasite will breach a crucial law of survival—the respect for host specificity—and may cross species barriers, thus attempting the "unspeakable" in order to gain unprecedented, if unpredictable, advantages.

In many instances, the outcome of such viral experimentation will be frustrated. Nevertheless, the large number of virions usually produced in productive infections allows for many failures during risky adventures into unproven territories. The survivors of the successful events in these interactions are viruses that we encounter as "evolving viruses" with quite uncharacteristic and unforeseeable properties, which can cause catastrophic problems in medical virology. Many of the truly lethal viruses of present medical practice have evolved through such strategies of still poorly understood switches in their host ranges. Among others, viruses that presently raise serious concerns in the medical community include HIV 1 and 2, coronavirus severe acute respiratory syndrome, Marburg virus, and avian influenza virus H5N1 *(1,2)*. These fatal disease-causing agents have alerted virologists to the potential of viruses that succeed in crossing species barriers and, in the course, acquire novel properties.

There are representatives of this class of viruses that have less immediate medical relevance but, nevertheless, have significance as models to study the essentials of nonproductive virus–host transitions. In my laboratory we have studied human adenovirus (Ad) type 12 and its interaction with cells from the Syrian hamster (*Mesocricetus auratus*) *(3–5)*. Ad12 undergoes a fully productive cycle in human cells in culture but induces undifferentiated tumors in newborn Syrian hamsters *(6–8)*. In crossing the barrier from human to hamster, Ad12 gains oncogenic potential. There is no evidence that human Ads are involved in human tumorigenesis *(9–12)*. When Ad12 infects human cells, all infected cells are killed in the course of this productive interaction. Hence, there are no surviving cells that could be transformed into tumor cells. Therefore, it is not surprising that Ad12 has not been identified as oncogenic in humans. It cannot be ruled out that Ad12 could be involved, however rarely, in a hit-and-run mechanism in human oncogenesis. This possibility becomes more plausible when one considers that, among hamster tumor cells, revertant cells have been isolated that have lost all of the previously integrated Ad12 genomes but have retained their oncogenic potential. Ad12 infection of hamster cells in culture is completely abortive; the replication block lies before viral DNA replication. The infected cells continue to replicate and show no cytopathic effect. Thus, in rare instances Ad12-infected hamster cells can go on to develop into tumor cells and form tumors in animals.

Among the factors contributing to the oncogenic potential of Ad12 in Syrian hamsters are the following:

1. The abortive mode of infection and the survival and continued growth of Ad12-infected hamster cells *(3)*.
2. The insertion of Ad12 DNA into the host genome *(13,14)*.
3. Alterations of cellular DNA methylation and transcription patterns *(15–17)*.
4. The Ad12 E1 and E4 gene products (reviewed in **ref. 18**).
5. Downregulation of cellular major histocompatibility complex genes by Ad12 *(19,20)*.
6. Downregulation of cellular defense genes by Ad12 *(20)*.
7. Many additional unknown factors.

In the first, major, part of this chapter, concepts for the study of the abortive Ad12–hamster cell system and of the tumor production by Ad12 in newborn hamsters will be described. The second part will present a summary of the techniques employed in these studies.

1.1. Main Features of the Abortive Ad12-Hamster Cell System

Primary hamster cells or the baby hamster kidney (BHK21) cell line are nonpermissive for infection with Ad12; the infection is abortive. However, the same cells are permissive for infection with Ad2, and this infection is productive *(3,21)*. The following steps in the virus–host interaction have been characterized:

1. Ad12 adsorbs to the hamster cell surface, although much less efficiently than to human cells *(22)*.
2. Ad12 particles enter the cytoplasm.
3. Ad12 DNA can be found in the nuclei of hamster cells *(13,22)*.
4. Ad12 DNA associates with the chromosomes *(23)*.
5. Ad12 DNA can integrate into the host chromosome *(13,24–26)*.
6. Some of the early Ad12 functions are transcribed, although inefficiently *(22,27)*.
7. By DNA array analyses, a limited number of changes in cellular transcription patterns can be documented upon the infection of BHK21 cells with Ad12 (Dorn, A. and Doerfler, W. unpublished results).
8. Ad12 DNA replication cannot be detected *(3,22,28)*.

9. Ad12 VA and L1 RNAs are not transcribed *(29)*.
10. The region downstream of the major late promoter (MLP) of Ad12 DNA carries a 33-bp-long mitigator element,* which inactivates the MLP in hamster cells *(30)*.
11. Late Ad12 genes are not detectably transcribed *(28,31)*.
12. *Per definitionem,* new Ad12 virions are not produced in the abortive system *(3)*.
13. There is no cytopathic effect in Ad12-infected hamster cells (the infected cells continue to replicate *[3]*).
14. As one of the biological consequences of the abortive infection of hamster cells, Ad12 induces undifferentiated tumors in newborn hamsters. (Ad2 fails to do so. The abortive mode of infection enables the Ad12 genome to become permanently fixed in the host genome and to survive for many cell generations. Productive infections would kill all infected cells.)

1.2. Attempts to Complement the Defect in Ad12 Replication in Hamster Cells

In the search for factors missing in hamster cells that would sustain Ad12 replication, we have tried to complement these defects in a stepwise manner. The nonpermissiveness of hamster cells for Ad12 replication seemed to be restricted to this particular serotype (since Ad2 underwent productive replication in hamster cells). Hence, there had to be a highly specialized barrier or barriers preventing Ad12 replication in hamster cells.

1.2.1. Ad12 in BHK21 Hamster Cells

BHK297-C131 hamster cells *(32)* constitutively express the E1 region of the integrated left 18.6% fragment of the Ad5 genome *(31,33)*. The characterization of these cells is as follows:

1. Ad12 DNA can replicate to a limited extent.
2. Ad12 late transcripts are synthesized.
3. Ad12 fiber mRNA is transcribed, and has the correct nucleotide sequence, tripartite leader and polyA tail.
4. However, fiber protein is not produced (raising the possibility of an additional translational block in this Ad5-E1-complemented hamster cell system).
5. Ad12 virions are not made.

We conclude that the availability of apparently sufficient concentrations of the Ad5 E1 functions facilitates the activation of Ad12 replication and transcription machinery. Although the late fiber mRNA seems to have most, if not all, properties of *bona fide* Ad12 fiber mRNA, it cannot be translated into fiber protein.

---

*This mitigator is not found in the genome of Ad2. Excision of the mitigator element from the Ad12 genome renders the MLP active in hamster cells and increases its activity in human cells.

1.2.2. Overexpression of the Ad12 pTp or E1A Genes or of the Ad2 pTP or E1A Genes: A New Role for pTP in Early Ad12 Infection?

The cloned pTP or E1A gene under the control of the human cytomegalovirus (HCMV) promoter was transfected into BHK21 cells, which were subsequently infected with Ad12. The results were *(34)*: (1) Ad12 DNA can replicate in these cells, although less efficiently than in permissive human cells; and (2) Ad12 late transcripts are not readily detectable (traces might be found late after infection). Although this approach was not particularly revealing for understanding the nature of the immediate block for Ad12 replication in BHK21 hamster cells, it yielded the unexpected result that overexpression of the Ad12 pTP gene alone, without enhanced transcription of the *E1A* genes, led to Ad12 DNA replication. It is conceivable that above-threshold concentrations of Ad12 pTP were capable of exerting previously unknown functions in preparing the hamster cells for nuclear import and replication of Ad12 DNA (even in the absence of large amounts of *E1A* gene products and beyond the well-known role that pTP can play in the initiation of Ad DNA replication). The data may suggest a novel role of Ad12 pTP in supporting Ad12 DNA nuclear import and replication in a nonpermissive cellular environment. What additional functions might pTP exert in controlling the uptake and transport of Ad12 DNA? Does the Ad12 DNA-bound TP in the infecting virion play a similar role?

1.2.3. Overexpression of the Human Coxsackie Ad Receptor in Hamster Cells

Following transfection, the human Coxsackie Ad receptor gene was transiently overexpressed in BHK21 hamster cells, which were subsequently infected with Ad12 at 8 h after transfection *(22)*. The results were: (1) these cells develop a cytopathic effect at approx 24 h p.i.; and (2) the transport and uptake of Ad12 DNA into the hamster cell nuclei is markedly enhanced and even more efficient than in Ad2-infected BHK21 cells.

In a previous study on Ad12 infection of human embryonic kidney cells, we demonstrated that the extracellular virions, which were isolated from the medium, had a 10-fold higher specific infectivity per virus particle than the intracellular virions *(35)*. By the then-available technology, gross physical differences could not be documented in the viral DNA and protein moieties between the intra- and extracellular Ad12 particles. It is conceivable that anteceding and/or during the release of the Ad12 virions from the cell, subtle alterations, perhaps the result of folding or processing modifications of virion proteins, bestow enhanced infectivity on the extracellular virions.

1.3. Synopsis of the Ad12 Hamster Tumor System

Ad12 can transform hamster cells in culture at extremely low efficiency. However, after injection into newborn Syrian hamsters (*Mesocricetus auratus*), tumors develop at high frequency *(6–8,36)*. Recently, we have reported on the consequences of intramuscular injection of Ad12 and the intraperitoneal dissemination of multiple tumors within about 30 d of virus application in 70–90% of the animals surviving injection *(8)*. Individual intraperitoneal tumors are of varying sizes and can exhibit identical or different integration profiles for the Ad12 genomes. The latter tumors are most likely of different origins; the former could be seed metastases within the animal's peritoneal cavity. Integration patterns of persisting Ad12 genomes have been determined by restriction cleavage of total intranuclear tumor cell DNA followed by Southern blotting and hybridization to intact Ad12 DNA or to the terminal fragments of the viral DNA (for methods, *see* **refs.** *8,24,25,37)*. Detailed analyses on more than 100 different tumors have documented the clonal derivation of Ad12-induced hamster tumors *(37)*.

1.3.1. On Ad Type 12 Oncogenesis

All tumor cells carry multiple copies of integrated, but no free, Ad12 DNA. Ad12 DNA usually inserts at only one chromosomal site. In rare instances (1/60), two insertion sites of Ad12 DNA in the hamster tumor genome have been observed *(37)*. The sites of viral DNA integration are different from tumor to tumor, except for tumors derived as intraperitoneal seed metastases. Ad12-induced tumors are clonal. In all cells derived from one tumor, integration sites are identical. In the Ad12-induced hamster tumor cells, only some of the early Ad12 genes are transcribed at low efficiency. Viral transcription profiles are very similar from tumor to tumor *(8)*. Late genes might be transcribed at extremely low levels, if at all. DNA array analyses of Ad12-induced tumor cells reveal both similarities and differences in cellular transcription patterns among different tumors *(8)*. In an Ad12-induced hamster tumor cell line, about 1 kb of the adjacent cellular DNA showed changes in its methylation profile. Apparently, the insertion of foreign DNA into an established mammalian genome can alter DNA methylation at the site of foreign DNA insertion *(38)*. In addition, foreign DNA insertion into an established genome can elicit alterations in DNA methylation and transcription patterns not only at the site of foreign DNA insertion but also at loci remote from the integration site *(15–17)*.

1.3.2. Consequences of Foreign DNA Insertion Into Established Mammalian Genomes

1.3.2.1. DE NOVO METHYLATION OF INTEGRATED FOREIGN DNA

When foreign DNA, such as Ad12 DNA in hamster cells, is genomically integrated into a mammalian genome, it frequently becomes *de novo* methylated *(24,39)* by an unknown mechanism. DNA methyltransferases 1 and 3a plus 3b seem to cooperate in this enzymatic reaction. The criteria for specific 5'-CG-3' dinucleotide selection in the methylation reaction are still unknown. *De novo* methylation is also important in embryonic development: patterns of DNA methylation are erased and subsequently *de novo* re-established. Because promoter methylation serves as a long-term signal for gene inactivation *(40)*, the introduction of 5-mC residues at specific sites has been considered an ancient defense against the activity of foreign genes *(41,42)*. The *de novo* methylation reaction of DNA has more generally important functions in many biological systems that are not directly related to virology. These include the following:

1. Gene inactivation in transgenic cells and organisms.
2. Genetic imprinting.
3. X-chromosome inactivation.
4. Altered transcription patterns during embryonal development.
5. Long-term silencing of retrotransposons.
6. Changes in transcription patterns in tumor cells.
7. Cloning of organisms.
8. Gene therapy.

1.3.2.2. INDIRECT, FAR-REACHING INSERTIONAL MUTAGENESIS AT SITES REMOTE FROM THE INTEGRATION LOCUS

The insertion of foreign (Ad12, lambda, or plasmid) DNA into established mammalian genomes can be associated with extensive alterations in DNA methylation patterns in repetitive DNA sequences, e.g., the retrotransposon genomes of the intracisternal A particles *(15)*. Changes in the methylation patterns of cellular genes can also be observed (major histocompatibility complex, IgCµ, serine protease, cytochrome P450) *(15)*. Thus, the integration of foreign DNA (several hundred kb to 1 Mb) on one chromosome, as in Ad12-induced tumor cells, might lead to structural perturbations of chromatin. This perturbation is hypothesized to be transmitted to neighboring chromosomes in interphase nuclei, possibly by interactions via the nuclear matrix. Methylation and transcription patterns can be altered *in cis* and *in trans*. The author pursues the concept that these changes are critically involved in causing the oncogenic phenotype *(43)*.

1.3.3. Loss of All Integrated Ad12 DNA Sequences From Ad12-Induced Hamster Tumor Cells Is Compatible With the Maintenance of the Oncogenic Phenotype

When Ad12-induced tumor cells are passaged continuously in culture, spontaneous revertants of these tumor cells occasionally arise that exhibit altered (usually more fibroblastic) morphology and have lost all or next to all integrated viral DNA sequences. These revertants still retain their oncogenic potential when reinjected into hamsters *(36,44)*. Hence, mechanisms other than those enacted by proteins encoded in the Ad12 E1 region must play a role at least in the maintenance of the Ad12-induced oncogenic phenotype. These findings have led me to study alterations in the methylation and transcription patterns of cellular genes and repetitive sequences in Ad12-transformed and Ad12-induced tumor cells as alternate important parameters in research on viral oncogenesis *(43)*.

1.3.3.1. CHARACTERISTICS OF THE REVERTANT TR12 DERIVED FROM THE AD12-TRANS-FORMED BHK21 HAMSTER CELL LINE T637

The Ad12-transformed hamster cell line T637 was derived by infecting cells of the BHK21 hamster fibroblast line *(45)* in culture with Ad12 *(46)*. Cell line T637 exhibits epitheloid morphology and differs distinctly from the parent fibroblastic BHK21 cells. Each T637 cell carries approx 15 copies of Ad12 DNA, which are integrated into the hamster cell genome *(14,15,24,25)* at a single chromosomal location in a "pearl necklace"-like array *(23)*. Upon continuous passage of T637 cells in culture, morphological revertants with a more fibroblastic cell morphology arise. Several of these revertants, designated TR, have been cloned and characterized *(47)*.

In the TR3 revertant of T637 cells, Ad12 DNA could no longer be detected in the cellular genome *(15,47)*. In the revertant TR12, only one copy plus a 3.9-kb fragment from the right terminus of a second copy of Ad12 DNA still reside integrated *(26)*, probably at the same chromosomal location *(48)*, which harbors the 15 copies of Ad12 DNA in cell line T637. The 15 copies of Ad12 DNA in cell line T637 are less intensely methylated than the persisting Ad12 DNA in the revertant cell line TR12 *(48)*. Detailed analyses of the patterns of methylation in the Ad12 integrates in cell line T637 and in its revertant TR12 are now being completed *(26)*. It is conceivable that the extent of DNA methylation of transgenic DNAs, e.g., of the Ad12 genomes in cell lines T637 and TR12, is related to their stability of integration in the recipient genome. Could hypermethylation stabilize foreign DNA integrated into an established mammalian genome like the single copy of Ad12 DNA in cell line TR12?

2. Materials

1. 40 m*M* Hydroquinone.
2. Sodium bisulfite stock: prepare by dissolving 8.1 g of sodium bisulfite in 16 mL of degassed water by gently inverting the tube; 1 mL of 40 m*M* hydroquinone is finally added and the solution is adjusted to pH 5.0 by adding 0.6 mL of freshly prepared 10 *M* NaOH.
3. Glassmilk (Gene Clean II Kit Bio 101 Inc., Nista, CA).
4. Qiaquick gel extraction kit (Qiagen, Hilden, Germany).
5. *Escherichia coli* XL1BlueMRF' (Stratagene, La Jolla, CA).
6. Applied Biosystems 377 DNA sequencer.
7. GeneAmp 9600 system (Perkin Elmer, Norwalk, CT).
8. Taq DNA polymerase.
9. Avian myeloblastosis virus reverse transcriptase.
10. Tfl DNA polymerase (RT Access Kit, Promega, Madison, WI).
11. 2% Agarose gels.
12. pGEM–T vector (Promega, Madison, WI).
13. 1% Igepal CA-630 (Sigma).
14. Pefabloc SC protease inhibitor (Roche).
15. Branson B-12 sonifier.
16. Normal rabbit immunoglobulin G.
17. Protein A–agarose conjugate (Santa Cruz Biotechnology).
18. Ad12-pTP rabbit antiserum.
19. Proteinase K.
20. ss-M13-DNA purification kit (Qiagen, Hilden, Germany).
21. Positively charged nylon membranes (Roche).
22. Macherey/Nagel kit (Düren, Germany).
23. α- or γ-^{32}P-dCTP (3000 Ci/mmol).
24. Superscript II reverse transcriptase (Invitrogen, Karlsruhe, Germany).
25. Sephadex G-50 columns (Roche).
26. DIG hybridization buffer (Roche).

3. Methods

This section will present an overview of the techniques used in our studies on the abortive interaction of Ad12 with hamster cells and on the analyses of Ad12-induced tumor cells. The details of some of these methods are described, unless they are well-known, routinely applied procedures used in molecular virology and biology. Among the more routine-based methods, which are described in detail in several of our referenced publications and will not be repeated here, are construction of expression vectors, expression clones, Southern blotting, analyses of newly synthesized viral/cellular DNA in transfected and infected cells, Western blot analysis, quantitative real-time reverse transcription-polymerase chain reaction (PCR) *(22)*, preparation of protein extracts

(22,34), Ad12 tumor induction in newborn Syrian hamsters *(8)*, DNA microarray technique: membranes, slides *(8,20)*, determination of DNA methylation patterns by methylation-sensitive restriction endonucleases like the HpaII–MspI system, HhaI a.o. *(15,24,48)*, arrays for the study of viral or cellular gene transcription *(20)*, studies on viral DNA integration *(8,14,24,25)*, and fluorescence *in situ* hybridization technique *(23)*.

3.1. The Bisulfite Protocol of the Genomic Sequencing Method

1. Prepare genomic DNAs from Ad12-infected cells or from Ad12-induced hamster tumor cells following standard protocols (*8,24*).
2. The genomic DNA is then alkali-denatured in 0.3 *M* NaOH for 15 min at 37°C and for 3 min at 95°C.
3. Subsequently, treat the DNA with sodium bisulfite. Mix the denatured DNA solution (66 µL) with 1.2 mL of the bisulfite solution (**Subheading 2.**, **item 3**), overlay with mineral oil, and incubate at 55°C for 16 h in a water bath in the dark.
4. From the above reaction mixture (**step 3**), purify the DNA using glassmilk.
5. For desulfonation, add 3 µL of 10 *M* NaOH and incubate for 10 min at 37°C.
6. The solution is thereupon neutralized, the DNA ethanol-precipitated, dried, and redissolved in 20 µL H$_2$O.
7. Amplify selected segments in the promoters by PCR with appropriate oligodeoxyribonucleotide (oligo) primers.
8. Clone reaction products into the pGEM–T vector (Promega, Madison, WI) after purification by using the Qiaquick gel extraction kit (Qiagen, Hilden, Germany) and transfect into *E. coli* XL1BlueMRF' (Stratagene, La Jolla, CA) applying standard procedures.
9. Isolate a number of clones and determine the nucleotide sequences with an Applied Biosystems 377 DNA sequencer.

The bisulfite reaction *(49,50)* converts all unmethylated C residues into U residues, which are copied during PCR amplification into T residues, whereas the 5-methyldeoxycytidine (5-mC) residues are refractory to the bisulfite conversion reaction. Thus, a C residue in the eventually determined nucleotide sequence proves the presence of a 5-mC residue in this position in the original genomic nucleotide sequence. All *bona fide* C residues score as Ts.

3.2. PCR Amplification

1. Perform PCR amplification in a reaction volume of 25 µL in a GeneAmp 9600 system by using 250 ng of bisulfite-treated genomic DNA, 10 m*M* Tris-HCl, pH 9.0, 50 m*M* KCl, 1.7 m*M* MgCl$_2$, 0.1% Triton X-100, 1 U of Taq DNA polymerase, 0.1 m*M* of each of the four dNTPs, and 250 ng of each primer. Primer sequences are chosen based on the published Ad12 DNA sequence *(51)*.
2. Amplification conditions are as follows: After the initial denaturation step at 94°C for 5 min, 35 cycles of denaturation at 94°C for 15 s, annealing for 15 s at tem-

peratures between 51°C and 57°C, and an extension reaction at 72°C for 30 s follows. A final extension step at 72°C for 5 min terminates the reaction.

3.3. Reverse Transcription-PCR

1. Reverse transcription reactions are performed in a one-step protocol, with 100–300 ng of DNase I-treated total RNA by using avian myeloblastosis virus reverse transcriptase and Tfl DNA polymerase following the manufacturer's guidelines.
2. The RT reaction is carried out at 48°C for 45 min, followed by PCR with 40 cycles of denaturation at 94°C for 30 s, annealing at 58°C for 1 min and elongation at 68°C for 2 min. Subsequently, a final elongation step at 68°C for 7 min is applied.
3. All reverse transcription (RT)-PCR reaction products are analyzed by electrophoresis on 2% agarose gels and are stained with ethidium bromide.
4. Products in the size range predicted from the published DNA sequence are purified from the gel, subcloned into the pGEM-T vector, and the DNA is sequenced.

3.4. Immunoprecipitation of the Ad12-pTP Complex

1. Transfect BHK21 cells with the cloned Ad12-pTP DNA by electroporation.
2. At 18 h after transfection, infect the cells with 100 plaque-forming units of Ad12 per cell.
3. At 28 h p.i., suspend about 10^7 cells in phosphate-buffered saline containing 1% Igepal CA-630, 0.1% sodium dodecyl sulfate (SDS), and 4 mM Pefabloc SC protease inhibitor.
4. The cells are disrupted by ultrasonic treatment for 2 min in a Branson B-12 sonifier. After removal of cell debris, treat cell extracts for 1 h at 4°C with 1.0 µg of normal rabbit immunoglobulin G and 20 µL of protein A–agarose conjugate.
5. Subsequently the beads are pelleted, and the lysate is incubated at 4°C overnight with 15 µL of Ad12-pTP rabbit antiserum and 30 µL of protein A–agarose.
6. Immunoprecipitates are collected by centrifugation for 5 min at 1000g and are washed four times with phosphate-buffered saline.
7. For the release of Ad12 DNA from this complex, suspend the beads in 50 µL of 10 mM Tris-HCl (pH 8.0), 2 mM ethylene diamine tetraacetic acid, 10 mM NaCl, 1% SDS, and treat for 3 h at 55°C with 1 µg of Proteinase K per mL.
8. This protocol is described in **ref. 34**. The DNA is then analyzed by standard restriction and blotting procedures.

3.5. Construction and Use of Single-Stranded DNA Arrays

1. Ad12-specific DNA segments corresponding to 28 different open reading frames of Ad12 DNA are amplified by PCR using appropriate primers and are cloned into M13mp18 vector DNA as described *(8)*.
2. Prepare single-stranded (ss) M13-Ad12 DNAs (we used 28 different clones) using the ss-M13-DNA purification kit.
3. Denature the ssDNAs in 0.4 M NaOH, 5 mM ethylene diamine tetraacetic acid, fix on positively charged nylon membranes, and hybridize to ^{32}P-labeled cDNA.

(We used amounts of 20 or 100 ng of Ad12-M13 DNA for HeLa·Ad12 or BHK21·Ad12 DNA array analysis, respectively.)

4. For reverse transcription and ^{32}P-labeling of cDNA, total RNA from mock- or Ad12-infected cells is isolated by using a Macherey/Nagel kit. In this reaction, 5 µg of RNA, 2 µg of oligo (dT) 12–18 primers, and 1 mM of each of the dNTPs are denatured at 65°C for 5 min, chilled on ice, then mixed with 5 µCi of (γ-^{32}P)-dCTP (3000 Ci/mmol), 10 mM dithiothreitol, superscript II reaction buffer, and 300 U of superscript II reverse transcriptase and are incubated at 42°C for 1.5 h.

5. Subsequently, the ^{32}P-labeled probes are purified by gel filtration on Sephadex G-50 columns and are hybridized to DNA arrays on membranes at 42°C for 20 h in DIG hybridization buffer.

6. Wash membranes three times for 20 min in 0.5X standard sodium citrate, 1% SDS at 65°C and expose for 14 h (HeLa·Ad12) or 48 h (BHK21·Ad12) to X-ray films.

Acknowledgments

For many years, research in the author's laboratory was supported by the Deutsche Forschungsgemeinschaft Bonn, Germany (through SFBs 74 and 274 and grant DO 165/17), by the Wilhelm Sander Foundation, Munich, by Amaxa GmbH, Cologne, and recently by the Institute for Clinical and Molecular Virology, University Erlangen–Nürnberg.

References

1. Avian flu special. (2005) *Nature* **435,** 399–409.
2. The next pandemic? (2005) *Foreign Affairs* **84/4,** 3–64.
3. Doerfler, W. (1969) Nonproductive infection of baby hamster kidney cells (BHK21) with adenovirus type 12. *Virology* **38,** 587–606.
4. Doerfler, W. (1991) Abortive infection and malignant transformation by adenoviruses: Integration of viral DNA and control of viral gene expression by specific patterns of DNA methylation. *Adv. Virus Res.* **39,** 89–128.
5. Hösel, M., Webb, D., Schröer, J., and Doerfler, W. (2003) The abortive infection of Syrian hamster cells with human adenovirus type 12. *Curr. Top. Microbiol. Immunol.* **272,** 415–440.
6. Trentin, J. J., Yabe, Y., and Taylor, G. (1962) The quest for human cancer viruses. A new approach to an old problem reveals cancer induction by human adenoviruses. *Science* **137,** 835–841.
7. Huebner, R., Rowe, W. P., and Lane, T. W. (1962) Oncogenic effects in hamsters of human adenovirus types 12 and 18. *Proc. Natl. Acad. Sci. USA* **48,** 2051–2058.
8. Hohlweg, U., Hösel, M., Dorn, A., et al. (2003) Intraperitoneal dissemination of Ad12-induced undifferentiated neuroectodermal hamster tumors: de novo methylation and transcription patterns of integrated viral and of cellular genes. *Virus Res.* **98,** 45–56.

9. Green, M., Mackey, J. K., Wold, W. S. M., and Ridgen, P. (1977a) Analysis of human cancer DNA for DNA sequences of human adenovirus type 4. *J. Natl. Cancer Inst.* **62,** 23–26.

10. Green, M., Mackey, J. K., Wold, W. S. M., and Ridgen, P. (1977b) Analysis of human cancer DNA for DNA sequences of human adenovirus serotypes 3, 7, 11, 14, 16 and 21 in group B. *Cancer Res.* **39,** 3479–3484.

11. Green, M., Wold, W. S. M., Mackey, J. K., and Ridgen, P. (1977) Analysis of human tonsil and cancer DNAs and RNAs for DNA sequences of subgroup C (serotypes 1, 2, 5 and 6) human adenoviruses. *Proc. Natl. Acad. Sci. USA* **76,** 6606–6610.

12. Mende, Y., Schneider, P. M., Baldus, S. E., and Doerfler, W. (2004) PCR-screening of human esophageal and bronchial cancers reveals absence of adenoviral DNA sequences. *Virus Res.* **104,** 81–85.

13. Doerfler, W. (1968) The fate of the DNA of adenovirus type 12 in baby hamster kidney cells. *Proc. Natl. Acad. Sci. USA* **60,** 636–643.

14. Doerfler, W., Gahlmann, R., Stabel, S., et al. (1983) On the mechanism of recombination between adenoviral and cellular DNAs: the structure of junction sites. *Curr. Top. Microbiol. Immunol.* **109,** 193–228.

15. Heller, H., Kämmer, C., Wilgenbus, P., and Doerfler, W. (1995) Chromosomal insertion of foreign (adenovirus type 12, plasmid, or bacteriophage λ) DNA is associated with enhanced methylation of cellular DNA segments. *Proc. Natl. Acad. Sci USA* **92,** 5515–5519.

16. Remus, R., Kämmer, C., Heller, H., Schmitz, B., Schell, G., and Doerfler, W. (1999) Insertion of foreign DNA into an established mammalian genome can alter the methylation of cellular DNA sequences. *J. Virol.* **73,** 1010–1022.

17. Müller, K., Heller, H., and Doerfler, W. (2001) Foreign DNA integration. Genome-wide perturbations of methylation and transcription in the recipient genomes. *J. Biol. Chem.* **276,** 14,271–14,278.

18. Endter, C. and Dobner, T. (2004) Cell transformation by human adenoviruses. *Curr. Top. Microbiol. Immunol.* **273,** 163–214.

19. Zantema, A. and van der Eb, A. J. (1995) Modulation of gene expression by adenovirus transformation. *Curr. Top. Microbiol. Immunol.* **199/III,** 1–23.

20. Dorn, A., Zhao, H., Granberg, F., et al. (2005) Identification of specific cellular genes up-regulated late in adenovirus type 12 infection. *J. Virol.* **79,** 2404–2412.

21. Strohl, W. A., Rouse, H., Teets, K., and Schlesinger, R. W. (1970) The response of BHK21 cells to infection with type 12 adenovirus. 3. Transformation and restricted replication of superinfecting type 2 adenovirus. *Arch. Ges. Virusforsch.* **31,** 93–112.

22. Webb, D., Hösel, M., Hochstein, N., Dorn, A., Auerochs, S., and Doerfler, W. (2005) hCAR-enhanced entry of adenovirus type 12 (Ad12) into non-permissive hamster cells fails to elicit viral replication. In revision.

23. Schröer, J., Hölker, I., and Doerfler, W. (1997) Adenovirus type 12 DNA firmly associates with mammalian chromosomes early after virus infection or after DNA

transfer by the addition of DNA to the cell culture medium. *J. Virol.* **71**, 7923–7932.

24. Sutter, D., Westphal, M., and Doerfler, W. (1978) Patterns of integration of viral DNA sequences in the genomes of adenovirus type 12-transformed hamster cells. *Cell* **14**, 569–585.

25. Knoblauch, M., Schröer, J., Schmitz, B., and Doerfler, W. (1996) The structure of adenovirus type 12 DNA integration sites in the hamster cell genome. *J. Virol.* **70**, 3788–3796.

26. Hochstein, N., Muiznieks, I., Mangel, L., Brondke, H., and Doerfler, W. (2006) Selection for transgenome stability and distinct hypermethylation: anatomy of a transgenome. Submitted.

27. Ortin, J., Scheidtmann, K.-H., Greenberg, R., Westphal, M., and Doerfler, W. (1976) Transcription of the genome of adenovirus type 12. III. Maps of stable RNA from productively infected human cells and abortively infected and transformed hamster cells. *J. Virol.* **20**, 355–372.

28. Schiedner, G. and Doerfler, W. (1996) Insufficient levels of NFIII and its low affinity for the origin of adenovirus type 12 (Ad12) DNA replication contribute to the abortive infection of BHK21 hamster cells by Ad12. *J. Virol.* **70**, 8003–8009.

29. Jüttermann, R., Weyer, U., and Doerfler, W. (1989) Defect of adenovirus type 12 replication in hamster cells: Absence of transcription of viral virus-associated and L1 RNAs. *J. Virol.* **63**, 3535–3540.

30. Zock, C. and Doerfler, W. (1990) A mitigator sequence in the downstream region of the major late promoter of adenovirus type 12 DNA. *EMBO J.* **9**, 1615–1623.

31. Schiedner, G., Schmitz, B., and Doerfler, W. (1994) Late transcripts of adenovirus type 12 DNA are not translated in hamster cells expressing the E1 region of adenovirus type 5. *J. Virol.* **68**, 5476–5482.

32. Visser, L., van Maarschalkerweerd, M. W., Rozijn, T. H., Wassenaar, A. D., Reemst, A. M., and Susenbach, J. S. (1980) Viral DNA sequences in adenovirus-transformed cells. *Cold Spring Harb. Symp. Quant. Biol.* **44**, 541–550.

33. Klimkait, T. and Doerfler, W. (1985) Adenovirus types 2 and 5 functions elicit replication and late expression of adenovirus type 12 DNA in hamster cells. *J. Virol.* **55**, 466–474.

34. Hösel, M., Webb, D., Schröer, J., Schmitz, B., and Doerfler, W. (2001) The overexpression of the adenovirus type 12 pTP or E1A gene facilitates Ad12 DNA replication in non-permissive BHK21 hamster cells. *J. Virol.* **75**, 16,041–16,053.

35. Brüggemann, U., Klenk, H.-D., and Doerfler, W. (1985) Increased infectivity of extracellular adenovirus type 12. *J. Virol.* **55**, 117–125.

36. Kuhlmann, I., Achten, S., Rudolph, R., and Doerfler, W. (1982) Tumor induction by human adenovirus type 12 in hamsters: loss of the viral genome from adenovirus type 12-induced tumor cells is compatible with tumor formation. *EMBO J.* **1**, 79–86.

37. Hilger-Eversheim, K. and Doerfler, W. (1997) Clonal origin of adenovirus type 12-induced hamster tumors: nonspecific chromosomal integration sites of viral DNA. *Cancer Res.* **57**, 3001–3009.

38. Lichtenberg, U., Zock, C., and Doerfler, W. (1988) Integration of foreign DNA into mammalian genome can be associated with hypomethylation at site of insertion. *Virus Res.* **11,** 335–342.

39. Sutter, D. and Doerfler, W. (1980) Methylation of integrated adenovirus type 12 DNA sequences in transformed cells is inversely correlated with viral gene expression. *Proc. Natl. Acad. Sci. USA* **77,** 253–256.

40. Doerfler, W. (1983) DNA methylation and gene activity. *Ann. Rev. Biochem.* **52,** 93–124.

41. Doerfler, W. (1991) Patterns of DNA methylation—evolutionary vestiges of foreign DNA inactivation as a host defense mechanism—a proposal. *Biol. Chem. Hoppe-Seyler* **372,** 557–564.

42. Yoder, J. A., Walsh, C. P., and Bestor, T. H. (1997) Cytosine methylation and the ecology of intragenomic parasites. *Trends Genet.* **13,** 335–340.

43. Doerfler, W. (2000) *Foreign DNA in Mammalian Systems.* Wiley-VCH, New York.

44. Pfeffer, A., Schubbert, R., Orend, G., Hilger-Eversheim, K., and Doerfler, W. (1999) Integrated viral genomes can be lost from adenovirus type 12-induced hamster tumor cells in a clone-specific, multistep process with retention of the oncogenic phenotype. *Virus Res.* **59,** 113–127.

45. Stoker, M. and Macpherson, I. (1964) Syrian hamster fibroblast line BHK21 and its derivatives. *Nature* **203,** 1355–1357.

46. Strohl, W. A., Rabson, A. S., and Rouse, H. (1967) Adenovirus tumorigenesis: Role of the viral genome in determining tumor morphology. *Science* **156,** 1631–1633.

47. Groneberg, J., Sutter, D., Soboll, H., and Doerfler, W. (1978) Morphological revertants of adenovirus type 12-transformed hamster cells. *J. Gen. Virol.* **40,** 635–645.

48. Orend, G., Knoblauch, M., and Doerfler, W. (1995) Selective loss of unmethylated segments of integrated Ad12 genomes in revertants of the adenovirus type 12-transformed cell line T637. *Virus Res.* **38,** 261–267.

49. Frommer, M., McDonald, L. E., Millar, D. S., et al. (1992) A genomic sequencing protocol that yields a positive display of 5-methylcytosine residues in individual DNA strands. *Proc. Natl. Acad. Sci. USA* **89,** 1827–1831.

50. Zeschnigk, M., Schmitz, B., Dittrich, B., Buiting, K., Horsthemke, B., and Doerfler, W. (1997) Imprinted segments in the human genome: different DNA methylation patterns in the Prader-Willi/Angelman syndrome region as determined by the genomic sequencing method. *Hum. Mol. Genet.* **6,** 387–395.

51. Sprengel, J., Schmitz, B., Heuss-Neitzel, D., Zock, C., and Doerfler, W. (1994) Nucleotide sequence of human adenovirus type 12 DNA: comparative functional analysis. *J. Virol.* **68,** 379–389.

15

Measurement of Natural-Killer Cell Lytic Activity of Adenovirus-Infected or Adenovirus-Transformed Cells

John M. Routes

Summary

Natural-killer (NK) cells are lymphocytes that do not express the CD3 T-cell receptor but do express the CD16 (FcγRIII) and CD56 (isoform of NCAM) in humans or NK1.1 antigen in certain strains of mice. NK cells display spontaneous lytic activity but do not exhibit immunological memory. NK cells are important mediators of antiviral and antitumor immunity. Standard NK cytolysis assays measure the ability of NK cells to kill certain target cells (tumor cells, virally infected cells) in short-term (usually 4–6 h) cytolysis assays. This chapter details the use of the NK cell cytolysis assay using polyclonal populations of human or rodent NK cells.

Key Words: Adenovirus; E1A; natural-killer cells.

1. Introduction

Natural-killer (NK) cells, a critical component of the innate immune response, are a lymphocyte subpopulation that do not express the CD3 antigen but do express the CD16 (FcγRIII) and CD56 (isoform of NCAM) in humans or NK1.1 antigen in certain strains of mice *(1,2)*. NK cells exhibit spontaneous lytic activity that is not major histocompatibility complex restricted. NK cells kill target cells predominantly via the perforin/granzyme pathway, although other death-inducing mechanisms (tumor necrosis factor [TNF]-related apoptosis-inducing ligand, TNF-α, and Fas ligand) are also utilized *(3)*. NK cells express a variety of activating and inhibitory receptors, and their ligands have in many cases been identified *(4)*. The balance of activating and inhibitory signals delivered via these NK cell receptors determines whether target cell lysis occurs *(5)*.

The expression of Ad serotype 2 or 5 (Ad2/5) *E1A* gene products sensitizes rodent and human cells to lysis by NK cells *(6,7)*. Ad2/5 *E1A* also sensitizes

From: *Methods in Molecular Medicine, Vol. 131:*
Adenovirus Methods and Protocols, Second Edition, vol. 2:
Ad Proteins, RNA, Lifecycle, Host Interactions, and Phylogenetics
Edited by: W. S. M. Wold and A. E. Tollefson © Humana Press Inc., Totowa, NJ

cells to apoptosis by perforin/granzyme, TNF-related apoptosis-inducing ligand, TNF-α, and Fas ligand *(6,8–11)*. This activity of *E1A* likely contributes to the ability of NK cells to selectively kill cells that express Ad2/5-*E1A* gene products. Using rodent models, NK cells have been shown to be an important component of the cellular immune response to Ad2/5-transformed cells and tumor cells that express *E1A (12–14)*. Based on these studies, it would appear likely that NK cells are also important in antiadenoviral immunity in humans.

1.1. Measurement of NK-Cell Lytic Activity

Standard NK cytolysis assays measure the ability of NK cells to kill certain target cells (tumor cells, virally infected cells) in short-term (usually 4–6 h) cytolysis assays. This assay can easily be adapted to the Ad system by the inclusion of appropriate Ad-infected or Ad-transformed target cells. NK-sensitive targets such as K562 (human leukemia cell line) or YAC-I (mouse lymphoma cell line) are usually included as positive controls in human and murine NK assays, respectively. However, NK-susceptible, Ad-infected, or Ad-transformed cell lines can be substituted as positive controls for K562 or YAC-I.

To measure the susceptibility of a target cell to NK-cell lysis, a constant number of target cells are labeled with ^{51}Cr and incubated with graded numbers of NK cells (effector cells). A sufficient period of time should be allowed for target-cell lysis, which is measured by radiolabel released into the supernatant. Nonradioactive methods of measuring NK-cell killing have been developed. However, the ^{51}Cr-release assay still is the most widely used and accepted method of quantifying NK-cell lytic activity.

Effector cells for NK cytolysis assays may include NK-cell clones, highly purified populations of NK cells, freshly isolated human peripheral-blood mononuclear cells (PBMCs), or rodent mononuclear cells—the latter usually obtained from the spleen. This section specifically covers the isolation of human and rodent mononuclear cells for use in NK assays. NK cytolysis assays should be performed at multiple effector-to-target cell (E:T) ratios, i.e., the ratio of the total number of effector cells to target cells. When using unpurified populations of mononuclear cells as a source of NK cells, effector cell number refers to the total number of mononuclear cells, not just NK cells. Results are commonly presented graphically as a percent of specific target-cell lysis at each E:T ratio. Another common method of presenting NK-killing data is by calculating lytic units, which are derived from the NK-killing curve. Lytic units are calculated by the formula: $LU_X = 10,000/E{:}TX$, where $E{:}T_X$ is the E:T ratio needed to lyse $X\%$ of target cells.

2. Materials

2.1. Isolation of Rodent Mononuclear Cells

1. RPMI-10% fetal calf serum (FCS) (RPMI-10): RPMI medium (GibcoI-BRL, Gaithersburg, MD), 10% FCS, heat-inactivated at 56°C for 1 h, 2 mM L-glutamine, penicillin G (1000 U/mL), and streptomycin (100 μg/mL).
2. Red blood cell (RBC)-lysing buffer: NH_4Cl (8.29 g/L), $KHCO_3$ (1.0 g/L), and ethylene diamine tetraacetic acid (0.0372 g/L). Add to 1000 mL, filter through a 0.2-μm filter, aliquot (85 mL per bottle), and refrigerate. Add 15 mL endotoxin-free, heat-inactivated FCS prior to use.
3. Hank's balanced buffer solution (HBBS).
4. Polyinosinic-polycytidylic acid (Poly I:C) (Sigma, St. Louis, MO): stock solution of 100 μg/mL in HBSS. Store in 2- to 3-mL aliquots at 100 μg/mL in HBSS in –20°C freezer.
5. Sterile scissors and forceps.
6. 100-Mesh wire sieve.
7. Freshly removed spleens from mice, hamsters, or rats (2–5 mo old).
8. Equipment and reagents for counting viable cells (e.g., 0.1% Trypan blue and hemocytometer).

2.2. Isolation of Human PBMC

1. RPMI-10, HBBS.
2. Ficoll-Hypaque solution (Ficoll-Paque, Pharmacia, Piscataway, NJ).
3. RosetteSep Enrichment Cocktail for human NK cells (StemCell Technologies, Vancouver, BC).
4. Heparin sulfate (1000 U/mL).
5. Equipment for drawing blood. **Caution:** Standard biosafety procedures must be followed when using human blood. Donors should be screened by serology for hepatitis B, hepatitis C, and HIV-1 prior to isolation of blood mononuclear cells.

2.3. NK Cytolysis Assay

1. Targets (Ad-infected or Ad-transformed cell lines): usually include YAC-I or K562 cells (positive controls) when performing rodent or human NK cytolysis assays, respectively.
2. RPMI-10.
3. Mononuclear cells.
4. 96-Well, tissue-culture-treated, flat-bottomed or V-bottom plates.
5. $Na_2{}^{51}CrO_4$: sodium chromate in normal sterile saline, 1 mCi/mL (NEN Research Products, Boston, MA).
6. 0.5% Sodium dodecyl sulfate (SDS).
7. Multichannel pipets (<200 mL).
8. ^{51}Cr-counting tubes.
9. γ-Scintillation counter.

3. Methods

3.1. Activate NK Cells (Mouse NK-Cell Assays)

1. Inject mice with 100 μg of poly I:C intraperitoneally 18 h prior to the assay. Poly I:C is an interferon inducer and activates NK cells in vivo. NK cells present in murine spleens have much lower resting NK-cell lytic activity than those from hamster or rat. Therefore, this step is necessary to enhance mouse NK-cell lytic activity.

3.2. Isolation of Mononuclear Cells From Spleen

1. Remove spleen under sterile conditions and place into a 60 × 15-mm Petri dish containing 5–10 mL of cold HBSS.
2. Cut spleen in half and puncture repeatedly with 19-gage needle. Remove splenocytes by repeatedly perfusing spleen with cold HBSS solution using a 6-mL syringe and 19-gage needle. The remainder of splenocytes may be removed by teasing and crushing with forceps. The process is complete when only predominately white fibrous tissue remains.
3. Centrifuge (200*g* for 10 min). Resuspend pellet in 5 mL cold RBC lysing buffer on ice for 5 min. Add 10 mL HBSS and centrifuge (200*g* for 10 min). Wash pellet twice with cold HBBS.
4. Resuspend pellet with 5–10 mL RPMI-10. Pass remaining cells through 100-mm mesh wire sieve (200-mm nylon mesh may be substituted), and count viable cells. Resuspend splenocytes at the following concentrations: mouse: 4×10^7 cells/mL; hamster or rat: 2×10^7 cells/mL (*see* **Note 1**).

3.3. Isolation of Mononuclear Cells From Human Peripheral Blood

1. Draw peripheral blood into a 30- to 60-mL syringe containing roughly 3–4 mL sterile heparin-sulfate solution (1000 U/mL) and mix.
2. To isolate purified (>90%) populations of polyclonal NK cells, add 50 μL RosetteSep Enrichment Cocktail per mL of whole blood to the heparinized blood, mix, and incubate for 20 min at room temperature. This step may be omitted if one elects to use PBMCs as the source of human NK cells.
3. Mix equal volumes of warm HBSS and heparinized blood in a sterile container.
4. Hold a 15-mL polystyrene centrifuge tube at a 45° angle and slowly pipet 10 mL HBSS/blood over 5 mL Ficoll-Hypaque solution. It is imperative that one maintains a sharp Ficoll/blood interface for optimal separation of mononuclear cells.
5. Centrifuge (room temperature) 30 min at 900*g*; brake off.
6. Carefully remove tubes from the centrifuge. There are four layers: the large top layer contains plasma and platelets, the second narrow layer contains mononuclear cells, followed by a larger Ficoll-Hypaque layer, and RBC/granulocytes at the bottom of the tube (*see* **Note 2**). If one uses RosetteSep (**step 2**), the second layer contains NK cells.
7. Remove the upper layer and discard. Carefully remove the thin mononuclear-cell layer and transfer to a centrifuge tube.

8. Wash cells three times with HBSS. In the initial wash, dilute mononuclear cells with at least threefold excess (v/v) of HBBS and spin at 500g for 10 min. The remainder of the centrifugations can be performed (at 200–300g for 10 min).
9. Resuspend mononuclear-cell pellet in an appropriate volume of RPMI-10 and count viable cells by Trypan blue exclusion (*see* **Note 1**). Resuspend cells at a final concentration of 1×10^7 cell/mL (mononuclear cells) or $12–25 \times 10^6$ cells/mL (purified NK cells).

3.4. Target-Cell Preparation

1. $1–2 \times 10^6$ target cells are passaged into fresh medium the night before the cytolysis assay. If one elects to use adenovirus-infected cells as target cells, infect the target cells the day before the cytolysis assay.
2. Prepare approx $1–3 \times 10^6$ target cells as a single cell suspension in RPMI-10.
3. Centrifuge at 200g for 10 min, pour off the supernatant, and gently resuspend cells (*do not vortex*) in remaining 100–200 µL of medium. Add 100 µL of heat-inactivated, lipopolysaccharide-free fetal bovine serum.
4. Add 100 µCi ^{51}Cr, gently resuspend, and incubate in 37°C water bath for 1 h. Gently resuspend every 15 min. Increased incubation times (1.5 h) and/or ^{51}Cr (up to 200 µCi) may be added if cells label insufficiently (*see* **Note 6**).
5. ^{51}Cr-labeled target cells are washed two times in HBSS (15 mL). Resuspend cell pellet in 10 mL of RPMI-10 and incubate an additional 30–45 min (mixing every 15 min) in 37°C water bath. (This additional purging step decreases the spontaneous release of radiolabel) (*see* **Note 5**).
6. Centrifuge, then resuspend pellet in complete medium; count viable cells by Trypan blue exclusion, and resuspend cells at a final concentration of 5×10^4 cells/mL. For most targets, 10^4 targets/well are used in cytolysis assays.

3.5. Cytolysis Assays

1. Unless limited by the number of mononuclear cells (effector cells) or target cells, one should generate killing curves using different E:T ratios. In general, make serial twofold dilutions of mononuclear cells prior to the addition of targets. Beginning and ending E:T ratios vary with the source of mononuclear cells. Typical killing curves use E:T ratios from 400:1 to 25:1 (mouse); 200:1 to 12:1 (hamster, rat), and 100:1 to 6:1 (human). When using purified populations of freshly isolated human NK cells, the E:T ratio typically starts at either 25:1 or 12:1 and ends at 1.5:1.
2. Add 0.2 mL of labeled target cells (1×10^4 cells/well) to replicate wells containing effector cells. Target cells added in triplicate to wells containing 100 µL of complete medium will be used to calculate the spontaneous release. Target cells added in triplicate along with 100 µL of 0.5% SDS will be used to calculate the total releasable counts.
3. Carefully centrifuge plates (200g) for 1 min (brake off) to enhance contact between target and effector cells. Incubate plates for 4–6 h at 37°C with 5% CO_2 in an incubator.

4. Using a multichamber pipet, carefully harvest 150 µL of supernatant and transfer to ^{51}Cr-counting tubes. Avoid disturbing cells at the bottom of the plate. For wells with 0.5% SDS, gently pipet several times before final harvest.
5. Count samples in γ-scintillation counter.
6. Calculate percent specific release (*see* **Notes 2, 4,** and **5**) for each E:T ratio using the arithmetic mean of triplicate cultures and the following formula:

$$\text{Percent specific release} = \frac{\text{experimental release} - \text{spontaneous release}}{\text{total release} - \text{spontaneous release}} \times 100$$

where total release = ^{51}Cr released from wells with 0.5% SDS, spontaneous release = ^{51}Cr released from wells with medium alone, and experimental release = ^{51}Cr released from wells with effector cells.
7. Calculate percent spontaneous release (*see* **Note 5**):

$$\text{Percent spontaneous release} = \frac{\text{spontaneous release}}{\text{total release}}$$

4. Notes

1. The following are average yields of mononuclear cells: human peripheral blood, 1×10^6 mononuclear cells per mL of blood; hamster, $4–6 \times 10^7$ mononuclear cells per spleen; immunocompetent mouse, $8–10 \times 10^7$ mononuclear cells per spleen; nude mouse, 5×10^7 mononuclear cells per spleen; immunocompetent rat, 1×10^8 mononuclear cells per spleen; nude rat, 4×10^8 mononuclear cells per spleen.
2. Percent-specific release (target-cell killing) of NK-susceptible cell lines (YAC-1 or K562) should be greater than 30% at the highest E:T ratio. However, NK activity from human blood mononuclear cells is donor variable. Common reasons for consistently low specific release of susceptible targets include not promptly removing mononuclear cells from Ficoll-Hypaque, extended incubation times in RBC lysing buffer, incubation of microtiter plates at room temperature, and incorrect counts. Under certain circumstances, assay time may need to be extended.
3. The total releasable counts calculated from each cell line should be at least 1000 cpm above background. If fewer counts are obtained, increase the amount of label or labeling time. Alternatively, one can increase the number of target cells added per well.
4. Percent spontaneous release values should be noted for each target-cell line. Values above 30% are unacceptable. The purge step should eliminate high spontaneous-release values.
5. The inability of a target-cell line to properly label or high spontaneous-release values may indicate improper cell-culture conditions prior to the assay or microbial contamination. Particular attention to maintaining healthy cell cultures is critical to the success of the assay. Cell lines should be routinely screened for contamination with *Mycoplasma*.

References

1. Papamichail, M., Perez, S. A., Gritzapis, A. D., and Baxevanis, C. N. (2004) Natural killer lymphocytes: biology, development, and function. *Cancer Immunol. Immunother.* **53,** 176–186.
2. Hamerman, J. A., Ogasawara, K., and Lanier, L. L. (2005) NK cells in innate immunity. *Curr. Opin. Immunol.* **17,** 29–35.
3. Smyth, M. J., Cretney, E., Kelly, J. M., et al. (2005) Activation of NK cell cytotoxicity. *Mol. Immunol.* **42,** 501–510.
4. Moretta, L. and Moretta, A. (2004) Unraveling natural killer cell function: triggering and inhibitory human NK receptors. *EMBO J.* **23,** 255–259.
5. Chiesa, S., Tomasello, E., Vivier, E., and Vely, F. (2005) Coordination of activating and inhibitory signals in natural killer cells. *Mol. Immunol.* **42,** 477–484.
6. Cook, J. L., May, D. L., Wilson, B. A., et al. (1989) Role of tumor necrosis factor-alpha in E1A oncogene-induced susceptibility of neoplastic cells to lysis by natural killer cells and activated macrophages. *J. Immunol.* **142,** 4527–4534.
7. Routes, J. M. and Cook, J. L. (1995) E1A gene expression induces susceptibility to killing by NK cells following immortalization but not adenovirus-infection of human cells. *Virology* **210,** 421–428.
8. Cook, J. L., Potter, T. A., Bellgrau, D., and Routes, B. A. (1996) E1A oncogene expression in target cells induces cytolytic susceptibility at a post-recognition stage in the interaction with killer lymphocytes. *Oncogene* **12,** 833–842.
9. Cook, J. L., Routes, B. A., Leu, C. Y., Walker, T. A., and Colvin, K. L. (1999) E1A oncogene-induced cellular sensitization to immune-mediated apoptosis is independent of p53 and resistant to blockade by E1B 19 kDa protein. *Exp. Cell. Res.* **252,** 199–210.
10. Routes, J. M., Ryan, S., Clase, A., et al. (2000) Adenovirus E1A oncogene expression in tumor cells enhances killing by TNF-related apoptosis-inducing ligand (TRAIL). *J. Immunol.* **165,** 4522–4527.
11. Duerksen-Hughes, P., Wold, W. S. M., and Gooding, L. R. (1989) Adenovirus E1A renders infected cells sensitive to cytolysis by tumor necrosis factor. *J. Immunol.* **143,** 4193–4200.
12. Cook, J. L. and Lewis, A. M., Jr. (1987) Immunological surveillance against DNA virus-transformed cells: correlations between natural killer cell cytolytic competence and tumor susceptibility of athymic rodents. *J. Virol.* **61,** 2155–2161.
13. Cook, J. L., Krantz, C. K., and Routes, B. A. (1996) Role of p300-family proteins in E1A oncogene induction of cytolytic susceptibility and tumor cell rejection. *Proc. Natl. Acad. Sci. USA* **93,** 13,985–13,990.
14. Routes, J. M., Ryan, S., Steinke, J., and Cook, J. L. (2000) Dissimilar immunogenicities of human papillomavirus E7 and adenovirus E1A proteins influence primary tumor development. *Virology* **277,** 48–57.

16

A Flow Cytometric Assay for Analysis of Natural-Killer Cell-Mediated Cytolysis of Adenovirus-Transformed Cells

Graham Bottley, Graham P. Cook, and G. Eric Blair

Summary

Natural-killer (NK) cells play an important role in recognizing and eliminating virally infected and transformed cells. To study this process, convenient assays for NK-cell function are required. Conventional NK-cell activity assays measure the release of ^{51}Cr from prelabeled target cells following membrane disruption. This chapter describes nonradiometric assays for NK-cell killing of adenovirus-transformed human cells that can be applied to multiple cell samples using flow cytometry.

Key Words: Human adenovirus; natural-killer cells; cell transformation; NK-cell-killing assay; nonradiometric killing assay.

1. Introduction

1.1. Natural-Killer Cells

Natural-killer (NK) cells are lymphocytes that play an important role in the control of viral infection and malignancy *(1)*. Antigenic peptides are presented to CD8$^+$ cytotoxic T-cells via major histocompatibility complex (MHC) class I molecules. It is known that many viruses downregulate MHC class I and hence are able to evade T-cell-mediated immunity *(2)*. The subversion of antigen presentation in adenovirus (Ad)-infected cells is mediated by a 19-kDa glycoprotein encoded by the E3 region *(3)*. In virally transformed cells, expression of adenovirus 12 (Ad12) E1A has been shown to lead to downregulation of MHC class I molecules on the surface of transformed rat and human cells, whereas cells expressing Ad2 or Ad5 E1A display normal or elevated MHC class I molecules *(4)*.

From: *Methods in Molecular Medicine, Vol. 131:*
Adenovirus Methods and Protocols, Second Edition, vol. 2:
Ad Proteins, RNA, Lifecycle, Host Interactions, and Phylogenetics
Edited by: W. S. M. Wold and A. E. Tollefson © Humana Press Inc., Totowa, NJ

However, NK cells are able to detect cells in which expression of MHC class I at the cell surface has been reduced and thus allow the recognition of infected or transformed cells that cannot be recognized by cytotoxic T-cells. NK cells express a repertoire of inhibitory receptors that bind to target cell MHC class I molecules. Reduced expression of MHC class I on the target cell results in a failure to trigger these inhibitory receptors, resulting in activation of the NK-cell cytolytic pathway. NK cytolysis (like that of cytotoxic T-cells) involves the induction of apoptosis in target cells, ultimately leading to membrane disruption and subsequent cell death *(5)*.

1.2. NK Cytolysis Assays

Conventional NK-cell activity assays measure the release of ^{51}Cr from prelabeled target cells following membrane disruption. The protocol requires the safe use and disposal of radioisotopes with associated problems, including potential health hazards. It is known that high spontaneous release can interfere with experimental results. This and other limitations have been previously described *(6,7)*. Therefore, assays have been developed for flow cytometric analysis of NK- and cytotoxic T-cell-specific lysis. The basic principle underlying these methods is to label the effector or target cells with a nontoxic fluorescent dye prior to co-culture. The cells are then stained with a DNA dye such as 7-aminoactinomycin-D (7-AAD) to identify cells with a disrupted membrane and are analyzed by flow cytometry. These assays show consistent reproducibility and have been found to correlate closely with the conventional ^{51}Cr release assay *(6–9)*. A range of labeling strategies has been used, including the green dyes PKH-1 *(6)*, PKH-2 *(10)*, PKH-26 *(11,12)*, F18 *(13)*, CFSE *(14)*, MitoTracker Green FM *(15)*, DiO *(9)*, effector-specific monoclonal antibody *(7,16)*, or EGFP-transfection of target K562 cells *(8)*.

The assay described here can be used to assess the susceptibility of a wide range of Ad-infected or Ad-transformed cells to NK cells. We provide protocols that are appropriate for either suspension or adherent target cells. We have used this protocol to analyze differences in susceptibility of Ad12- and Ad5-transformed human cells to NK-cell killing *(17)*.

2. Materials
2.1. NK-Cell Isolation, Enrichment, and Activation

1. Vacutainer–ethylene diamine tetraacetic acid (EDTA) tubes (Becton Dickinson, Cowley, UK).
2. Dulbecco's phosphate-buffered saline (PBS) A (Sigma, Poole, UK).
3. Lymphoprep™ (Nycomed, Norway).
4. NK Cell Isolation Kit II (Miltenyi Biotec, Bisley, UK).

5. NK-cell culture medium: Dulbecco's modified Eagle's medium (Sigma), 5% fetal calf serum (FCS) (Sigma), 10% human AB serum (Harlan Seralabs, Loughborough, UK), and 50 U/mL human rIL-2 (Sigma).
6. Sterile 15-mL conical tubes (Becton Dickinson).
7. Hemacytometer (Fisher Scientific, Loughborough, UK).

2.2. NK Cytolysis Assay

1. 24-Well tissue culture plates (Corning BV, The Netherlands).
2. Vybrant DiI membrane dye (Molecular Probes, Cambridge, UK).
3. Complete RPMI growth medium: RPMI medium (Sigma) supplemented with 10% FCS (Sigma) and 4 mM glutamine (Sigma).
4. K562: a human erythroleukaemic cell line (ATCC No. CCL-243).
5. PBS.
6. 1X Trypsin-EDTA solution (Sigma).
7. 7-AAD (Sigma).
8. Flow cytometer with a 488-nm laser and 585-nm/670-nm detectors (e.g., a Becton Dickinson FACSCalibur).
9. Sterile 15-mL conical tubes (Becton Dickinson).

3. Methods

3.1. NK-Cell Isolation, Enrichment, and Activation

1. Peripheral blood should be collected using EDTA tubes. NK cell numbers depend on a number of factors; therefore, volume of blood collected should be optimized according to the isolation protocol used.
2. Dilute the blood 1:1 with sterile, room temperature PBS and mix gently.
3. Prepare an appropriate conical tube with a volume of room temperature Lymphoprep equivalent to the volume of the diluted blood.
4. Carefully overlay the diluted blood onto the Lymphoprep, ensuring that the separate layers are maintained (*see* **Note 1**).
5. Centrifuge at 400g at room temperature for 25 min, with no brake.
6. Remove the interface between the layers with a sterile Pasteur pipet and pool in a fresh conical tube.
7. Add 20 mL of sterile, room temperature PBS to the interface material.
8. Centrifuge at 800g at room temperature for 5 min, with braking.
9. Completely remove supernatant and resuspend pellet in 5 mL of sterile prechilled PBS. Cells should then be placed on ice.
10. Count cells using a hemacytometer.
11. NK cells are enriched in our laboratory with the NK Cell Isolation Kit II from Miltenyi Biotec, following the manufacturer's instructions. This system is a negative selection process using magnetic beads (*see* **Note 2**).
12. Analyze NK cells by flow cytometry for expression of the NK-cell marker CD56 and the pan-T-cell marker CD3 to confirm identity. NK cells are CD56^{+ve} and CD3^{-ve}.

Table 1
Cell Numbers Required

Well	Target cells	NK cells
Target control	0.5×10^5	—
1:1 E:T ratio	0.5×10^5	0.5×10^5
2:1 E:T ratio	0.5×10^5	1×10^5
5:1 E:T ratio	0.5×10^5	2×10^5
10:1 E:T ratio	0.5×10^5	5×10^5
20:1 E:T ratio	0.5×10^5	10×10^5
NK control	—	1×10^5

NK, natural-killer; E:T, effector-to-target cell.

13. Culture enriched NK cells in NK-cell culture medium for 7–10 d to ensure proliferation and activation.

3.2. NK-Cell Labeling and Assay Setup

1. The killing assay is set up with the cell numbers shown in **Table 1**. Plate target cells in 24-well plates at a density of 5×10^4 cells per well in 1 mL of appropriate growth medium and allow to adhere. Plate three wells of each target cell line as controls. 1.95×10^6 NK cells are required for each target cell repeat. K562 (a human erythroleukemic cell line that grows in suspension) should be used as a positive control for NK-cell activity.
2. Effector cells (NK cells) should be fluorescently labeled using the Vybrant DiI system according to the manufacturer's instructions and preincubated for 30 min before addition to target cells (*see* **Notes 3** and **4**). Add effector cells in 1 mL of complete RPMI growth medium at an appropriate concentration (**Table 1**; *see* **Note 5**).

3.3. Cytotoxicity Assay

1. Incubate co-cultures at 37°C and 5% CO_2 for 6 h or as appropriate.
2. *Suspension target cells (e.g., K562):* each cell suspension should be transferred from the cell culture plate directly to a microcentrifuge tube and placed on ice. The control effector cells should also be transferred to a microcentrifuge tube on ice.
   Adherent target cells: remove and retain the overlaying medium of adherent target cells and wash each well gently with 100 µL PBS. Trypsinize the target cells using 100 µL of trypsin–EDTA, neutralizing the trypsin with the retained medium from the appropriate well. The cells should then be transferred to microcentrifuge tubes on ice.
3. Wash all cells once in PBS. Resuspend pellet in 300 µL of 20 µg/mL 7-AAD (in PBS) and incubate on ice for 20 min (*see* **Note 6**).

Table 2
Sample FACSCalibur Settings

Parameter	Setting
FSC	E-1/7.37
SSC	420
FL2 (585 nm)	435
FL3 (670 nm)	475
FL2-FL3 compensation	15.2%
FL3-FL2 compensation	12.7%

FSC, forward scatter; SSC, side scatter.

4. Transfer all cell suspensions to appropriate flow cytometry tubes and analyze with appropriate settings. Suggested settings for the Becton Dickinson FACSCalibur are given in **Table 2**. Data should be saved for all events, and the sample should be run until 10,000 unlabeled target events have been acquired (*see* **Note 7**).

3.4. Data Analysis

1. Plot forward scatter (FSC) against side scatter (SSC) on a dot plot and gate out debris (Gate 1) as shown in **Fig. 1** (*see* **Note 8**).
2. Plot Gate 1 events on an SSC against DiI fluorescence (FL2) dot plot. Apply a further gate (Gate 2) around the DiI^{-ve} events (target cells) as shown in **Fig. 2** (*see* **Note 9**).
3. Display all events corresponding to Gate 1 and Gate 2 on a 7-AAD (FL3) histogram and apply a marker to the 7-AAD-positive cells (the dead cells) as shown in **Fig. 3**.
4. Calculate the percentage of dead cells for each sample (*see* **Note 10**). Specific lysis is calculated by subtracting lysis in the absence of NK cells (spontaneous lysis) from lysis in their presence. The results are expressed as a percentage of the target population killed. The data from the three control repeats should be averaged to determine spontaneous lysis.

4. Notes

1. When layering the diluted blood onto the Lymphoprep, it is important to maintain a slow and steady flow rate to avoid mixing the layers. It is also very important to centrifuge the blood/Lymphoprep tube without a brake. When removing the interface containing the cells, care must be taken to avoid contamination with platelets because of excessive plasma from above the interface or polymorphonuclear cell contamination in Lymphoprep from below the interface.

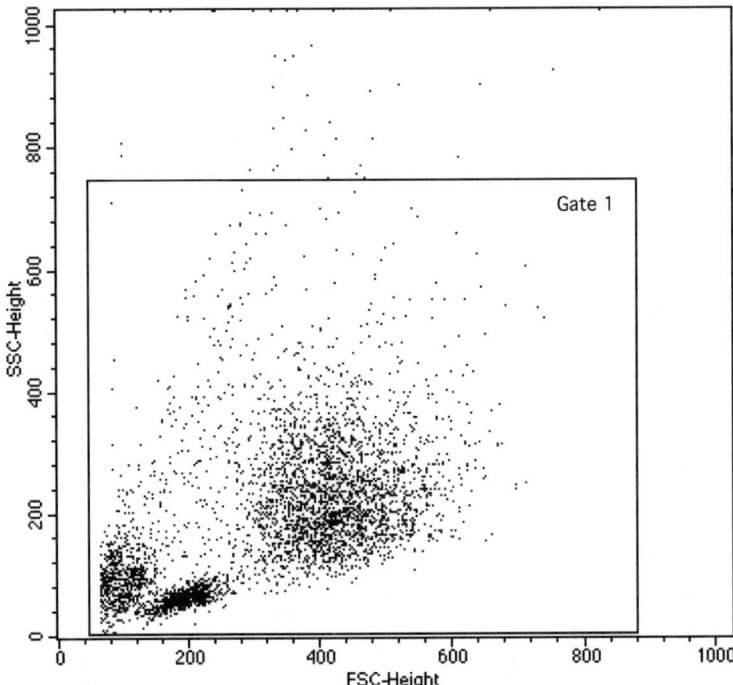

Fig. 1. Forward scatter/side scatter (FSC/SSC) dot-plot: Gate 1 is placed to exclude cells with very high FSC or SSC characteristics, indicative of cellular aggregates or debris.

2. Numerous alternative NK-enrichment strategies are commercially available. All of these methods have advantages and disadvantages in terms of time, cost, and purity of isolated cells. The method of choice will depend mainly on the resources and preferences of the individual researcher.

3. When labeling target or effector cells, great care must be taken to avoid transferring free dye to the subsequent co-culture by using stringent wash steps. This prevents unintentional labeling. If incorrect labeling is seen despite careful washing, it is often helpful to transfer the cells into a fresh tube during the washing steps.

4. If one NK-cell sample is used to test the sensitivity of multiple targets, it is usually easier to label the NK cells. Similarly, if multiple NK cell samples are to be tested against a single target, it is easier to label the target cells.

5. If the assay is performed with a large number of target cells and/or a long incubation period, it is possible that the varying number of cells in the same volume over the range of effector-to-target cell (E:T) ratios may affect the results. Experiments requiring longer incubation times may benefit from having an increased volume of medium in each well to offset medium-related effects.

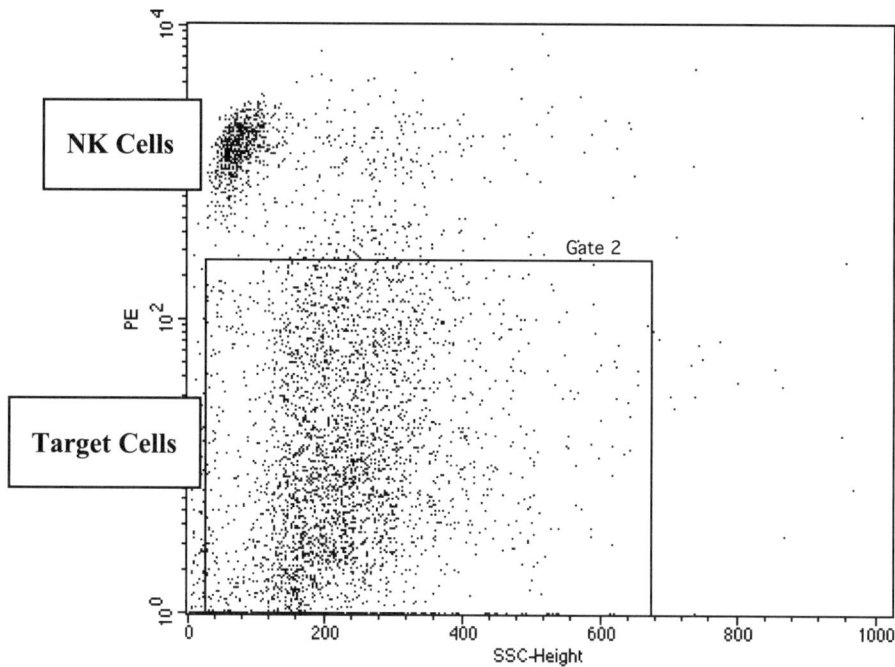

Fig. 2. Side scatter/DiI fluorescence dot-plot: Only gated events from **Fig. 1** are plotted. Gate 2 is located on target (unlabeled) cells.

6. 7-AAD is used preferentially in this protocol to avoid spectral overlap with the DiI membrane dye. If a green membrane dye is used, propidium iodide can be used as the DNA stain instead of the 7-AAD.

7. Although 10,000 target events is optimal, an absolute minimum acquisition number for gated target cells should be 1000 events. In cases of high effector activity, very few target cells can remain, especially at high E:T ratios. If it is only possible to acquire fewer than 1000 events, the results should be treated with caution. In these cases it may be necessary to reduce the time scale or E:T ratio of the experiment or scale the experiment up and start with more target cells.

8. When gating on the FSC/SSC dot-plot, care should be taken not to exclude very small, nongranular events (bottom left corner) because dead cells may be present in this population.

9. Gate 2 should be placed to ensure exclusion of effector cells using the effector-cell-only dot-plot. It is not uncommon for target cells to take up some cell dye from the labeled effector cells. Therefore, Gate 2 should not be placed using control target cell dot-plots.

10. Target cells should be healthy and actively dividing at the time of plating to ensure accurate results, and spontaneous cell death should be less than 10%. Most cell lines exhibit less than 5% spontaneous cell death.

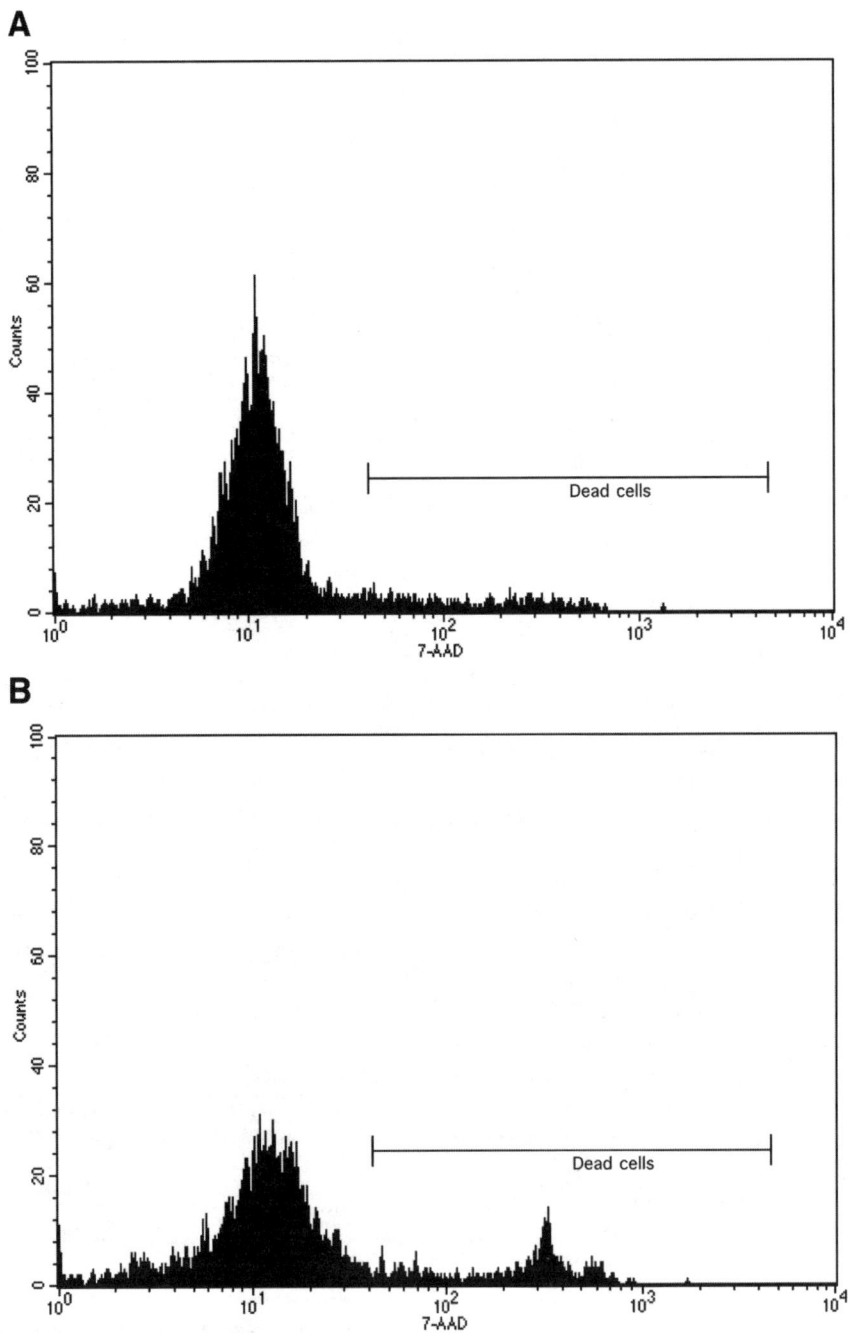

Fig. 3. 7-Aminoactinomycin-D histograms: Control K562 **(A)** and K562/PBNK co-culture at 5:1 ratio **(B)** showing dead cell enumeration (marker region).

Acknowledgments

We would like to thank Jo Quinton and Josie Meade for comments on the manuscript. Work from the authors' laboratory is funded by Yorkshire Cancer Research and the West Riding Medical Research Trust.

References

1. Lanier, L. L. (2005) NK cell recognition. *Annu. Rev. Immunol.* **23,** 225–274.
2. Hewitt, E. W. (2003) The MHC class I antigen presentation pathway: strategies for viral immune evasion. *Immunology* **110,** 163–169.
3. Burgert, H. G., Ruzsics, Z., Obermeier, S., Hilgendorf, A., Windheim, M., and Elsing, A. (2002) Subversion of host defense mechanisms by adenoviruses. *Curr. Top. Microbiol. Immunol.* **269,** 273–318.
4. Blair, G. E. and Blair-Zajdel, M. E. (2004) Evasion of the immune system by adenoviruses. *Curr. Top. Microbiol. Immunol.* **273,** 3–28.
5. Russell, J. H. and Ley, T. J. (2002) Lymphocyte-mediated cytotoxicity. *Annu. Rev. Immunol.* **20,** 323–370.
6. Slezak, S. E. and Horan, P. K. (1989) Cell-mediated cytotoxicity. A highly sensitive and informative flow cytometric assay. *J. Immunol. Methods* **117,** 205–214.
7. Goldberg, J. E., Sherwood, S. W., and Clayberger, C. (1999) A novel method for measuring CTL and NK cell-mediated cytotoxicity using annexin V and two-color flow cytometry. *J. Immunol. Methods* **224,** 1–9.
8. Kantakamalakul, W., Jaroenpool, J., and Pattanapanyasat, K. (2003) A novel enhanced green fluorescent protein (EGFP)-K562 flow cytometric method for measuring natural killer (NK) cell cytotoxic activity. *J. Immunol. Methods* **272,** 189–197.
9. Piriou, L., Chilmonczyk, S., Genetet, N., and Albina, E. (2000) Design of a flow cytometric assay for the determination of natural killer and cytotoxic T-lymphocyte activity in human and in different animal species. *Cytometry* **41,** 289–297.
10. Johann, S., Blumel, G., Lipp, M., and Forster, R. (1995) A versatile flow cytometry-based assay for the determination of short- and long-term natural killer cell activity. *J. Immunol. Methods* **185,** 209–216.
11. Lee-MacAry, A. E., Ross, E. L., Davies, D., et al. (2001) Development of a novel flow cytometric cell-mediated cytotoxicity assay using the fluorophores PKH-26 and TO-PRO-3 iodide. *J. Immunol. Methods* **252,** 83–92.
12. Flieger, D., Gruber, R., Schlimok, G., Reiter, C., Pantel, K., and Riethmuller, G. (1995) A novel non-radioactive cellular cytotoxicity test based on the differential assessment of living and killed target and effector cells. *J. Immunol. Methods* **180,** 1–13.
13. Radosevic, K., Garritsen, H. S., Van Graft, M., De Grooth, B. G., and Greve, J. (1990) A simple and sensitive flow cytometric assay for the determination of the cytotoxic activity of human natural killer cells. *J. Immunol. Methods* **135,** 81–89.
14. Marcusson-Stahl, M. and Cederbrant, K. (2003) A flow-cytometric NK-cytotoxicity assay adapted for use in rat repeated dose toxicity studies. *Toxicology* **193,** 269–279.

15. Vizler, C., Nagy, T., Kusz, E., Glavinas, H., and Duda, E. (2002) Flow cytometric cytotoxicity assay for measuring mammalian and avian NK cell activity. *Cytometry* **47,** 158–162.

16. Godoy-Ramirez, K., Franck, K., and Gaines, H. (2000) A novel method for the simultaneous assessment of natural killer cell conjugate formation and cytotoxicity at the single-cell level by multi-parameter flow cytometry. *J. Immunol. Methods* **239,** 35–44.

17. Bottley, G., Cook, G. P., Meade, J. L., Holt, J. R., Hoeben, R. C., and Blair, G. E. (2005) Differential expression of LFA-3, Fas and MHC Class I on Ad5- and Ad12-transformed human cells and their susceptibility to lymphokine activated killer (LAK) cells. *Virology* **338,** 297–308.

17

Large-Scale Purification and Crystallization of Adenovirus Hexon

John J. Rux and Roger M. Burnett

Summary

This chapter provides a protocol for the large-scale purification of adenovirus type 2 and 5 virions and the soluble major coat protein hexon. The purified virus particles remain intact and are suitable for vector, vaccine, or structural studies and can also be used as seed stock for further rounds of infection. The hexon may be used to produce crystals suitable for high-resolution X-ray crystallographic studies. Briefly, virus is propagated in HeLa cell suspension cultures. The infected cells are lysed, virions and hexon are separated by centrifugation, and the protein is then further purified by anion exchange chromatography. The entire purification procedure takes approx 1 wk and typically yields 10^{13} virus particles and 10–20 mg of highly purified hexon.

Key Words: Adenovirus; hexon; protein purification; crystallization; X-ray diffraction.

1. Introduction

Adenovirus (Ad), shortly after its discovery in 1953 *(1)*, was one of the first biological entities to be imaged using the electron microscope (EM) *(2)*. Since then, it has been under continuous scrutiny by structural biologists (reviewed in **ref. 3**). The characteristic shape of the Ad virion, with its projecting fibers, made it a fascinating object for early EM studies, which focused on how its more than 10 structural proteins are organized to form the virion. The size of the virion, with a diameter of 914 Å at the fivefold vertices *(4)* and a mass of more than 150×10^6 Daltons *(5)*, presents a formidable challenge for X-ray crystallographic studies.

An alternative approach to determining the virion structure was facilitated by the property that cells infected with Ad contain a 10- to 100-fold excess of most structural proteins in soluble form compared with those incorporated into

From: *Methods in Molecular Medicine, Vol. 131:*
Adenovirus Methods and Protocols, Second Edition, vol. 2:
Ad Proteins, RNA, Lifecycle, Host Interactions, and Phylogenetics
Edited by: W. S. M. Wold and A. E. Tollefson © Humana Press Inc., Totowa, NJ

virions *(6)*, which permits the isolation of these proteins for structural studies. The major coat protein, hexon, was the first animal virus protein to be crystallized *(7)*, and its three-dimensional structure has since been determined in progressively increasing detail *(8–12)*. Because hexon represents more than 60% of the total particle weight *(5)*, its initial low-resolution structure could be used to develop a model for the Ad capsid *(8)*. Further studies using cryo-electron microscopy and image reconstruction *(4)* revealed the complete virion to 35 Å resolution. Subsequent work combined the crystallographic and EM images and revealed details of the minor proteins *(13)*. This approach, incorporating the crystal structure of the penton base *(14)*, has now provided a quasi-atomic model of the entire capsid at approx 10 Å resolution *(15)*.

This chapter describes a protocol for the large-scale purification of virions of Ad types 2 and 5 and the soluble major coat protein, hexon. It follows our original protocol *(16)*, but has been updated with new resources. The virus particles can be used as seed stock for further rounds of infection, for EM studies on the intact particle or its fragments *(15,17,18)*, or in gene replacement, oncolytic, or vaccine vector investigations *(19–23)*. The hexon may be used to produce crystals suitable for high-resolution X-ray crystallographic studies, as described here. An extension of this protocol to other Ad types depends on the ease of viral growth and the quantity of soluble hexon found in the infected cell. Several members of subgroup C (types 1, 2, 5, and 6), simian Ad 25 (adC68), and avian CELO hexon have been crystallized from soluble protein *(24–28)*, but the fastidious subgroup F viruses (types 40 and 41), which grow only in 293 cells, yield little hexon (Hay, R. T., personal communication, 1989). Structural studies on other virion proteins such as the fiber *(29,30)*, its receptor-binding knob *(31–34)*, penton base *(14)*, and the complete penton *(35)*, which are present in low quantities in the infected cell, are being facilitated by expression systems. These studies have been difficult to extend to hexon, as this protein has a complicated folding pathway requiring the presence of the Ad-coded nonstructural 100 K protein *(36)*, but co-expression of the two genes can produce correctly folded hexons *(37)*.

The hexon purification scheme using anion-exchange chromatography has been developed in our laboratory over several years and is based on a variety of other published protocols *(38–43)* as well as our own experience *(16)*. The protocol has been optimized to provide the large amount of protein required for structural studies in as short a time as possible. Once preparations have been made, the procedure requires approx 1 wk to obtain 10–20 mg of purified hexon (*see* **Note 1**).

2. Materials

The materials listed are those routinely used to perform the Ad hexon purification protocol. Substitutions could be made with equivalent items as available.

2.1. Cells and Virus

Human Ad types 2 and 5 are routinely propagated in suspension cultures of HeLa S3 cells as described; human Ad types 2 (American Type Culture Collection [ATCC] VR-846) and 5 (ATCC VR-5) and HeLa S3 cells (ATCC CCL 2.2) are available from the ATCC.

2.2. Stock Solutions

All stock buffer solutions are filtered through 0.22-μm bottle top filters (Corning, 430624).

1. Joklik's solution (20 L): 2 × (10 L) minimal essential medium powder (Joklik-modified) (Invitrogen, 22300), 40 g NaHCO$_3$ (Sigma-Aldrich), 20 L distilled/deionized water. Dissolve powder in the water, and filter through 0.2-μm culture capsule (Pall-Gelman) to sterilize. Store in 500-mL bottles in the dark at 4°C (stable for at least 6 mo). Warm to room temperature before use (*see* **Note 2**).
2. Complete medium (545.5 mL): 500 mL Joklik's solution, 40 mL horse serum (Hyclone, SH30074.02), 5 mL 200 m*M* L-glutamine (J.R.H. Biosciences, 59202), 0.5 mL (50 mg/mL) gentamicin sulfate (Cambrex/BioWhittaker, 17-518Z). Make fresh as needed, keep solution sterile and in the dark at room temperature.
3. Incomplete medium: Same as complete medium without the horse serum.
4. Tris-NaCl (500 mL): 20 m*M* Tris-HCl, pH 8.1, 0.5 *M* NaCl (Sigma-Aldrich).
5. Phosphate-buffered saline (PBS) (500 mL; Sigma-Aldrich).
6. Tris-CsCl (2 × 100 mL): to 100 mL 20 m*M* Tris-HCl, pH 8.1, add 58.84 g CsCl (Sigma-Aldrich) to obtain a density of 1.435 (solution refractive index 1.3750), and to another 100 mL 20 m*M* Tris-HCl, pH 8.1, add 45.42 g CsCl to obtain a density of 1.336 (solution refractive index 1.3657).
7. BTP (7 L): 10 m*M* *bis-tris*-propane (pH 7.0) (USB).
8. BTP-NaCl (2 L): 10 m*M* BTP (pH 7.0), 0.1 *M* NaCl (Sigma-Aldrich).
9. Hexon storage buffer (100 mL): 10 m*M* Sodium phosphate (pH 7.0), 0.02% Sodium azide (Sigma-Aldrich).
10. Crystallization buffer (500 mL): 1 *M* Sodium citrate (pH 3.20) (Sigma-Aldrich; *see* **Note 3**).
11. Leupeptin (1 mL): 2 mg leupeptin (Sigma-Aldrich) in water, store at –20°C. Make freshly for each preparation.
12. Pepstatin (1 mL): 1 mg pepstatin (Sigma-Aldrich) in 1 mL (200 proof) ethanol; store at –20°C. Make freshly for each preparation.
13. Phenylmethylsulfonyl fluoride (PMSF) (1 mL): 17 mg PMSF (Sigma-Aldrich) in 1 mL (200 proof) ethanol; store at –20°C. Make freshly for each preparation. **Caution:** Both the powder and solution are highly toxic; gloves are required.

14. Trypan blue stain, 0.4% (100 mL; Invitrogen/Gibco). Store at 25°C.
15. Virus assay buffer: 20 mM Sodium phosphate (pH 7.2) (Sigma-Aldrich), 0.5% Sodium dodecyl sulfate (SDS) (BioRad).
16. Dimethylsulfoxide (Sigma-Aldrich).
17. Freon: 1,1,2-trichloro-1,2,2-trifluoroethane, $Cl_2FCCClF_2$ (Mallinckrodt). Alternatively, chloroform: $CHCl_3$ or hexanes (85% n-Hexane): $CH_3(CH_2)_4CH_3$ can be substituted (Mallinckrodt).
18. High-vacuum grease (Dow Corning).
19. Disinfectant cleaning solution: Amphyl (Spectrum Chemical).

2.3. Equipment

1. Spinner flasks: 2 × 50-, 100-, 250-, and 500-mL, 1- and 3-L (Bellco).
2. Conical-bottom centrifuge bottles: 4 × 750 mL (Bellco).
3. 37°C CO_2 incubator or warm room (*see* **Note 4**); two heavy-duty magnetic stir plates (Bell Stir, Bellco).
4. Laminar flow hood.
5. Hemacytometer with improved Neubauer ruling (Fisher); sterile disposable 1-mL pipets; and microscope.
6. Eppendorf Pipetteman pipets: 1000-, 100-, and 10-µL tips.
7. Disposables: 1- and 5-mL syringes (Beckton Dickinson); 50-mL tubes (Falcon); 0.45-µm syringe filters; Amicon Ultra-15 Centrifugal filter unit, 10,000 nominal molecular-weight limit (NMWL) (Millipore, UFC901008); Amicon Ultra-4 10,000 NMWL (Millipore, UFC801008); 24-well tissue culture plates (Linbro); 22-mm siliconized cover slips (Hampton Research).
8. Refractometer (Fisher).
9. SW28 swinging-bucket and Ti-50 ultracentrifuge rotors; Ultra-Clear and Quick-Seal centrifuge tubes; Quick-Seal tube sealer (Beckman).
10. Dialysis membrane, molecular-weight cutoff 50,000 (Spectra/Por, 132130).
11. Chromatography: fast performance liquid chromatography (FPLC) apparatus; diethylamino ethanol (DEAE)-Sephadex Fast Flow packed medium (100 mL) packed in HR 16/50 column; Mono-Q 10/10 Column (Pharmacia). Alternatively, prepacked HiTrap DEAE Sepharose FF or ANX Sepharose FF columns (GE/ Amersham) can replace the manually packed DEAE-Sephadex column.
12. Electrophoresis: PhastSystem, 7.5% homogeneous SDS–polyacrylamide gel electrophoresis (PAGE) gels, 8/1 applicators, Silver Stain Kit (Pharmacia). Alternatively, NuPAGE 7% Tris-acetate gels, 1.0 mm, 15-well (Invitrogen, cat. no. EA03555BOX) with Tris-acetate SDS running buffer (Invitrogen, cat. no. LA0041), SilverQuest Silver Staining kit (Invitrogen, cat. no. LC6070).

3. Methods

3.1. Preparations for Hexon Purification

3.1.1. Prepare Solutions

The Joklik's solution and complete medium, which are required for HeLa cell culture, should be prepared first. To test the sterility of the Joklik's solution, it is prepared at least 2 wk in advance and then closely examined before use to make sure the solution is still clear. If the solution is cloudy or contains suspended particulates, it is contaminated and must be discarded. The remaining solutions may be prepared after the HeLa cell culture is underway. The density of the Tris-CsCl solutions should be checked by measuring their refractive indices and adding either distilled/deionized water or CsCl to the solutions as necessary to obtain the correct values.

3.1.2. HeLa Cell Culture

A starting culture of S3 HeLa cells must be obtained from either ATCC or another laboratory. The cells are then maintained in a suspension culture as described below (*see* **Notes 4–9**). All work with cells is done in the laminar flow hood to maintain sterile conditions (*see* **Note 6**). If the cell culture becomes contaminated, it must be discarded and another culture started.

1. Thoroughly clean and autoclave the spinner flasks, letting them cool to room temperature.
2. If a starter culture is available in suspension, skip **steps 3–6**.
3. Warm a 500-mL bottle of complete medium in a 37°C water bath.
4. Thaw the frozen cells in a 37°C water bath for 1 min.
5. Within the laminar flow hood, transfer cells to a sterile 15-mL tube and add warm complete medium dropwise over a period of 10 min; slowly bring the volume up to 15 mL.
6. Sediment the cells at $1000g$ for 10 min, decant off medium, add fresh complete medium, and resuspend cells by gently and repeatedly drawing the solution into a sterile 25-mL pipet and releasing. Sufficient volume is needed to cover the stirbar in the spinner flask (e.g., 25-mL culture for a 50-mL spinner flask or 40 mL for a 100-mL spinner flask) to keep the cells growing well.
7. Remove a 1.0-mL fraction from the culture and count the number of cells (*see* **Notes 7–9**). Adjust the cell concentration to $2–5 \times 10^5$ cells/mL; cells can be sedimented at $1000g$ for 10 min and excess medium decanted if necessary.
8. Log-phase HeLa cells have a doubling time of 24 h. The suspension culture must be split regularly to maintain a cell concentration of $2–10 \times 10^5$ cells/mL. Cells are split every 2–3 d by discarding a portion of the culture and replacing the discarded volume with complete medium.
9. It is easier and less expensive to maintain a small volume of cells in suspension and to scale up only when necessary.

10. To scale up the cell culture volume, do not split the culture but increase its volume with fresh complete medium to bring the cell concentration to $2–5 \times 10^5$ cells/mL.
11. Begin to scale up the volume of the suspension culture from approx 100 mL to 2.5 L 5–6 d before the cells are to be infected.
12. For long-term storage, the cells can be concentrated to 10^7 cells/mL, frozen overnight at –70°C in a mixture of 50% Joklik's solution, 40% horse serum, and 10% dimethylsulfoxide, and then stored under liquid nitrogen.

3.1.3. Assay of Virus Stock

The titer of the virus stock is conveniently estimated with a spectrophotometric assay, which is based on the absorbance of light at a wavelength of 260 nm (A_{260nm}). Using a viral plaque assay, Challberg and Ketner *(44)* determined the appropriate conversion factor (3.5×10^9 PFU/mL). To estimate the virus titer, add 10 μL purified virus stock to 990 μL virus assay buffer in a quartz cuvet and measure the absorbance.

$$\text{Virus titer (PFU/mL)} = A_{260nm} \times 100 \text{ (dilution factor)} \times 3.5 \times 10^9 \text{ PFU/mL}$$

3.1.4. Chromatography

Column chromatography is conveniently performed using a Pharmacia FPLC within a chromatography refrigerator. The chromatography solvents, columns, and samples are kept at 4°C. The DEAE column must first be packed (*see* **Note 10**). Then, both the DEAE and Mono-Q columns must be equilibrated with column buffer by flushing with five column volumes of 10 m*M* BTP (pH 7.0) before use. If an automated system like the FPLC is to be used, methods for sample loading, column flushing, and elution should be programmed before starting (*see* **Note 11**). These methods can be quickly modified (e.g., based on sample volume) as necessary just before use.

3.2. Infection of Cells With Ad

Before infecting the cells with Ad, the cells must be removed from the complete medium containing horse serum, washed, and resuspended into incomplete medium (**steps 1–9**, below). This prevents an interaction of serum proteins with the virus. After the cells have been infected during the incubation step, they are returned to complete medium (**steps 10–15**, below).

1. Place the 3.0-L spinner flask containing the 2.5-L HeLa cell culture into the laminar flow hood. Remove 1 mL culture. Count the number of cells (*see* **Note 9**).
2. Pour the 2.5-L culture into the four sterile glass 750-mL centrifuge bottles; close 3-L stir flask to maintain sterile conditions.
3. Balance the solution levels in the centrifuge bottles using a sterile 25-mL pipet.

4. Sediment the cells at 1000*g* for 10 min.
5. Calculate the amount of virus needed to infect at 10 PFU per cell.
6. Decant the supernatant medium back into the 3-L spinner flask under sterile conditions (*see* **Note 12**).
7. Wash cell pellets with incomplete medium: add 25 mL incomplete medium, mix cells by swirling, and gently draw the cell suspension up and down twice with a 25-mL pipet.
8. Rinse the sterile 250-mL spinner flask with incomplete medium.
9. Pipet cells into the rinsed 250-mL spinner flask, rinse centrifuge bottles with 25 mL incomplete medium, add rinse to spinner flask, and adjust the final volume to 250 mL with incomplete medium.
10. Remove virus stock from freezer; place in laminar flow hood.
11. Use 1000-µL Pipetteman sterilized with ethanol to add virus into the 250-mL cell culture.
12. Incubate 250-mL spinner flask at 37°C for 1 h.
13. Place virus-contaminated vials, tips, etc., into a beaker containing disinfectant.
14. After 1-h incubation, pour infected cells back into the 3-L spinner flask containing the complete medium from **step 6**.
15. Incubate the cells in a 3-L spinner culture for 40 h at 37°C.

3.3. Harvest of Cells, Virus, and Proteins

All further steps in the purification protocol are carried out at 4°C. The cells are harvested, washed, and resuspended in a basic buffer solution before they are disrupted (**steps 1–15**, below). The basic pH of the buffer and the addition of inhibitors reduce the activity of cellular proteases. Lipids are extracted from the soluble fraction (**steps 16–24**, below), and then the soluble antigens (including hexon) are separated from the virus particles on a CsCl cushion (**steps 25–32**, below).

1. Isolate infected cells by centrifugation as described above (**Subheading 3.2., steps 2–4**).
2. Place contaminated glass pipets into 250-mL cylinder containing disinfectant.
3. Decant medium into a large spinner flask, leaving enough medium to cover 80% of the conical bottom of the four 750-mL centrifuge bottles.
4. Mix the cells and medium remaining in the bottles by slowly swirling the solution. Gently draw the cells up and down with a 25-mL pipet to fully resuspend the cells without breaking them.
5. Transfer the cell suspension to 50-mL Falcon tubes on ice (three tubes with approx 30 mL/tube).
6. Rinse bottles sequentially with 20 mL medium from the 3-L spinner flask, and divide the 20-mL rinse into the three Falcon tubes.
7. Sediment the cells at 1000*g* for 10 min.
8. Aspirate the supernatant into a flask containing disinfectant.

9. Resuspend the cells into cold PBS with a 25-mL pipet. Use 10 mL PBS per 1 L of initial culture volume (e.g., 25 mL PBS for 2.5-L culture).
10. Sediment the cells at 1000g for 10 min at 4°C.
11. Aspirate the supernatant into a flask containing disinfectant.
12. Resuspend the cells into cold 20 mM Tris-HCl + 0.5 M NaCl with a 25-mL pipet. Use 10 mL Tris-NaCl/L of initial culture volume.
13. Add protease inhibitors: 50 µL leupeptin, 100 µL pepstatin, and 100 µL PMSF per 100 mL of cell suspension.
14. Lyse cells by repeating five freeze–thaw cycles:
    a. Freeze cell suspension in an ethanol dry-ice bath for 45 min.
    b. Thaw cells in 1000-mL beaker of warm water; mix the cell suspension by swirling the Falcon tubes in the warm water bath.
    c. Forcibly pipet the cell suspension in each tube up and down 20 times with a 25-mL pipet for each tube to help break up cells and debris. If the color of the solution begins to turn brown, add more Tris-NaCl buffer to each tube to keep the solution basic. After freezing cells for the fifth time, the solution can be stored frozen at –20°C overnight.
15. Sediment cell debris.
    a. Finish the fifth freeze–thaw cycle.
    b. Pellet the DNA by centrifugation at 3000g for 10 min.
16. Decant supernatant into three new Falcon tubes; store on ice.
17. Resuspend DNA pellets in 10 mL Tris-NaCl buffer.
18. Pellet DNA by centrifugation at 3000g for 10 min.
19. Combine supernatants, mix them by pouring solution from tube to tube, and divide solution into four 50-mL tubes (approx 15 mL per tube).
20. Add 15 mL Freon (*see* **Note 13**) to each tube, cap tubes tightly, and seal them with Parafilm.
21. Shake tubes vigorously by hand for 5 min each.
22. Separate the organic from the aqueous phase by centrifugation at 2000g for 10 min.
23. Pipet off the top layer (aqueous phase) and place it in two new Falcon tubes.
24. Re-extract the organic phase with an additional 2 mL Tris-NaCl buffer, and combine with the aqueous layer from the previous extraction.
25. Cool the ultracentrifuge and the SW-28 rotor to 5°C.
26. Place 5-mL cushions of 1.43 g/mL Tris-CsCl solution into two ultracentrifuge tubes.
27. Layer the virus solution onto the CsCl cushion carefully to avoid mixing the layers. Fill the tubes completely to the top, adding more Tris-NaCl buffer if required. The total virus solution volume must be low enough to fill only two tubes; otherwise, the concentration of the virus will be too low for it to be visualized as a band during the final CsCl density gradient step.
28. Sediment the virus solution at 136,000g for 1 h at 5°C.
29. Check that the refractive index of the 1.34 g/mL Tris-CsCl solution is 1.3660, adjusting the solution density with water or CsCl as necessary.
30. Carefully place the centrifuge tubes onto a rack in the laminar flow hood, taking care not to mix the layers.

31. Transfer the clear top layers containing the soluble proteins into clean Falcon tubes on ice.
32. Place the centrifuge tubes with the bottom layers containing the virus on ice.

3.4. Purification of Virus

The virus is purified by CsCl density gradient centrifugation (**steps 1–6,** below), then isolated, pooled, and stored (**steps 7–13,** below).

1. Bring virus solution on ice to the refractometer. Check the refractive index of the virus solution and, if necessary, adjust it to 1.3660 by adding a few drops of the 1.43 g/mL CsCl stock solution to increase, or water to decrease, the value.
2. Use a Pasteur pipet to fill Quick-Seal tubes with virus solution. Tubes must be filled completely to their top line. Fill the first tube completely before starting the second tube. Use the 1.34 mg/mL CsCl solution to ensure that the last tube is completely full.
3. Weigh the tubes, and balance to within ±0.05 g.
4. Seal tubes with tube sealer following the manufacturer's instructions, add tube caps, and place into Ti-50 rotor.
5. Spin the samples at 136,000g for a minimum of 18 h at 5°C to form the CsCl density gradient.
6. Dialyze soluble protein antigen solution (top layer from CsCl cushion; **Subheading 3.3., step 31**) overnight into 3–4 L 10 m*M* BTP (pH 7.0) using 50,000 molecular-weight cutoff membrane. The Spectra/Por dialysis tubing has a volume/cm of approx 3.7 mL/cm.
7. After the gradient has formed, allow rotor to stop with the BRAKE OFF (approx 30 min).
8. Carefully place tubes containing virus in a rack in the laminar flow hood; **do not disturb** the gradient.
9. Two translucent white bands containing virus will be visible in each tube if it is illuminated from the side in the darkened hood.
10. Remove the lower band, which contains the intact virus particles, as follows:
    a. Make a hole in the sealed tube anywhere at the top with a 28-gage needle.
    b. Place a small amount of high-vacuum grease on the side of the tube just below the lower virus band.
    c. Puncture the tube through the grease with an 18-gage needle attached to a 3-mL syringe with the needle bevel facing up (toward the virus band).
    d. Raise the tip of the 18-gage needle into the virus band by gently lowering the barrel of the syringe, and carefully draw out the virus band without disturbing the gradient.
11. Pool the virus-containing fractions.
12. Add glycerol to a final concentration of 20% (v/v).
13. Mix solution, place 100-µL aliquots into labeled cryotubes, and store them in a –20°C freezer.

3.5. Purification of Hexon

The purification consists of three separate column chromatography steps. The first step uses a DEAE-Sepharose fast-flow anion-exchange column that separates the hexon from crude cell lysate (*see* **Note 10**). The hexon is then further purified by anion-exchange chromatography with Mono-Q columns (*see* **Note 14**).

3.5.1. DEAE-Sepharose Fast-Flow Chromatography

Figure 1 depicts typical results for the purification of Ad hexon by DEAE-Sepharose fast-flow chromatography using a salt gradient. The protocol is as follows:

1. Remove sample from dialysis tubing.
2. Sediment sample at 5000*g* for 20 min at 4°C (*see* **Note 15**).
3. Decant supernatant containing hexon into a clean Falcon tube; place sample in chromatography refrigerator.
4. Dilute the sample 1:1 with column buffer to reduce the sample viscosity.
5. Prepare solvents:
   Solvent A: 10 m*M* BTP (pH 7.0).
   Solvent B: 10 m*M* BTP (pH 7.0) + 1.0 *M* NaCl.
6. If an automated system like the FPLC is to be used, warm up the absorbance and conductivity detectors, then program a method to equilibrate the DEAE-Sepharose fast-flow column (30 × 1.6 cm, 60 mL) with solvent A at 1.0 mL/min until the specific conductance has stabilized.
7. Load the diluted sample at 1.0 mL/min.
8. Step the gradient to 10% B and flush the column with two column volumes (120 mL) at 1.0 mL/min. Save the solvent flow-through in case the column fails to bind hexon.
9. After proteins that do not bind to the column have eluted, begin collecting fractions of 4 mL/tube, and start a gradient from 10 to 100% B over 400 mL.
10. Step the gradient back to 0% B and wash for five column volumes to re-equilibrate the column with solvent A.

3.5.2. Assay Fractions by SDS-PAGE

To determine which chromatography fractions contain hexon, we routinely analyze them by SDS-PAGE using the Pharmacia PhastSystem because of its speed and convenience (**Fig. 1**). The samples are run on 7.5% homogeneous PhastGels and are stained using the Pharmacia silver-stain kit according to the manufacturer's instructions. A convenient alternative to the PhastSystem is to use minigels, such as the NuPAGE 7% Tris-Acetate gels and silver staining kits from Invitrogen.

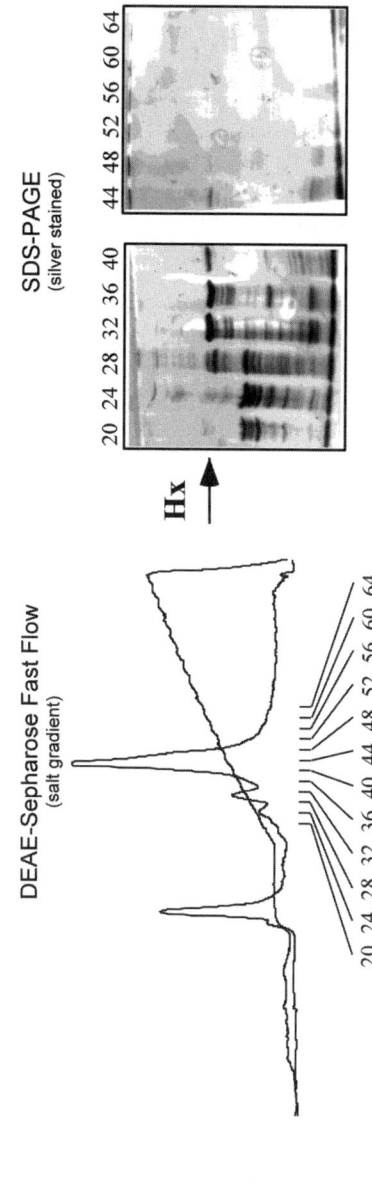

Fig. 1. First column: Purification of adenovirus hexon by DEAE-Sepharose fast flow-chromatography and analysis by sodium dodecyl sulfate–polyacrylamide gel electrophoresis (SDS-PAGE). The chromatogram is a tracing of A_{280nm}, and the gradient is a tracing of specific conductivity. Fractions spanning the hexon peak are labeled and are analyzed by silver-stained SDS-PAGE. The position on the gels that corresponds to the hexon band is labeled Hx.

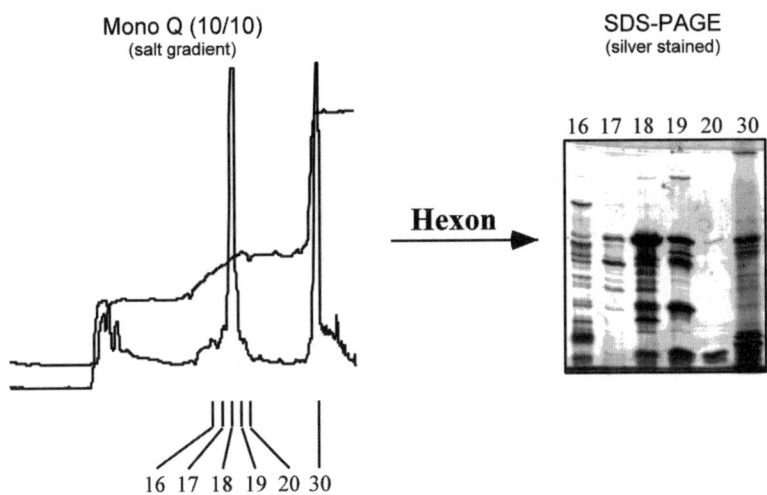

Fig. 2. Second column: Purification of adenovirus hexon by Mono-Q (10/10) chromatography and analysis by sodium dodecyl sulfate–polyacrylamide gel electrophoresis. Labeled corresponding to **Fig. 1**.

3.5.3. Mono-Q HR Chromatography

Figure 2 depicts typical results for the purification of Ad hexon on a Mono-Q HR 10/10 column. The protocol is as follows:

1. Either pool and dialyze the fractions containing hexon, as described above (**Subheading 3.4., step 6**), or concentrate the protein and exchange its buffer as described below:
   a. Place up to 15 mL of the fractions into an Amicon Ultra-15 Centrifugal Filter Unit; add the fractions containing the least amount of hexon first.
   b. Spin at 3000g for 30 min at 4°C.
   c. Collect the filtrate, add more fractions; repeat the concentration and sample addition cycle until the sample volume is less than 15 mL.
2. Dilute the sample 1:1 with 10 mM BTP (pH 7.0) to reduce its ionic strength and viscosity.
3. Prepare solvents:
   Solvent A: 10 mM BTP (pH 7.0).
   Solvent B: 10 mM BTP (pH 7.0) + 1.0 M NaCl.
4. If an automated system like the FPLC is to be used, program a method to equilibrate the Mono-Q column (10 × 1.0 cm, 8 mL) with solvent A at 2.0 mL/min until the specific conductance has stabilized.
5. Load the diluted sample at 2.0 mL/min.
6. Flush the column with 30 mL solvent A at 2.0 mL/min. Save the solvent flowthrough in case the column fails to bind hexon.

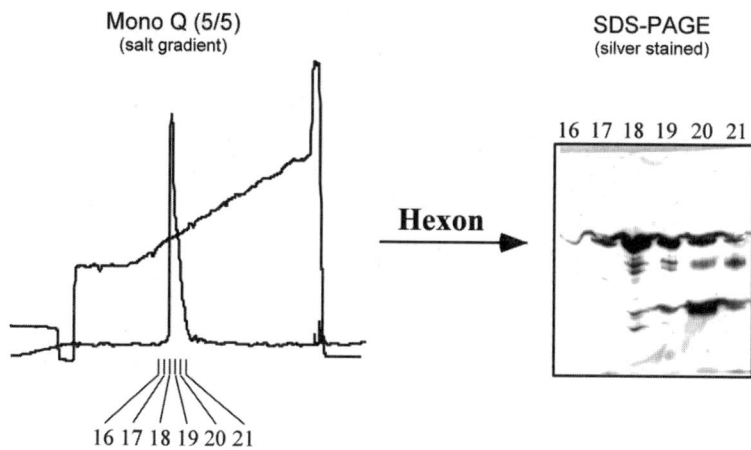

Fig. 3. Third column: Purification of adenovirus hexon by Mono-Q (5/5) chromatography and analysis by sodium dodecyl sulfate–polyacrylamide gel electrophoresis. Labeled corresponding to **Fig. 1**.

7. Step the gradient to 25% B, and flush the column with 30 mL at 2.0 mL/min.
8. Begin to collect fractions of 4 mL/tube, and start a gradient from 25% to 40% B over 50 mL.
9. Hold the gradient at 40% B for 12 min (24 mL), step to 100% B, and flush column with 8 mL.
10. Step the gradient back to 0% B, and wash with five column volumes to re-equilibrate the column with solvent A.
11. Assay the fractions by SDS-PAGE as before (**Subheading 3.5.2.**).

After anion-exchange chromatography with the Mono-Q (10/10) column, the hexon-containing fractions may still be contaminated with significant amounts of other proteins, as judged by SDS-PAGE analysis. If further purification is required, the Mono-Q chromatography step can be repeated with a smaller Mono-Q column as shown in **Fig. 3** (*see* **Note 16**).

3.5.4. Concentration of Hexon and Buffer Exchange

1. In preparation for crystallization trials, the purified protein must be concentrated and exchanged into the hexon-storage buffer. Concentrate the hexon fractions containing the least amount of hexon first. Add the hexon fractions to Amicon Ultra-15 10,000 NMWL Centrifugal Filter Units, with up to 15 mL per concentrator.
2. Spin at 3500*g* for 30 min at 4°C.
3. Collect and save the filtrate; add more sample to concentrators; repeat the concentration and sample addition cycle until the total volume is small enough to fit into a single Ultra-4 Centrifugal Filter Unit (4 mL).

Fig. 4. Adenovirus type 5 hexon crystal (approx 0.5 mm) grown in a 10-µL hanging drop containing 5 µL (10.6 mg/mL) hexon and 5 µL of the 1 mL (1.0 *M* sodium citrate pH 3.2) reservoir solution.

4. Combine the concentrated samples into a single concentrator, carefully rinse the previously used concentrators with a small amount of storage buffer, combine the rinse with the concentrated hexon sample, and continue to concentrate.
5. Dilute the concentrated sample with storage buffer and concentrate again; repeat this process until the sample has been exchanged 99% or more into the hexon-storage buffer.

The final hexon concentration should be 10 mg/mL, which can be checked spectrophotometrically. A 1.0 mg/mL concentration in hexon storage-buffer has an absorbance of 1.43 at 280 nm with a 1-cm path length. Dilute 10 µL concentrated hexon solution to 1000 µL and determine the absorbance.

Hexon concentration (mg/mL) = $A_{280nm} \times 100$ (dilution factor) $\times$ 1.43 mg /mL

3.6. Crystallization of Hexon

After concentration and buffer exchange, the freshly purified hexon is used immediately for crystallization trials. Hexon is crystallized (**Fig. 4**) using the vapor-diffusion hanging drop method *(45)* (*see* **Note 17**). A drop of protein solution is suspended over a reservoir of precipitation solution and allowed to equilibrate. The gradual increase in precipitant concentration within the drop under the appropriate conditions causes crystal formation. The steps for hexon crystallization are as follows:

1. Heat a small amount of high-vacuum grease in a 25-mL beaker on a hot plate until melted.
2. Place 1.0 mL of 1.0 *M* sodium citrate (pH 3.20) in the well of a 24-well tissue-culture tray (*see* **Note 3**).
3. Place a thin layer of melted high-vacuum grease around the rim of the well.
4. Place a 5-μL drop of 10 mg/mL hexon solution on a 22-mm-diameter siliconized cover slip.
5. Use a 20-μL Pipetteman to mix 5 μL precipitation buffer from the reservoir with the 5-μL hexon drop. Take care not to produce bubbles when mixing the solutions.
6. Place the cover slip over the well with forceps; gently push down on the rim of the cover slip to ensure a complete seal.
7. Repeat **steps 2–6** for each crystallization trial.
8. Store the crystallization tray at 25°C where it will not be exposed to large temperature fluctuations, vibrations, or other disturbances that could upset the equilibrium of the drops (*see* **Note 18**).

4. Notes

1. Time table (the procedure should be performed as quickly as possible):
   Preparation: **Subheading 3.1.**
   Day 1: **Subheading 3.2.**
   Day 2: (free)
   Day 3: **Subheading 3.3., steps 1–14.**
   Day 4: **Subheading 3.3., step 15 to Subheading 3.4., step 6.**
   Day 5: **Subheading 3.4., step 7 to Subheading 3.5.1.**
   Day 6: **Subheading 3.5.2. to Subheading 3.5.3., step 2.**
   Day 7: **Subheading 3.5.3., steps 3–11.**
2. Although not always feasible, it is best to use freshly prepared medium for cell culture (*46*). A key indicator of expired medium is that the doubling time for the cells will increase from 24 h to 2 or more days.
3. Hexon crystallization is highly sensitive to solution pH. It can be difficult to consistently prepare the crystallization buffer with precisely the correct pH for crystal growth. To ensure that a suitable crystallization solution is available, prepare a series of 50-mL 1 *M* sodium citrate solutions that vary in pH between 3.15 and 3.25 in 0.02 pH unit increments. Set up two to three crystallization trials with each solution to determine which yields the best crystals; then use this subsequently.
4. The HeLa cell suspension cultures grow well in a 37°C incubator with a 5% CO_2 atmosphere; however, the culture can also be maintained in an incubator or warm room without the added CO_2.

5. The spinner flask must be placed on a heavy-duty magnetic stirrer that does not get hot with prolonged use. The Bellco magnetic stirrer is set to position 3, which is fast enough to maintain the cells in suspension yet not so fast as to damage the cells. One magnetic stirrer is placed in the incubator (or warm room) and the other is placed in the laminar flow hood.

6. Maintaining sterile conditions while handling the cells and solutions is of particular importance. Take care that sterile pipets do not become contaminated by touching the sides of flasks or the hood. Sterilize the hood with disinfectant before use, and disinfect each item before placing it in the hood. Do not forget to wipe your gloves with disinfectant and to change them frequently.

7. Starting a suspension culture from frozen HeLa cells can be difficult. Often the frozen culture will contain many dead cells. To help distinguish live from dead cells, add an equal volume of Trypan blue stain to the cell fraction before counting the cells. The dead cells take up the blue dye, while the live cells remain pink. Over time, the live cells will divide and eventually outnumber the dead ones. After splitting the culture several times, very few dead cells will remain.

8. Cells often grow slowly immediately after being transferred into a new container. Ensure that the cells are growing well before infecting them with virus.

9. Cells rapidly settle to the bottom of the spinner flask when not stirred. Take care that cells are well suspended before removing a fraction for counting, and mix the fraction before examining cells under the microscope.

10. The crude cell lysate can contaminate FPLC columns resulting in increased backpressure and decreased efficiency. The DEAE column is used before the Mono-Q columns to act as a prefilter because the material is inexpensive and the DEAE column can be repacked with fresh material as needed. Use of the DEAE column extends the lifetime of the Mono-Q columns and so can save both time and money. Alternatively, prepacked disposable columns such as the HiTrap DEAE Sepharose FF or ANX Sepharose FF columns, which conveniently attach to the FPLC, may be used. Sample solutions must be filtered (see **Note 15**).

11. If using an automated system such as the Pharmacia FPLC, care must be taken to monitor the solvent backpressure. The pressure limits on the pumps should be set to 2 MPa for the DEAE-Sepharose column and 4 MPa for the Mono-Q columns to avoid column-bed compression, which could damage the medium. If the solvent pressure exceeds these values during a run, decrease the flow rate and continue.

12. To avoid contamination when decanting the culture medium, do not let the two flasks touch each other. Also, the cell pellet tends to slide out of the flask when pouring out the last portion of the solution. Remove the remaining medium from cells with a sterile Pasteur pipet by aspiration.

13. Because of the limited availability of Freon, substitution with another organic solvent such as chloroform, $CHCl_3$, or hexanes (85% n-hexane), $CH_3(CH_2)_4CH_3$, may be necessary.

14. Anion-exchange chromatography often yields chromatograms with a trailing shoulder in the hexon peak (**Fig. 1**). Blanche et al. *(47)* showed that this shift in retention time is a result of the formation of isoaspartate residues by deamidation

of surface-exposed asparagines. To optimize the stability of Ad vectors for clinical use, appropriate modifications should be made to accessible Asn residues engaged in Asn-Gly pairs *(12,47)*.

15. Samples to be loaded onto the FPLC must not contain particulates that could clog its valves, lines, and columns. Particulates may be removed either by filtration with a 0.45-μm membrane or by sedimentation. Crude hexon solutions can be difficult to filter, so sedimentation is the preferred method.

16. The protocol for the Mono-Q (5/5) column is analogous to that described for the Mono-Q (10/10) column. The following modifications should be made. Use 1 mL/min flow rate throughout. Change gradient profile after loading sample to: step gradient to 25% B; flush with 40 mL; start gradient from 25% B to 62.5% B over 125 mL; step gradient to 100% B and flush column with 8 mL; step gradient back to 0% B, and wash with five column volumes to re-equilibrate the column with solvent A.

17. Hexon crystals have also been grown in capillary tubes *(8)* and from bulk solution as a final purification step *(41)*.

18. Crystal growth should be apparent within hours of setup time. The crystals reach 0.3–0.5 mm maximum diameter in 3–5 d and grow no further after 2 wk. For optimal results, avoid handling the crystallization trays for the first 3–5 d after they have been set up.

Acknowledgments

We thank the many pioneers who enabled our work by developing Ad protocols over the years. In our own laboratory, Drs. Markus Grütter, Janice L. White, Jan van Oostrum, Donatella Pascolini, Paula R. Kuser, and Susan L. Pichla have all made contributions. Dr. Robert Ricciardi, Dental School, and the Vector Core Facility of the Gene Therapy Program at the University of Pennsylvania have been generous in providing us with Ad. We are also indebted to Dr. Louise Showe for helpful discussions regarding HeLa cell culture and to Dr. Ann H. Rux for her critical evaluation of the protocol and recommendations for current resources. This work has been supported by the National Institute of Allergy and Infectious Diseases (AI-17270), the National Science Foundation (MCB-0094577), the Wistar Cancer Center (CA 10815), and the Pennsylvania Department of Health.

References

1. Rowe, W. P., Huebner, R. J., Gillmore, L. K., Parrott, R. H., and Ward, T. G. (1953) Isolation of a cytopathogenic agent from human adenoids undergoing spontaneous degeneration in tissue culture. *Proc. Soc. Exp. Biol. Med.* **84,** 570–573.

2. Horne, R. W., Brenner, S., Waterson, A. P., and Wildy, P. (1959) The icosahedral form of an adenovirus. *J. Mol. Biol.* **1,** 84–86.

3. Rux, J. J. and Burnett, R. M. (2004) Adenovirus structure. *Hum. Gene Ther.* **15,** 1167–1176.

4. Stewart, P. L., Burnett, R. M., Cyrklaff, M., and Fuller, S. D. (1991) Image reconstruction reveals the complex molecular organization of adenovirus. *Cell* **67**, 145–154.

5. van Oostrum, J. and Burnett, R. M. (1985) Molecular composition of the adenovirus type 2 virion. *J. Virol.* **56**, 439–448.

6. White, D. O., Scharff, M. D., and Maizel, J. V., Jr. (1969) The polypeptides of adenovirus. III. Synthesis in infected cells. *Virology* **38**, 395–406.

7. Pereira, H. G., Valentine, R. C., and Russell, W. C. (1968) Crystallization of an adenovirus protein (the hexon). *Nature* **219**, 946–947.

8. Burnett, R. M., Grütter, M. G., and White, J. L. (1985) The structure of the adenovirus capsid. I. An envelope model of hexon at 6 Å resolution. *J. Mol. Biol.* **185**, 105–123.

9. Roberts, M. M., White, J. L., Grütter, M. G., and Burnett, R. M. (1986) Three-dimensional structure of the adenovirus major coat protein hexon. *Science* **232**, 1148–1151.

10. Athappilly, F. K., Murali, R., Rux, J. J., Cai, Z., and Burnett, R. M. (1994) The refined crystal structure of hexon, the major coat protein of adenovirus type 2, at 2.9 Å resolution. *J. Mol. Biol.* **242**, 430–455.

11. Rux, J. J. and Burnett, R. M. (2000) Type-specific epitope locations revealed by X-ray crystallographic study of adenovirus type 5 hexon. *Mol. Therapy* **1**, 18–30.

12. Rux, J. J., Kuser, P. R., and Burnett, R. M. (2003) Structural and phylogenetic analysis of adenovirus hexons by use of high-resolution X-ray crystallographic, molecular modeling, and sequence-based methods. *J. Virol.* **77**, 9553–9566.

13. Stewart, P. L., Fuller, S. D., and Burnett, R. M. (1993) Difference imaging of adenovirus: Bridging the resolution gap between X-ray crystallography and electron microscopy. *EMBO J.* **12**, 2589–2599.

14. Zubieta, C., Schoehn, G., Chroboczek, J., and Cusack, S. (2005) The structure of the human adenovirus 2 penton. *Mol. Cell* **17**, 121–135.

15. Fabry, C. M. S., Rosa-Calatrava, M., Conway, J. F., et al. (2005) A quasi-atomic model of human adenovirus type 5 capsid. *EMBO J.* **24**, 1645–1654.

16. Rux, J. J., Pascolini, D., and Burnett, R. M. (1999) Large scale purification and crystallization of adenovirus hexon, in *Adenovirus Methods and Protocols* (Methods in Molecular Medicine, vol. 21) (Wold, W. S. M., ed.), Humana Press, Totowa, NJ, pp. 259–275.

17. Furcinitti, P. S., van Oostrum, J., and Burnett, R. M. (1989) Adenovirus polypeptide IX revealed as capsid cement by difference images from electron microscopy and crystallography. *EMBO J.* **8**, 3563–3570.

18. Saban, S. D., Nepomuceno, R. R., Gritton, L. D., Nemerow, G. R., and Stewart, P. L. (2005) CryoEM structure at 9Å resolution of an adenovirus vector targeted to hematopoietic cells. *J. Mol. Biol.* **343**, 526–537.

19. Kozarsky, K. F. and Wilson, J. M. (1993) Gene therapy: Adenovirus vectors. *Curr. Opin. Genet. Dev.* **3**, 499–503.

20. Seth, P. (ed.) (1999) *Adenoviruses: From Basic Research to Gene Therapy Applications*, R.G. Landes, Austin, TX.

21. Doerfler, W. and Böhm, P. (eds.) (2003) *Adenoviruses: Model and Vectors in Virus Host Interactions* (Current Topics in Microbiology and Immunology, vols. 272 & 273), Springer-Verlag, Berlin.

22. Everts, M. and Curiel, D. T. (2004) Transductional targeting of adenoviral cancer gene therapy. *Curr. Gene Ther.* **4**, 337–346.

23. Tatsis, N. and Ertl, H. C. J. (2004) Adenoviruses as vaccine vectors. *Mol. Ther.* **10**, 616–629.

24. Franklin, R. M., Petterson, U., Åkervall, K., Strandberg, B., and Philipson, L. (1971) Structural proteins of adenovirus. V. Size and structure of the adenovirus type 2 hexon. *J. Mol. Biol.* **57**, 383–395.

25. Cornick, G., Sigler, P. B., and Ginsberg, H. S. (1971) Characterization of crystals of type 5 adenovirus hexon. *J. Mol. Biol.* **57**, 397–401.

26. Döhner, L. and Hudemann, H. (1972) Untersuchungen zur Kristallisation von Hexonen der Adenoviren der Gruppe III. *Arch. Gesamte Virusforsch.* **38**, 279–289.

27. Xue, F. and Burnett, R. M. (2006) Capsid-like arrays in crystals of chimpanzee adenovirus hexon. *J. Struct. Biol.* **154**, 217–221.

28. Xu, L., Benson, S. D., and Burnett, R. M. (2006) Nanoporous crystals of Chicken Embryo Lethal Orphan (CELO) adenovirus major coat protein, hexon. *J. Struct. Biol.* doi: 10.1016/j.jsb.2006.08.017.

29. Devaux, C., Adrian, M., Berthet-Colominas, C., Cusack, S., and Jacrot, B. (1990) Structure of adenovirus fibre. I. Analysis of crystals of fibre from adenovirus serotypes 2 and 5 by electron microscopy and X-ray crystallography. *J. Mol. Biol.* **215**, 567–588.

30. van Raaij, M. J., Mitraki, A., Lavigne, G., and Cusack, S. (1999) A triple beta-spiral in the adenovirus fibre shaft reveals a new structural motif for a fibrous protein. *Nature* **401**, 935–938.

31. Xia, D., Henry, L. J., Gerard, R. D., and Deisenhofer, J. (1994) Crystal structure of the receptor-binding domain of adenovirus type 5 fiber protein at 1.7 Å resolution. *Structure* **2**, 1259–1270.

32. van Raaij, M. J., Louis, N., Chroboczek, J., and Cusack, S. (1999) Structure of the human adenovirus serotype 2 fiber head domain at 1.5 A resolution. *Virology* **262**, 333–343.

33. Bewley, M. C., Springer, K., Zhang, Y. B., Freimuth, P., and Flanagan, J. M. (1999) Structural analysis of the mechanism of adenovirus binding to its human cellular receptor, CAR. *Science* **286**, 1579–1583.

34. Durmort, C., Stehlin, C., Schoehn, G., et al. (2001) Structure of the fiber head of Ad3, a non-CAR-binding serotype of adenovirus. *Virology* **285**, 302–312.

35. Schoehn, G., Fender, P., Chroboczek, J., and Hewat, E. A. (1996) Adenovirus 3 penton dodecahedron exhibits structural changes of the base on fibre binding. *EMBO J.* **15**, 6841–6846.

36. Cepko, C. L. and Sharp, P. A. (1982) Assembly of adenovirus major capsid protein is mediated by a nonvirion protein. *Cell* **31**, 407–415.

37. Molinier-Frenkel, V., Lengagne, R., Gaden, F., et al. (2002) Adenovirus hexon protein is a potent adjuvant for activation of a cellular immune response. *J. Virol.* **76**, 127–135.

38. Pettersson, U., Philipson, L., and Höglund, S. (1967) Structural proteins of adenoviruses. I. Purification and characterization of the adenovirus type 2 hexon antigen. *Virology* **33,** 575–590.
39. Dowdle, W. R., Lambriex, M., and Hierholzer, J. C. (1971) Production and evaluation of a purified adenovirus group-specific (hexon) antigen for use in the diagnostic complement fixation test. *Appl. Microbiol.* **21,** 718–722.
40. Boulanger, P. A. and Puvion, F. (1973) Large-scale preparation of soluble adenovirus hexon, penton and fiber antigens in highly purified form. *Eur. J. Biochem.* **39,** 37–42.
41. Grütter, M. and Franklin, R. M. (1974) Studies on the molecular weight of the adenovirus type 2 hexon and its subunit. *J. Mol. Biol.* **89,** 163–178.
42. Jörnvall, H., Pettersson, U., and Philipson, L. (1974) Structural studies of adenovirus type-2 hexon protein. *Eur. J. Biochem.* **48,** 179–192.
43. Siegel, S. A., Hutchins, J. E., and Witt, D. J. (1987) Purification of adenovirus hexon by high performance liquid chromatography. *J. Virol. Meth.* **17,** 211–217.
44. Challberg, S. S. and Ketner, G. (1981) Deletion mutants of adenovirus 2: Isolation and initial characterization of virus carrying mutations near the right end of the viral genome. *Virology* **114,** 196–209.
45. McPherson, A. (1982) *Preparation and Analysis of Protein Crystals,* Wiley, New York.
46. Owens, J., Walthall, B., and Murphy, T. (1993) The effects of freshly prepared cell culture medium vs stored cell culture medium on high-density cell culture. *Am. Biotech. Lab.* **11,** 64–66.
47. Blanche, F., Cameron, B., Somarriba, S., Maton, L., Barbot, A., and Guillemin, T. (2001) Stabilization of recombinant adenovirus: Site-directed mutagenesis of key asparagine residues in the hexon protein. *Anal. Biochem.* **297,** 1–9.

18

Synthesis and Assay of Recombinant Adenovirus Protease

Joseph M. Weber

Summary

All adenoviruses (Ads) sequenced so far encode a single endopeptidase of the cysteine class, named adenain. The Ad2 adenain is a 204-residue, nearly inactive monomer, which is activated during virus maturation by an 11-residue cleavage fragment of capsid protein pVI. This chapter describes the synthesis, purification, activation, and assay of recombinant human Ad type adenain.

Key Words: Adenain; adenovirus endopeptidase; cysteine protease.

1. Introduction

Adenoviruses (Ads) encode a cysteine endopeptidase, adenain, synthesized late in virus infection, which is essential for virion maturation and infectivity *(1,2)*. The enzyme is encapsidated (approx 20 molecules per virion) and may also have a role during decapsidation *(3–5)*. Although there are approx 100 Ad serotypes known to infect vertebrates, so far only the human Ad type 2 (Ad2) enzyme has been studied. The recombinant protein expressed in *Escherichia coli* and insect cells has been purified and characterized *(3,6–8)*. The enzyme is a 204-residue monomer of 24,838 Daltons with a pI of 10.59 and optimal activity at 45°C and pH 8.0. The recombinant enzyme is stimulated by an 11-amino-acid C-terminal cleavage fragment from viral capsid protein pre-VI: GVQSLKRRRCF. This peptide, termed pVI-CT, regulates enzyme activity in vivo during virus infection. It is bound to Cl04 on the enzyme via a disulfide bridge *(9)*. Adenain is similarly activated by a cleavage fragment from actin *(10)*. Mutational analysis and X-ray crystallography identified the active-site triad as H54-C122-E71 *(8,9,11,12)*. The substrate specificity of the enzyme is

From: *Methods in Molecular Medicine, Vol. 131:*
Adenovirus Methods and Protocols, Second Edition, vol. 2:
Ad Proteins, RNA, Lifecycle, Host Interactions, and Phylogenetics
Edited by: W. S. M. Wold and A. E. Tollefson © Humana Press Inc., Totowa, NJ

(M,I,L)XGG-X or (M,I,L)XGX-G *(13)*. In the course of virus maturation, adenain cleaves the viral structural proteins pre-IIIa, pre-VI, pre-VIII, the core proteins pre-VII, pre-TP, pre-X (also called μ), the nonstructural protein L1-52K, and the cytoskeletal proteins actin and cytokeratin 18. Adenain may deubiquitinate cellular proteins *(14)*. Alkylating agents and E64 inhibit the protease and virus infection as expected *(15)*. Specific inhibitors are not yet available, although the pVI-CT peptide reduces infectious virus yield by premature activation of adenain *(16)*. To date, 35 protease genes have been sequenced from a wide range of different virus serotypes. The translated amino acid sequences range from 201 to 214 residues and show both variable and highly conserved regions.

2. Materials

1. Buffer A: 10 mM Tris-HCl, pH 8.5, 1 mM ethylene diamine tetraacetic acid, 5 mM mercaptoethanol, 10% glycerol.
2. Buffer B: 10 mM phosphate buffer, pH 6.7, 1 mM ethylene diamine tetraacetic acid, 5 mM mercaptoethanol, 10% glycerol.
3. The fluorescent peptide substrate (Leu-Arg-Gly-Gly-NH)$_2$-rhodamine, called R110, was available from Molecular Probes (Eugene, OR). An alternative fluorescent peptide, z-Leu-Arg-Gly-Gly-AMC, is available from Bachem.
4. Ad2 or other serotypes can be obtained from individual investigators or from the American Type Culture Collection (Rockville, MD). The virus is easily grown to high titer in human cell lines such as HeLa.
5. 12- to 14-kDa Cutoff dialysis membrane.
6. 300 mM Phosphate buffer.
7. MES buffer: 10 mM 2-(N-morpholino)ethanesulfonic acid (Sigma, St. Louis, MO).
8. Diethylamino ethanol (DEAE)-Sephacell (Pharmacia Biotech, Uppsala, Sweden).

3. Methods

3.1. Synthesis of Recombinant Protease in E. coli

The Ad2 protease coding sequence has been cloned into several types of vectors for expression in insect cells or *E. coli (3,6–8,17)*. For practical reasons expression in *E. coli* is preferable (*see* **Note 1**). The following protocol was used with the pLPV construct in *E. coli*. Cloning was described elsewhere *(6,7)*.

1. Grow a 3-mL culture of AR120 cells containing pLPV in the recommended medium for approx 4 h and then add to 1 L prewarmed medium and incubate at 37°C overnight (approx 16 h).
2. Centrifuge out the cells at 2510g for 20 min.
3. Resuspend cells in 8 mL of buffer A and transfer into a 50-mL tube and pellet again at 3000 rpm (2510g) for 30 min.

4. Remove supernatant. Resuspend pellet in 5 mL buffer A and freeze–thaw five times, followed by sonication by 10×5 s bursts with a probe at a median setting. This is done in an ice bath and at intervals to avoid overheating the probe.

5. Centrifuge at $2510g$ for 30 min. The supernatant may contain variable amounts (5–20% of the total) of soluble protease and may be purified separately.

3.2. Chromatographic Purification of Protease

1. Resuspend the pellet from **Subheading 3.1., step 5** in 5 mL of saturated urea in buffer A and vortex it vigorously until the pellet is completely resuspended.

2. Equilibrate DEAE-Sephacell with buffer A to make 15 mL of slurry. Wash the DEAE-Sephacell with buffer A until the pH of the slurry reaches 8.5. This step is time-consuming but very important.

3. Combine the 5 mL suspension (from **Subheading 3.2., step 1**) with the 15 mL DEAE-Sephacell and rotate at 4°C (cold room) for 6 h or overnight.

4. Centrifuge at $400g$ in a benchtop centrifuge and wash the pellet three times with 15 mL each of buffer A. Combine the four supernatants. To remove all traces of DEAE particles, centrifuge the washes at $3000g$ in a Sorvall RC-5B superspeed, SS34 rotor.

5. Dialyze against buffer B in a 12- to 14-kDa cutoff membrane. If the resulting volume is too great for chromatography, concentrate it by covering the dialysis bag with polyvinylpyrrolidone or polyethylene glycol crystals.

6. Pack a K16/20 column in the cold room with buffer B pre-equilibrated hydroxyapatite. Connect the outlet of the column to a peristaltic pump and fraction collector set for 50-drop fractions.

7. Load the dialyzed protein and collect the flow-through. Check this later; if the column is functioning correctly the flow-through should not contain protease.

8. Wash column with 10 mL of buffer B. Never allow column to run dry.

9. Elute the protease with a gradient by adding seven column volumes (100–120 mL) of buffer B to the mixing container and the same volume of 300-mM phosphate buffer to the second container of the gradient former.

10. Assay the gradient fractions for the protease. Some fractions should achieve between 60 and 90% purity. Greater purity requires an additional SP-Sepharose chromatography step.

11. Combine the protease containing fractions and dialyze against 10 mM MES buffer, pH 7.0, and 5 mM β-mercaptoethanol.

12. Pre-equilibrate with MES buffer a K16/20 column packed with SP-Sepharose. Load the dialyzed protease and elute with a gradient of 10 mM MES, 0.5 M NaCl.

3.3. Protease Assays

1. The protease migrates as a 23-kDa band under denaturing conditions in sodium dodecyl sulfate-polyacrylamide gel electrophoresis. Purified Ad2 virions contain approx 20 molecules of protease and can serve as a convenient marker. Detection is by immunoblotting and staining with antiprotease serum *(7)*. The protease runs between viral proteins VI and pre-VII.

2. Protease activity (*see* **Notes 2** and **3**) can be detected by the cleavage of viral precursor proteins, particularly the abundant pre-VII, actin, or ovalbumin. In a denatured state, to expose the cleavage site, any protein carrying the consensus sites (M,I,L)XGG-X or (M,I,L)XGX-G should be cleavable. Viral precursors can be produced by labeling of the proteins for 1 h with ^{35}S-methionine at 24 h after infection with Ad. After the pulse, the cells are washed, boiled for 3 min to inactivate endogenous protease, and a low-speed supernatant can serve as substrate (the 20-kDa viral protein pre-VII).

3. The fluorescent peptide substrate R110 gives rapid quantitative assays at 1–5 μ*M* concentration *(18,19)*. The excitation and emission wavelengths are 492 and 523 nm, respectively, both with a 5-nm slit width. Enzyme activity results in increased fluorescence.

4. Notes

1. The protease has been expressed as a fusion protein with N-terminal tags of protein A or hexahistidine to facilitate purification. These tags did not appear to interfere with enzyme activity.

2. Preparations of recombinant protease require the addition of a peptide to acquire full enzyme activity *(4,8,18)*. The enzyme should be incubated (15 min at 37°C) with 200-fold molar excess of preoxidized peptide GVQSLKRRRCF. Some enzyme preparations lose activity after 1 wk. Addition of 1 m*M* β-mercaptoethanol was found to restore some activity.

3. A variety of enzyme reaction buffers have been used successfully, such as 20 m*M* phosphate buffer and 10 m*M* Tris-HCl. The pH should be slightly alkaline, between 7.5 and 8.0, and the optimal temperature is 45°C. Addition of 1 m*M* dithiothreitol or β-mercaptoethanol or 25 m*M* NaCl may improve activity.

References

1. Weber, J. M. (1995) The adenovirus endopeptidase and its role in virus infection, in *Molecular Repertoire of Adenoviruses* (Doerfler, W. and Petra Bohm, P., eds.), *Curr. Topics Microbiol. Immunol.* **199/I**, 227–235.

2. Weber, J. M. and Tihanyi, K. (1994) Adenovirus endopeptidases. *Methods Enzymol.* **244D**, 595–604.

3. Anderson, C. W. (1990) The proteinase polypeptide of adenovirus serotype 2 virions. *Virology* **177**, 259–272.

4. Cotten, M. and Weber, J. M. (1995) The adenovirus protease is required for virus entry into host cells. *Virology* **213**, 494–502.

5. Greber, U. F., Webster, P., Weber, J., and Helenius, A. (1996) The role of the adenovirus protease in virus entry into cells. *EMBO J.* **15**, 1766–1777.

6. Houde, A. and Weber, J. M. (1990) Adenovirus proteinases: comparison of amino acid sequences and expression of the cloned cDNA in *Escherichia coli. Gene* **88**, 269–273.

7. Tihanyi, K., Bourbonnière, M., Houde, A., Rancourt, C., and Weber, J. M. (1993) Isolation and properties of the adenovirus type 2 proteinase. *J. Biol. Chem.* **268**, 1780–1785.

8. Webster, A., Hay, R. T., and Kemp, G. (1993) The adenovirus protease is activated by a virus-coded disulphide-linked peptide. *Cell* **72,** 97–104.

9. Ding, J., McGrath, W. J., Sweet, R. M., and Mangel, W. F. (1996) Crystal structure of the human adenovirus proteinase with its 11 amino acid cofactor. *EMBO J.* **15,** 1778–1783.

10. Brown, M. T., BcBride, K. M., Baniecki, L. M., Reich, N. C., Marriott, G., and Mangel, W. F. (2002) Actin can act as a cofactor for a viral proteinase in the cleavage of the cytoskeleton. *J. Biol. Chem.* **277,** 46,298–46,303.

11. Grierson, A. W., Nicholson, R., Talbot, P., Webster, A., and Kemp, G. (1994) The protease of adenovirus serotype 2 requires cysteine residues for both activation and catalysis. *J. Gen. Virol.* **75,** 2761–2764.

12. Rancourt, C., Tihanyi, K., Bourbonnière, M., and Weber, J. M. (1994) Identification of active-site residues of the adenovirus endopeptidase. *Proc. Natl. Acad. Sci. USA* **91,** 844–847.

13. Webster, A., Russell, W. C., and Kemp, G. D. (1989) Characterization of the adenovirus proteinase: development and use of a specific peptide assay. *J. Gen. Virol.* **70,** 3215–3223.

14. Balakirev, M. Y., Jaquinod, M., Haas, A. L., and Chroboczek, J. (2002) Deubiquitinating function of adenovirus proteinase. *J. Virol.* **76,** 6323–6331.

15. Sircar, S., Keyvani-Amineh, H., and Weber, J. M. (1996) Inhibition of adenovirus infection with protease inhibitors. *Antiviral Res.* **30,** 147–153.

16. Ruzindana-Umunyana, A., Sircar, S., and Weber, J. M. (2000) The effect of mutant peptide cofactors on adenovirus protease activity and virus infection. *Virology* **270,** 173–179.

17. Mangel, W. F., McGrath, W. J., Toledo, D. L., and Anderson, C. W. (1993) Viral DNA and a viral peptide can act as cofactors of adenovirus virion proteinase activity. *Nature* **361,** 274–275.

18. Diouri, M., Geoghegan, K. F., and Weber, J. M. (1995) Functional characterization of the adenovirus proteinase using fluorogenic substrates. *Protein Pept. Lett.* **6,** 363–370.

19. Mangel, W. F., Toledo, D. L., Brown, M. T., Martin, J. H., and McGrath, W. J. (1996) Characterization of three components of human adenovirus proteinase activity *in vitro. J. Biol. Chem.* **271,** 536–543.

19

Assay for the Adenovirus Proteinase

Purification of the Enzyme and Synthesis of a Fluorogenic Substrate

Walter F. Mangel and William J. McGrath

Summary

Human adenovirus proteinase (AVP), the first member of a new class of cysteine protein-ases, is required for the synthesis of infectious virus. As such, it is an attractive target for proteinase inhibitors that act as antiviral agents. However, before potential inhibitors can be screened, a quick, sensitive, and quantitative assay for the enzyme is required. Here, methods for purification of a recombinant AVP expressed in *Escherichia coli* are presented and a fluorogenic substrate is designed, synthesized, and purified and then used in the development of a quick, sensitive, and quantitative assay for the enzyme. The reporting group in the sub-strate is Rhodamine 110, possibly the most detectable compound known. The substrate con-tains the proteinase consensus cleavage sequence (Leu-Arg-Gly-Gly). The synthesis and purification of (Leu-Arg-Gly-Gly-NH$_2$)-Rhodamine is described. It is then used to develop assays with AVP and its various cofactors. The resultant assays are quite sensitive; enzyme activity at low nanomolar concentrations can readily be detected.

Key Words: Adenovirus proteinase; fluorogenic substrate; Rhodamine; cysteine protein-ase; cofactors; enzyme assay; recombinant enzyme; Michaelis–Menten kinetics; enzyme puri-fication; fluorescence.

1. Introduction

Human adenovirus encodes a proteinase (AVP) whose activity is essential for the production of infectious virus *(1)*. Properties of a temperature-sensitive mutant (ts-1) indicate that one of the functions of the proteinase during virion assembly is to cleave six major virion precursor proteins to the mature counter-parts found in wild-type virus *(2,3)*. There are approx 70 molecules of AVP per virion *(4)*, and they must cleave at 3200 sites to render a virus particle infec-tious. In addition, AVP may play roles in both viral entry into cells *(5–7)* and the release of progeny virus *(8,9)*. The ts-1 mutation mapped to the L3 23K

From: *Methods in Molecular Medicine, Vol. 131:*
Adenovirus Methods and Protocols, Second Edition, vol. 2:
Ad Proteins, RNA, Lifecycle, Host Interactions, and Phylogenetics
Edited by: W. S. M. Wold and A. E. Tollefson © Humana Press Inc., Totowa, NJ

gene *(1,10)*. This gene was cloned and expressed in *Escherichia coli (11–13)* and baculovirus-infected insect cells *(14)*; the resultant 204-amino-acid, 23-kDa protein was purified *(11,14–16)*.

1.1. Enzyme Requires Cofactors

AVP is unusual in that it requires cofactors for maximal activity (k_{cat}/K_m). Proteinase activity can be observed in wild-type virus, but not in ts-1 virus *(11,14)*. Proteinase activity was not observed using purified recombinant AVP. However, recombinant AVP in vitro complemented the mutation in ts-1 virions, restoring proteinase activity when mixed together. This implied that a cofactor may be required, and two were identified. One viral cofactor, pVIc, is the 11-amino-acid peptide (GVQSLKRRRCF) derived from the C terminus of virion precursor protein pVI. Preceding the pVIc sequence is the AVP consensus cleavage sequence IVGL-G, in which cleavage occurs between the L and G. The second cofactor identified is the viral DNA *(11,17)*. More recently, a third, cellular cofactor for AVP has been discovered. The cytoskeletal protein actin has been shown to complement AVP activity *(9,18)*.

1.2. Proteinase Substrates

Synthetic substrates are useful not only in assaying proteinases, but also in determining a proteinase's specificity and selectivity. An ideal substrate for a proteinase would be sensitive, specific, and selective. Sensitivity refers to the detectability of the reporting group for enzymatic activity. Specificity refers to the efficiency (k_{cat}/K_m) with which an enzyme-catalyzed reaction occurs. Selectivity in a substrate refers to how well it interacts with one specific enzyme and not with others. Protein and small peptide substrates are not particularly suitable for quantitative enzyme studies because they can present multiple sites of cleavage and because the peptide products are, in many cases, a constantly changing mixture of secondary substrates and inhibitors. In contrast, synthetic substrates have a well-defined chemical structure, and the kinetics of hydrolysis is simplified by cleavage at a single, uniform position.

1.3. Fluorogenic Substrates for AVP

A specific, sensitive, and quantitative assay for AVP is described based on the observation that proteinase activity in wild-type virus will cleave small peptides with sequences that correspond to the amino-terminal side of cleavage sites in virion precursor proteins *(19)*. The assay uses as a substrate a *bis*-substituted Rhodamine110 compound **(Fig. 1)**. *Bis*-substituted Rhodamine110 compounds contain tetrapeptides corresponding to sequences cleaved in adenovirus precursor proteins that are covalently linked to the two nitrogen atoms of Rhodamine110 *(20–23)*. The *bis*-substituted substrates are virtually

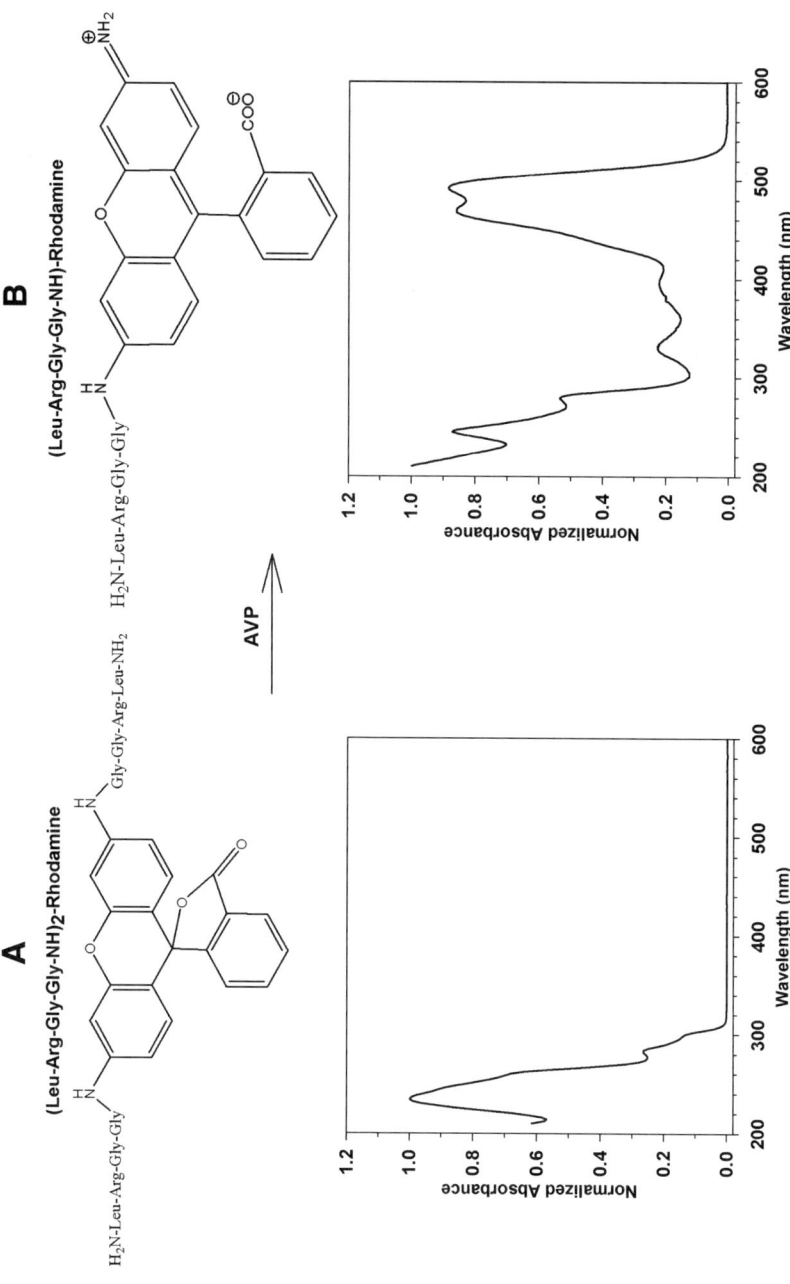

Fig. 1. Chemical structure of (**A**) (Leu-Arg-Gly-Gly-NH)$_2$-Rhodamine, (**B**) Leu-Arg-Gly-Gly-NH-Rhodamine, and their spectral properties. The substrate Leu-Arg-Gly-Gly-NH)$_2$-Rhodamine is drawn in the lactone state; it does not absorb at 490 nm. Hydrolysis of one of the amide bonds next to Rhodamine110 by a cofactor-activated adenovirus proteinase yields the product Leu-Arg-Gly-Gly-NH-Rhodamine, which is drawn in the quinone state; it absorbs strongly at 492 nm and emits most of the light as fluorescence; the quantum yield is greater than 0.9.

nonfluorescent, because the fluorophore is in the lactone state. Upon cleavage by an endoproteinase of one of the two peptide bonds adjacent to the Rhodamine moiety, the resultant mono-substituted Rhodamine moiety switches to the quinone state concomitant with a 3500-fold increase in fluorescence intensity. Conversion from the lactone to the quinone state results in a high degree of conjugation and hence stability, which is one reason why the susceptible bonds in these substrates are so unusually reactive. These substrates are easily prepared with high yields. Compared with the corresponding 7-amino-4-methylcoumarin-based analogs, peptidyl-Rhodamine substrates exhibit an increase in sensitivity of 50- to 300-fold. The Rhodamine-based substrate assay is extremely useful in the study of proteinases (*11,17,20,23–26*).

2. Materials

1. M9-TBY broth: 1.0% (w/v) bacto-tryptone, 0.5% (w/v) NaCl, 0.2% (w/v) yeast extract in 0.1% (w/v) NH_4Cl, 0.3% (w/v) KH_2PO_4, 0.6% (w/v) Na_2HPO_2, 0.4% (w/v) dextrose, and 1 mM $MgSO_4$, supplemented with ampicillin to 50 µg/mL.
2. 0.1 M Isopropyl-thio-β-galactoside (IPTG) stock solution.
3. Lysis buffer: 50 mM Tris-HCl, pH 8.0, 0.015 M NaCl, 0.05% (v/v) Triton X-100, and 5 mM β-mercaptoethanol.
4. 2.5 × 15-cm column TSK-diethylamino ethyl (DEAE) Sepharose (Toyopearl; Montgomery, PA).
5. 1.5 × 20-cm column S-Sepharose Fast Flow (Pharmacia; Piscataway, NJ).
6. 1.5 × 4-cm column Chelating Sepharose (Pharmacia; Piscataway, NJ) charged with zinc ions. Column is washed with water, then charged with zinc by passing two column volumes of 0.2 M $ZnCl_2$ in 5 mM HCl through the column. The column is next washed with five column volumes water and, finally, equilibrated in lysis buffer minus reducing agent.
7. Rhodamine 110 is from Exciton (Dayton, OH). All solvents (dimethylformamide [DMF], dimethylsulfoxide, ethyl acetate, ether) are high-performance liquid chromatography-grade or higher and are from Aldrich (Milwaukee, WI); they are kept neat by storage with molecular sieves (Alfa-Products; Danvers, MA). Cbz-amino acids, 4 M HBr/acetic acid solution, and IPTG are from Sigma (St. Louis, MO). Analytical TLC silica strips are from Macherey-Nagel (Easton, PA). Analytical TLC solvent is 8:1:1, methyl ethyl ketone:acetone:water. All assays for fluorescence are performed in four-sided clear, plastic disposable cuvets (polystyrene or acrylic) from Sarstedt (Newton, NC).
8. Alcoholic base contains 1 kg of KOH pellets, 2 L of 100% alcohol, and 4 L of water.

3. Methods

3.1. Purification of Recombinant AVP Expressed in E. coli

1. Grow BL21(DE3) cells transformed with a pET 3a construct containing the L3-23k gene under control of the T7 operator system to an optical density of 0.4 –0.6 in M9-TBY medium at 37°C (*see* **Note 1**).
2. Add IPTG to 0.5 mM, lower the temperature to 30°C, and allow the cells to grow overnight.
3. Harvest the cells by centrifugation at 7500g for 10 min. Store the bacterial cell pellets at –20°C until needed.
4. Suspend the bacterial cell pellet in 0.05 volume of lysis buffer. Initiate lysis by the addition of egg white lysozyme to 100 µg/mL, and incubate at 4°C for 45 min.
5. Subject the cells to three cycles of quick freezing and thawing.
6. Digest DNA by addition of MgCl$_2$ to 3 mM and DNase to 25 µg/mL, and incubate at 4°C for 45 min.
7. Sonicate the cell suspension for three 30-s bursts.
8. Centrifuge at 10,000g for 15 min, collect supernatant. Wash pellet with 0.25 initial volume lysis buffer and centrifuge at 10000g for 10 min. Combine supernatants.
9. Load the combined supernatants onto a DEAE-Sepharose column equilibrated with lysis buffer and collect fractions.
10. Wash the column with at least five column volumes of lysis buffer.
11. Strip the column with 50 mL lysis buffer supplemented with 1 M NaCl.
12. Fractions containing AVP are identified by activity assays and sodium dodecyl sulfate–polyacrylamide gel electrophoresis (SDS-PAGE) and are usually located in the flow-through.
13. Pool the fractions containing AVP.
14. Load the DEAE pool on the SP-Sepharose Fast Flow column equilibrated with lysis buffer and collect fractions.
15. Wash the column with at least 15 column volumes of lysis buffer lacking both reducing agent and detergent.
16. Elute AVP with a linear gradient of 0.015–0.4 M NaCl in 50 mM Tris-HCl, pH 8.0.
17. Collect fractions and identify AVP by activity assays and SDS-PAGE. AVP usually elutes between 0.1 and 0.2 M NaCl.
18. Pool the fractions containing AVP.
19. Apply the S-Sepharose AVP pool to the chelating Sepharose column that has been charged with zinc and previously equilibrated with lysis buffer lacking reducing agents. Collect fractions.
20. Wash the column with five column volumes of each of the following buffers: 50 mM Tris-HCl, pH 8.0, 0.1 M NaCl; 50 mM Tris-HCl, pH 8.0, 1 M NaCl; 50 mM Tris-HCl, pH 8.0, 0.1 M NaCl; 50 mM Tris-HCl, pH 8.0, 0.1 M NaCl, 0.035 M imidazole; and 50 mM Tris-HCl, pH 8.0, 0.1 M NaCl.
21. Elute the AVP with 50 mM Tris-HCl, pH 8.0, 0.1 M NaCl, 0.015 M ethylene diamine tetraacetic acid (EDTA).

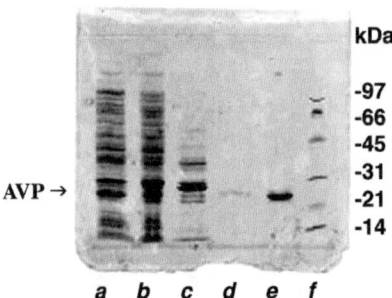

Fig. 2. Purification of the adenovirus proteinase (AVP). Samples were separated by sodium dodecyl sulfate–polyacrylamide gel electrophoresis and stained with Coomassie Brilliant Blue: lysate from bacteria after induction by isopropyl-thio-β-galactoside (lane a), 10,000*g* supernatant of the lysate (lane b), flow-through from diethylamino ethyl chromatography (lane c), pool that contains AVP activity from S-Sepharose chromatography (lane d), pool that contains AVP activity from zinc–iminodiacetic acid–Sepharose column (lane e), and molecular mass markers (lane f).

22. Fractions containing AVP are identified by activity assays and SDS-PAGE and usually elute with the 0.015 *M* EDTA buffer (**step 21**).
23. Pool AVP-containing fractions and dialyze overnight at 4°C against ≥50 vol 5 m*M* Tris-HCl, 0.005 *M* NaCl, 0.1 m*M* EDTA.
24. Store pure AVP in aliquots at –80°C at around 10 mg/mL (*see* **Note 2**).
25. An SDS-polyacrylamide gel showing the state of purification after various steps in the purification process is shown in **Fig. 2**.

3.2. Synthesis and Purification of the AVP Substrate (Leu-Arg-Gly-Gly-NH)$_2$-Rhodamine

The fluorogenic substrate (Leu-Arg-Gly-Gly-NH)$_2$-Rhodamine is synthesized and purified by a procedure that is a modification of published procedures for other Rhodamine-based substrates *(20–24)*.

1. (Cbz-Gly-Gly-NH)$_2$-Rhodamine is synthesized by adding 23.2 mmol Cbz-Gly-Gly (FW 266.3) to 154 mL neat DMF:pyridine (1:1, v/v) at 4°C with stirring. After Cbz-Gly-Gly dissolves or until no further Cbz-Gly-Gly goes into solution, 20.8 mmol 1-(3-dimethylaminopropyl)-3-ethylcarbodiimide is added. After 5 min, 0.82 mmol Rhodamine 110 in 6 mL neat DMF:pyridine (1:1, v/v) is added and the reaction allowed to proceed for 2 d until all the Rhodamine 110 is *bis*-substituted (*see* **Note 3**). The extent of the reaction is judged by analytical TLC using methyl ethyl ketone:acetone:water (8:1:1). The positions of Rhodamine 110 and Cbz-Gly-Gly-NH-Rhodamine can be visualized with a black lamp (long wavelength) because these compounds fluoresce green when excited by 490-nm light. Upon dipping the TLC strip in alcoholic KOH (**Subheading 2.**, **item 8**), the spot containing (Cbz-Gly-Gly-NH)$_2$-Rhodamine appears pink.

2. The product is purified by precipitation. To 20 mL of reaction is added 150 mL ethyl acetate. After centrifugation at 9000*g*, the precipitate in one centrifuge tube is dissolved in 20 mL neat DMF. Then, that 20 mL is used to dissolve the precipitate in a second centrifuge tube. This process is repeated until all the pellets are dissolved in a total volume of 20 mL DMF.

3. The Cbz blocking group is removed by treatment with 4 *M* HBr/acetic acid *(20,23)*. First, the 20 mL of (Cbz-Gly-Gly-NH)$_2$-Rhodamine is precipitated by adding 150 mL ethyl acetate. After centrifugation at 9000*g*, the precipitate in the centrifuge bottle is dissolved in 24 mL 4 *M* HBr/acetic acid. A stirring bar is added, and the centrifuge bottle is sealed by surrounding its top with parafilm. The solution is stirred at room temperature for 1 h.

4. (Gly-Gly-NH)$_2$-Rhodamine is purified by adding 60 mL ethyl ether and centrifugation at 9500*g* for 15 min. The pellet is dissolved in 10 mL DMF and the precipitation process repeated twice more. The final pellet is dissolved in 10 mL neat DMF:pyridine (1:1, v/v). The extent of deblocking (Cbz-Gly-Gly-NH)$_2$-Rhodamine is assessed by analytical TLC using methyl ethyl ketone: acetone:water (8:1:1) as the solvent. The positions of (Cbz-Gly-Gly-NH)$_2$-Rhodamine and (Gly-Gly-NH)$_2$-Rhodamine can be visualized with a black lamp (long wavelength); they will appear as pink spots upon dipping the TLC strip in alcoholic KOH. The spot containing (Cbz-Gly-Gly-NH)$_2$-Rhodamine will be near the solvent front, whereas the spot containing (Gly-Gly-NH)$_2$-Rhodamine will remain on the origin.

5. (Cbz-Arg-Gly-Gly-NH)$_2$-Rhodamine is synthesized by adding 23.2 mmol Cbz-Arg (FW 266.3) to 150 mL neat DMF:pyridine (1:1, v/v) at 4°C. After Cbz-Arg dissolves or when no further Cbz-Arg goes into solution, 20.8 mmol 1-(3-dimethylaminopropyl)-3-ethylcarbodiimide is added. After 5 min, the 10 mL of (Gly-Gly-NH)$_2$-Rhodamine in neat DMF:pyridine (1:1, v/v) synthesized in **step 4** is added and the reaction is allowed to proceed for 2 d until all the (Gly-Gly-NH)$_2$-Rhodamine is converted to (Cbz-Arg-Gly-Gly-NH)$_2$-Rhodamine. The course of the reaction is monitored by TLC as described in **Subheading 3.2., step 1**.

6. (Cbz-Arg-Gly-Gly-NH)$_2$-Rhodamine is purified by precipitation (*see* **Note 4**). First, place 20-mL aliquots of the reaction into centrifuge bottles and add 150 mL ethyl acetate. After centrifugation at 9000*g*, dissolve the precipitate in one centrifuge tube in 20 mL neat DMF. Then, use that 20 mL to dissolve the precipitate in a second centrifuge tube. Repeat this process until all the pellets are dissolved in a total volume of 20 mL DMF. Next, add 150 mL of 1.2 *N* HCl and, after centrifugation at 9000*g*, dissolve the precipitate in 20 mL DMF. Finally, add 150 mL of ethyl acetate. After centrifugation at 9000*g*, the precipitate is ready for deblocking.

7. Remove the Cbz blocking group as described for (Cbz-Gly-Gly-NH)$_2$-Rhodamine in **step 3**.

8. Purify (Arg-Gly-Gly-NH)$_2$-Rhodamine as described in **step 4**.

9. Synthesize (Cbz-Leu-Arg-Gly-Gly-NH)$_2$-Rhodamine by adding 23.2 mmol Cbz-Leu (FW 265) to 150 mL neat DMF:pyridine (1:1, v/v) at 4°C. After Cbz-Leu

dissolves or until no further Cbz-Leu goes into solution, add 20.8 mmol 1-(3-dimethylaminopropyl)-3-ethylcarbodiimide. After 5 min, add the 10 mL of (Arg-Gly-Gly-NH)$_2$-Rhodamine in neat DMF:pyridine (1:1, v/v) synthesized in **step 8** and allow the reaction to proceed for 2 d until all the (Arg-Gly-Gly-NH)$_2$-Rhodamine is converted to (Cbz-Leu-Arg-Gly-Gly-NH)$_2$-Rhodamine. Monitor the course of the reaction by TLC as described in **step 1**.

10. Purify (Cbz-Leu-Arg-Gly-Gly-NH)$_2$-Rhodamine as described in **step 6**.
11. Deblock (Cbz-Leu-Arg-Gly-Gly-NH)$_2$-Rhodamine as described **step 3**.
12. Purify (Leu-Arg-Gly-Gly-NH)$_2$-Rhodamine as described in **step 4**.
13. Dissolve the final precipitate in dimethyl sulfoxide and store in aliquots at −80°C. The identity, purity, and amount of each derivative is determined by quantitative amino acid analysis after complete acid hydrolysis.

3.3. Assay for Proteinase Activity

1. Assays are carried out in four-sided clear plastic disposable cuvets (*see* **Note 5**) at 37°C in a total volume of 1 mL containing 10 mM Tris-HCl, pH 8, 5 mM octylglucoside, AVP, and one or more cofactors: pVIc, DNA, polyE, or actin. For example, to assay the stimulation of AVP by pVIc, the assay mixture also contained 20 nM AVP, 20 mM NaCl, and 40 μM pVIc (*see* **Note 6**). No NaCl is present in assays with DNA, polyE, or actin (*see* **Note 7**).
2. Preincubate the assay mixture without substrate for 5 min at 37°C.
3. Add the substrate (Leu-Arg-Gly-Gly-NH)$_2$-Rhodamine. For example, adding substrate to 2 μM in an assay of AVP and pVIc will achieve half-maximal velocity, because the K_m is 2 μM (*see* **Note 8**).
4. Monitor the increase in fluorescence as a function of time in a spectrofluorometer. The excitation wavelength is 492 nm, and the emission wavelength is 523 nm, both with a 4-nm slit width.
5. Enzyme activity is defined as the rate of fluorescence increase of the assay containing AVP with cofactors minus the rate of fluorescence increase of an identical assay but without cofactor.
6. The rate of fluorescence increase is converted to units of moles of product per second using a standard curve. The standard curve is a plot of fluorescence vs the molar concentration of Leu-Arg-Gly-Gly-NH-Rhodamine.
7. Assays for measuring the Michaelis–Menten constants K_m and k_{cat} contain AVP saturated with ligand(s) and substrate concentrations ranging from one-fifth to five times the K_m (*see* **Notes 9** and **10**).

4. Notes

1. The induction and purification protocol generally results in tens of milligrams of pure AVP per liter of bacterial culture (*16*).
2. The concentration of AVP is determined with a calculated molar extinction coefficient at 280 nm of 26,510 M^{-1}cm^{-1} (*27*).
3. Each synthetic reaction must be driven to completion before the product is purified. The purification procedures are able only to separate reactants and products,

not the mono-substituted intermediates such as Cbz-Gly-Gly-NH-Rhodamine, which is highly fluorescent.

4. Once arginine is attached to a Rhodamine-based substrate, it can readily be purified from unincorporated blocked amino acids by precipitation with 1.2 N HCl.
5. Expensive quartz cuvets are not required because Rhodamine 110 compounds absorb maximally in the 490 nm range, not in the ultraviolet range.
6. The 20 mM NaCl is present to prevent nonspecific absorption of protein to the walls of the cuvet.
7. NaCl is absent because the binding of DNA, actin, or polyE *(28)* to AVP is sensitive to ionic strength *(9,29)*.
8. During the assay, no more than 5% of the substrate should be hydrolyzed. That way only mono-substituted product will form. If more than 5% of the substrate is hydrolyzed, the kinetics of substrate hydrolysis will not be linear, because there will be two different hydrolysis products: mono-substituted Rhodamine and Rhodamine.
9. If the K_m is very high, too much fluorescence may be generated during an assay. In this case, the rate of substrate hydrolysis can be determined by measuring the rate of increase in absorbance at 492 nm.
10. There are many commercially available computer programs that will calculate K_m and k_{cat}.

Acknowledgments

Research supported by the Office of Biological and Environmental Research of the U.S. Department of Energy under Prime Contract No. DE-AC0298CH 10886 with Brookhaven National Laboratory and by National Institutes of Health Grant AI41599.

References

1. Weber, J. (1976) Genetic analysis of adenovirus type 2 III. Temperature- sensitivity of processing of viral proteins. *J. Virol.* **17,** 462–471.
2. Hannan, C., Raptis, L. H., Dery, C. V., and Weber, J. (1983) Biological and structural studies with adenovirus type 2 temperature-sensitive mutant defective for uncoating. *Intervirology* **19,** 213–223.
3. Mirza, A. and Weber, J. (1980) Infectivity and uncoating of adenovirus cores. *Intervirology* **13,** 307–311.
4. Brown, M. T., McGrath, W. J., Toledo, D. L., and Mangel, W. F. (1996) Different modes of inhibition of human adenovirus proteinase, probably a cysteine proteinase, by bovine pancreatic trypsin inhibitor. *FEBS Lett.* **388,** 233–237.
5. Greber, U. F., Webster, P., Weber, J., and Helenius, A. (1996) The role of the adenovirus protease in virus entry into cells. *EMBO J.* **15,** 1766–1777.
6. Cotten, M. and Weber, J. M. (1995) The adenovirus proteinase is required for entry into host cells. *Virology* **213,** 494–502.
7. Greber, U. F. (1998) Virus assembly and disassembly: the adenovirus cysteine protease as a trigger factor. *Rev. Med. Virol.* **8,** 213–222.

8. Chen, P. H., Ornelles, D. A., and Shenk, T. (1993) The adenovirus L3 23-kilo-dalton proteinase cleaves the amino-terminal head domain from cytokeratin 18 and disrupts the cytokeratin network of HeLa cells. *J. Virol.* **67,** 3507–3514.

9. Brown, M. T., McBride, K. M., Baniecki, M. L., Reich, N. C., Marriott, G., and Mangel, W. F. (2002) Actin can act as a cofactor for a viral proteinase in the cleavage of the cytoskeleton. *J. Biol. Chem.* **277,** 46,298–46,303.

10. Yeh-Kai, L., Akusjarvi, G., Alestrom, P., Pettersson, U., Tremblay, M., and Weber, J. (1983) Genetic identification of an endopeptidase encoded by the adenovirus genome. *J. Mol. Biol.* **167,** 217–222.

11. Mangel, W. F., McGrath, W. J., Toledo, D. L., and Anderson, C. W. (1993) Viral DNA and a viral peptide can act as cofactors of adenovirus virion proteinase activity. *Nature* **361,** 274–275.

12. Tihanyi, K., Bourbonniere, M., Houde, A., Rancourt, C., and Weber, J. M. (1993) Isolation and properties of adenovirus type 2 proteinase. *J. Biol. Chem.* **268,** 1780–1785.

13. Anderson, C. W. (1993) Expression and purification of the adenovirus proteinase polypeptide and of a synthetic proteinase substrate. *Protein Express. Purif.* **4,** 8–15.

14. Webster, A., Hay, R. T., and Kemp, G. (1993) The adenovirus protease is activated by a virus-coded disulphide-linked peptide. *Cell* **72,** 97–104.

15. Webster, A. and Kemp, G. (1993) The active adenovirus protease is the intact L3 23K protein. *J. Gen. Virol.* **74,** 1415–1420.

16. Mangel, W. F., Toledo, D. L., Brown, M. T., Martin, J. H., and McGrath, W. J. (1996) Characterization of three components of human adenovirus proteinase activity *in vitro*. *J. Biol. Chem.* **271,** 536–543.

17. McGrath, W. J., Baniecki, M. L., Li, C., et al. (2001) Human adenovirus proteinase: DNA binding and stimulation of proteinase activity by DNA. *Biochemistry* **40,** 13,237–13,245.

18. Brown, M. T. and Mangel, W. F. (2004) Interaction of actin and its 11-amino acid C-terminal peptide as cofactors with the adenovirus proteinase. *FEBS* **563,** 213–218.

19. Webster, A., Russell, S., Talbot, P., Russell, W. C., and Kemp, G. D. (1989) Characterization of the adenovirus proteinase: substrate specificity. *J. Gen. Virol.* **70,** 3225–3234.

20. Leytus, S. P., Melhado, L. L., and Mangel, W. F. (1983) Rhodamine-based compounds as fluorogenic substrates for serine proteases. *Biochem. J.* **209,** 299–307.

21. Mangel, W. F., Leytus, S. P., and Melhado, L. L. (1987) Novel Rhodamine derivatives as fluorogenic substrates. U.S. Patent 4,640,893 (February 1987).

22. Mangel, W. F., Leytus, S. P., and Melhado, L. L. (1985) Rhodamine derivatives as fluorogenic substrates for proteinases. U.S. Patent 4,557,862 (December 1985).

23. Leytus, S. P., Patterson, W. L., and Mangel, W. F. (1983) New class of sensitive, specific, and selective substrates for serine proteinases: fluorogenic, amino acid peptide derivatives of Rhodamine. *Biochem. J.* **215,** 253–260.

24. Leytus, S. P., Toledo, D. L., and Mangel, W. F. (1984) Theory and experimental method for determining individual kinetic constants for fast-acting, irreversible, protease inhibitors. *Biochim. Biophys. Acta* **788,** 74–86.
25. McGrath, W. J., Abola, A. P., Toledo, D. L., Brown, M. T., and Mangel, W. F. (1996) Characterization of human adenovirus proteinase activity in disrupted virus particles. *Virology* **217,** 131–138.
26. Baniecki, M. L., McGrath, W. J., McWhirter, S. M., et al. (2001) Interaction of the human adenovirus proteinase with its eleven amino-acid cofactor pVIc. *Biochemistry* **40,** 12,349–12,356.
27. Gill, S. G. and von Hippel, P. H. (1989) Calculation of protein extinction coefficients from amino acid sequence data. *Anal. Biochem.* **182,** 319–326.
28. Bajpayee, N. S., McGrath, W. J., and Mangel, W. F. (2005) Interaction of the adenovirus proteinase with protein cofactors with high negative charge densities. *Biochemistry* **44,** 8721–8729.
29. McGrath, W. J., Ding, J., Sweet, R. M., and Mangel, W. F. (1996) Preparation and crystallization of a complex between human adenovirus serotype 2 proteinase and its 11-amino-acid cofactor pVIc. *J. Struct. Biol.* **117,** 77–79.

20

Cofactors of the Adenovirus Proteinase

Measuring Equilibrium Dissociation Constants and Stoichiometries of Binding

Walter F. Mangel and William J. McGrath

Summary

Human adenovirus proteinase (AVP) is required for the synthesis of infectious virus. AVP is synthesized in an inactive form; it is unusual in that it requires cofactors for activation of enzyme activity. Inside nascent virions, an 11-amino-acid peptide and the viral DNA are cofactors for activation; this enables the enzyme to cleave virion precursor proteins, rendering the virus particle infectious. In the cytoplasm, actin is a cofactor for activation, and an actin–AVP complex can cleave cytokeratin 18 and actin itself; this may prepare the infected cell for lysis. Experimental protocols are presented to determine stoichiometries of binding and equilibrium dissociation constants, K_d values, for the binding of pVc, DNA, or actin to AVP by changes in enzyme activity. Techniques are also presented for measuring stoichiometries of binding and K_d values for the binding of various lengths of DNA to AVP by changes in fluorescence polarization. Finally, the binding of different size classes of polymers of glutamic acid to AVP, the K_d values, and stoichiometries of binding are characterized by fluorescence polarization in an indirect assay involving competition with fluorescein-labeled DNA.

Key Words: Adenovirus proteinase; cofactors; equilibrium dissociation constant; stoichiometry of binding; viral precursor proteins; cell lysis; actin; Michaelis–Menten constants; fluorescence polarization; fluorogenic substrate.

1. Introduction

The adenovirus proteinase (AVP) is essential for reproduction of the virus. AVP is synthesized in an inactive form *(1)*. Inside nascent virions, it becomes activated to cleave virion precursor proteins to render the virus particle infectious. Activation of AVP inside nascent virions occurs upon binding of cofactors to the enzyme; cofactors include viral DNA and pVIc, the 11-amino-acid

From: *Methods in Molecular Medicine, Vol. 131:*
Adenovirus Methods and Protocols, Second Edition, vol. 2:
Ad Proteins, RNA, Lifecycle, Host Interactions, and Phylogenetics
Edited by: W. S. M. Wold and A. E. Tollefson © Humana Press Inc., Totowa, NJ

peptide pVIc derived from the C terminus of the precursor to viral protein VI (2–4). Binding of AVP to DNA is independent of DNA sequence and is a reflection of AVP binding to polymers of high negative charge density; for example, AVP binds to polyE (5) as well as it does to DNA (4). A third cofactor, actin, has recently been identified (6,7).

The two viral cofactors increase the k_{cat} for substrate hydrolysis and decrease the K_m. The predominant effect of pVIc on AVP is on k_{cat}, which increases 107-fold from 0.002/s, whereas the K_m decreases only 10-fold. The predominant effect of DNA on AVP is on K_m, which decreases 28-fold from 94 μM, while the k_{cat} decreases only 11-fold. Together, there is a synergistic effect of the cofactors. The k_{cat} increases 1206-fold, and the K_m decreases 28-fold. Other proteinases bind to DNA (8), but only the enzymatic activity of AVP is stimulated by being bound to DNA.

Extensive analysis of the interactions of AVP and its cofactors using the methodology described here has led us to propose a model for the regulation of AVP by its cofactors (4,9,10). AVP is synthesized as a relatively inactive enzyme. The K_m for (Leu-Arg-Gly-Gly-NH)$_2$-Rhodamine is 95 μM, and the k_{cat} is 0.002/s. If AVP were synthesized as an active enzyme, it would probably cleave virion precursor proteins before virion assembly, thereby preventing the formation of immature virus particles. Consistent with this hypothesis is the observation that if exogenous pVIc is added to cells along with Ad5 virus, the synthesis of infectious virus in those cells is severely diminished (2,11,12).

AVP quite possibly enters empty capsids bound to the viral DNA and remains bound during the maturation of the virus particle. It enters empty capsids bound to the viral DNA because the K_d for the binding of AVP to 12-mer ds-DNA is quite low—63 nM. Inside the young virion, the viral DNA is positioned next to the C terminus of virion protein pVI. Protein VI is a DNA-binding protein (13). AVP is partially activated by being bound to the viral DNA. The K_m decreases 10-fold and the k_{cat} increases 11-fold compared with AVP alone. Thus, AVP–DNA complexes can cleave pVI at the proteinase consensus cleavage site preceding the amino acid sequence of pVIc. The liberated pVIc can then bind either to the viral DNA, $K_{d(app.)}$ of 693 nM, to AVP molecules in solution, K_d of 4400 nM, or, most likely, to the AVP–DNA complex that liberated it, K_d of 90 nM (2). Once pVIc is bound to AVP, the penultimate cysteine in pVIc forms a disulfide bond with Cys104 of AVP (14). Now AVP is permanently activated. Compared with AVP alone, with pVIc–AVP–DNA, the K_m decreases 28-fold and the k_{cat} increases 1206-fold.

How can 70 fully activated proteinases bound to the viral DNA inside the virion (15) cleave precursor proteins 3200 times to render a virus particle infectious? For this to occur, either enzyme or substrate must move inside young virions. Perhaps the proteinase moves along the viral DNA searching for processing sites on precursor proteins like the *E. coli* RNA polymerase holoenzyme moves along DNA searching for a promoter. AVP–pVIc and RNA polymerase both exhibit an appreciable, non-sequence-specific affinity for DNA, K_d = 60 nM, in nucleotide base pairs for AVP–pVIc and 100 nM in nucleotide base pairs for RNA polymerase *(16)*.

RNA polymerase binds to DNA via a two-step mechanism. Initially RNA polymerase binds to any place on DNA via free diffusion in three-dimensional space. Next, it dissociates from the DNA to a point where, although free to move, it is still near the original binding site. This enables it, with high probability and within a short period of time, to reassociate with the same or a nearby binding site. Once it locates a promoter, it exhibits an enormous affinity for that sequence (K_d = 10 fM in nucleotide base pairs) *(16)*.

Because of "nonspecific" binding, the search process for the promoter in the second step occurs in reduced dimensionality or volume. That is how RNA polymerase can reach a promoter at a rate faster than a diffusion-controlled rate; diffusion is occurring in one dimension.

In the case of AVP, perhaps the viral DNA serves as a scaffold next to which reside the 3200 processing sites that must be cleaved. The 70 AVP–pVIc complexes could then move along the viral DNA via one-dimensional diffusion, using the DNA as a guidewire in cleaving precursor proteins. By using the viral DNA as a guidewire, AVP could quickly (via one-dimensional diffusion) and efficiently (by the alignment of the cleavage sites near the DNA and by moving along the DNA) process the numerous virion precursor proteins.

2. Materials

1. All assays for fluorescence are performed in four-sided clear plastic disposable cuvettes (polystyrene or acrylic) from Sarstedt (Newton, NC; *see* **Note 1**).
2. For assays in a spectrofluorometer, the excitation wavelength is 492 nm, and the emission wavelength is 523 nm, both with a 4-nm slit width.
3. Assay buffer in the absence of DNA (*see* **Note 2**): 10 mM Tris-HCl, pH 8.0, 2 mM octylglucoside, 20 or 100 mM NaCl, and 15 μM (Leu-Arg-Gly-Gly-NH)$_2$-Rhodamine.
4. Assay buffer in the presence of DNA (*see* **Note 3**): 10 mM Tris-HCl, pH 8.0, 2 mM octylglucoside, and 15 μM (Leu-Arg-Gly-Gly-NH)$_2$-Rhodamine.
5. AVP–pVIc complexes are preformed by incubating 50 μM AVP with 50 μM pVIc overnight at 4°C. During the incubation, a covalent bond is formed between Cys10' of pVIc and Cys104 of AVP *(3)*.
6. Rhodamine substrates are synthesized and purified as described in Chapter 19. Some Rhodamine substrates are available from commercial sources such as Molecular Probes (Eugene, OR), Promega (Madison, WI), Beckman-Coulter (Fullerton, CA), and Hoffman-LaRoche (Indianapolis, IN).
7. 5'-Fluorescein-labeled DNA was obtained from Invitrogen (Carlsbad, CA).

3. Methods

3.1. Cofactor Equilibrium Dissociation Constants Measured by Enzyme Activity Assays

The theoretical basis of the assay is that AVP is inactive. Three cofactors (pVIc, DNA, and actin) bind to AVP and in doing so stimulate enzyme activity. Thus, the progress of a binding reaction can be monitored by changes in enzyme activity.

1. Assays are performed in assay buffer. For assays of pVIc binding to AVP, 20 mM NaCl is present. For assays containing DNA, polyE or actin, no NaCl is present. The AVP concentration in assays in the absence of DNA, polyE or actin is 50 nM; in their presence, it is 20 nM.
2. For assays measuring the binding of DNA, polyE or actin to AVP–pVIc complexes, the AVP–pVIc complexes are preformed overnight as described in **Subheading 2.5.**
3. Increasing concentrations of ligand are added to a fixed concentration of enzyme and the reactions incubated at 37°C until binding equilibrium is reached, usually 5 min.
4. Substrate, (Leu-Arg-Gly-Gly-NH)$_2$-Rhodamine, is added to a concentration of 15 μM (*see* **Note 4**).
5. The increase in fluorescence as a function of time is monitored in an ISS PC-1 spectrofluorometer (Urbana, IL).
6. Enzyme activity is defined as the rate of fluorescence increase in the assay containing AVP with cofactors minus the rate of fluorescence increase in an identical assay but without the cofactor whose binding is being measured.
7. The rate of fluorescence increase is converted to units of moles of product per second by using a standard curve. The standard curve is a plot of fluorescence vs the molar concentration of Leu-Arg-Gly-Gly-NH-Rhodamine.
8. The K_d is calculated as follows: a graph is drawn of the rate of substrate hydrolysis vs the ligand concentration. The data points are best fitted to a hyperbola. From the graph, the concentration of ligand bound, [ligand]$_b$, at any initial ligand concentration [Ligand]$_i$ is:

$$[\text{ligand}]_b = [\text{AVP}]_o \, [R_i/R_{max}]$$

where [AVP]$_o$ is the AVP concentration, R_i is the rate of substrate hydrolysis at a specific initial concentration [Ligand]$_i$, and R_{max} is the rate of substrate hydrolysis at infinite ligand concentration. The concentration of ligand free, [Ligand]$_f$, is:

$$[\text{Ligand}]_f = [\text{Ligand}]_i - [\text{ligand}]_b$$

From a plot of [ligand]$_b$ vs [Ligand]$_f$, the K_d can be obtained by standard methods. This analysis is valid only if the stoichiometry of binding of ligand to AVP is 1:1; otherwise the K_d obtained by this method is an apparent K_d, $K_{d(\text{app.})}$.

3.2. Stoichiometry of Binding by Enzyme Activity Assays

The stoichiometry of binding of a cofactor to AVP is determined under tight-binding conditions, i.e., under conditions where one of the two components of the binding reaction, either AVP or the ligand, is present at a constant concentration at least five times greater than its K_d. The K_d for the binding of AVP to pVIc is 4.4 μM *(2)*; to 12-mer double-stranded DNA it is 63 nM *(4)*; and to actin it is 4.2 nM *(7)*. The K_d for the binding of AVP–pVIc complexes to 1.3-kDa polyE is 56.2 nM *(5)*.

1. Assays are performed in assay buffer. For assays of pVIc binding to AVP, 20 mM NaCl is present. For assays containing DNA, polyE, or actin, no NaCl is present. The AVP concentration in each assay is fivefold greater than its K_d for the ligand.
2. For assays measuring the binding of DNA, polyE, or actin to AVP–pVIc complexes, the AVP–pVIc complexes are preformed overnight as described in **Subheading 2.5.**
3. Increasing concentrations of ligand are added to a fixed concentration of enzyme and the reactions incubated at 37°C until binding equilibrium is reached, usually 5 min.
4. Substrate, (Leu-Arg-Gly-Gly-NH)$_2$-Rhodamine, is added to a concentration of 15 μM (*see* **Note 4**).
5. The increase in fluorescence as a function of time is monitored in an ISS PC-1 spectrofluorometer (Urbana, IL).
6. Enzyme activity is defined as the rate of fluorescence increase of the assay containing AVP in the presence of the cofactor whose binding is being measured minus the rate of fluorescence increase of an identical assay but without the cofactor whose binding is being measured.
7. The data are graphed as rate of substrate hydrolysis vs ligand concentration **(Fig. 1)**. At concentrations of ligand lower than that of AVP, there will be no free ligand and the rate of substrate hydrolysis will be directly proportional to the concentration of ligand. At concentrations of ligand that are greater than that of AVP, AVP will be saturated and the rate of substrate hydrolysis will be independent of the ligand concentration. The data in the graph can be characterized by two straight lines. The line characterizing the rate of substrate hydrolysis at low ligand concentrations goes through the origin; the line characterizing the rate of substrate hydrolysis at higher ligand concentrations is parallel to the abscissa. The two lines intersect at the point that reflects the minimum amount of ligand required to saturate all of the AVP in the assay.

3.3. Cofactor Equilibrium Dissociation Constants Measured by Fluorescence Polarization

The theoretical basis of the assay is that fluorescence polarization measurements provide information on molecular rotation. When a ligand binds to a

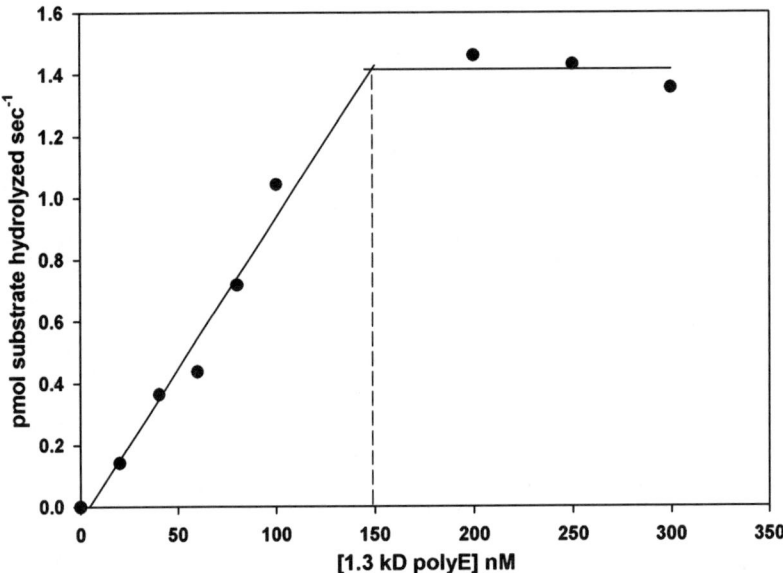

Fig. 1. Stoichiometry of binding of adenovirus proteinase (AVP)–pVIc to 1.3-kDa polyE. The assays were carried out as described in **Subheading 3.2**. Increasing concentrations of polyE were incubated with 150 nM AVP–pVIc. After 5 min at 37°C, (Leu-Arg-Gly-Gly-NH)$_2$-Rhodamine was added and the rate of substrate hydrolysis measured. The line parallel to the abscissa represents the average of the data points. The other line was derived by least-squares regression analysis of the initial data points.

protein, its rate of rotation decreases. Thus, fluorescence polarization measurements can be used to monitor the binding of a ligand to a protein.

3.3.1. Direct Binding Method

1. In the direct method, a fluorophore is attached to a ligand, and binding of the ligand is monitored by changes in anisotropy. This can be used to obtain directly the K_d of the labeled ligand.
2. Fluorescence anisotropy experiments to characterize the binding of ligand to 5'-fluorescein-labeled DNA are performed at 25°C in the thermostated sample compartment of an ISS PC-1 spectrofluorometer. An Oriel 530-nm bandpass filter is inserted before the emission photomultiplier. The excitation wavelength is 490 nm. All slit widths are set to 8 nm. The anisotropy of a solution is obtained by measuring the fluorescence emission intensities parallel and perpendicular to the incident light.
3. Assays are carried out in an initial volume of 1.8 mL containing 10 nM 5'-fluorescein-labeled 12-mer ssDNA in 10 mM Tris-HCl, pH 8.0, and 2 mM octylglucoside.

4. The anisotropy of the solution is determined prior to the addition of any binding species.
5. A small aliquot of AVP or of preformed complexes of AVP–pVIc is added, and the binding reaction is allowed to equilibrate for 90 s prior to measuring the change in anisotropy.
6. The process in **step 5** is repeated until binding saturation is reached.
7. Binding saturation is reached when for three consecutive additions of AVP or AVP–pVIc the overall change in polarization is less than 1%. At this stage, the overall change in the volume of the assay should not have exceeded the original volume by more than 10%.
8. The K_d is calculated from the fluorescence anisotropy data as follows: the anisotropy, r, is defined as:

$$r = (I_\| - GI_\perp)(I_\| + 2GI_\perp)^{-1}$$

where $I_\|$ and $I_\perp$ are the fluorescence emission intensities with polarizers parallel and perpendicular, respectively, to that of the exciting beam. The instrument correction factor G is determined for each experiment (*see* **Note 5**). The anisotropy when the DNA is saturated with ligand, r_{max}, is determined empirically; increasing concentrations of ligand are added to a constant concentration of labeled DNA until further addition of the ligand results in no further change in anisotropy (**Fig. 2A**). The fraction of ligand bound, f_b, at any ligand concentration is:

$$f_b = (r_{obs} - r_o)/(r_{max} - r_o) \text{ and } f_f = 1 - f_b$$

where r_o is the initial anisotropy of the DNA in the absence of other components and r_{obs} is the anisotropy at the specific ligand concentration. The concentration of free ligand, L_f, and bound ligand, L_b, are:

$$L_f = f_f \text{ [DNA] and } L_b = f_b\text{[DNA]}$$

where [DNA] is the molar concentration of DNA molecules. From a plot of L_b vs L_f, the K_d is determined by standard methods. This analysis is valid only if the stoichiometry of binding of ligand to AVP is 1:1, otherwise the K_d obtained by this method is an apparent K_d—$K_{d(app.)}$.

3.3.2. Lifetimes of the Excited State

For the conclusions in the fluorescence anisotropy experiments to be valid, the lifetime of the excited state of the fluorophore must not change during the assay (*see* **Note 6**).

1. The lifetime of the excited state of the fluorescein moiety is measured in an SLM 4800 Spectrofluorometer using an excitation wavelength of 490 nm and an Oriel 520-nm bandpass filter before the emission photomultiplier.
2. The phase-shift method uses a frequency modulator tank filled with 19% ethanol to generate sinusoidally varying excitation light (*17*).

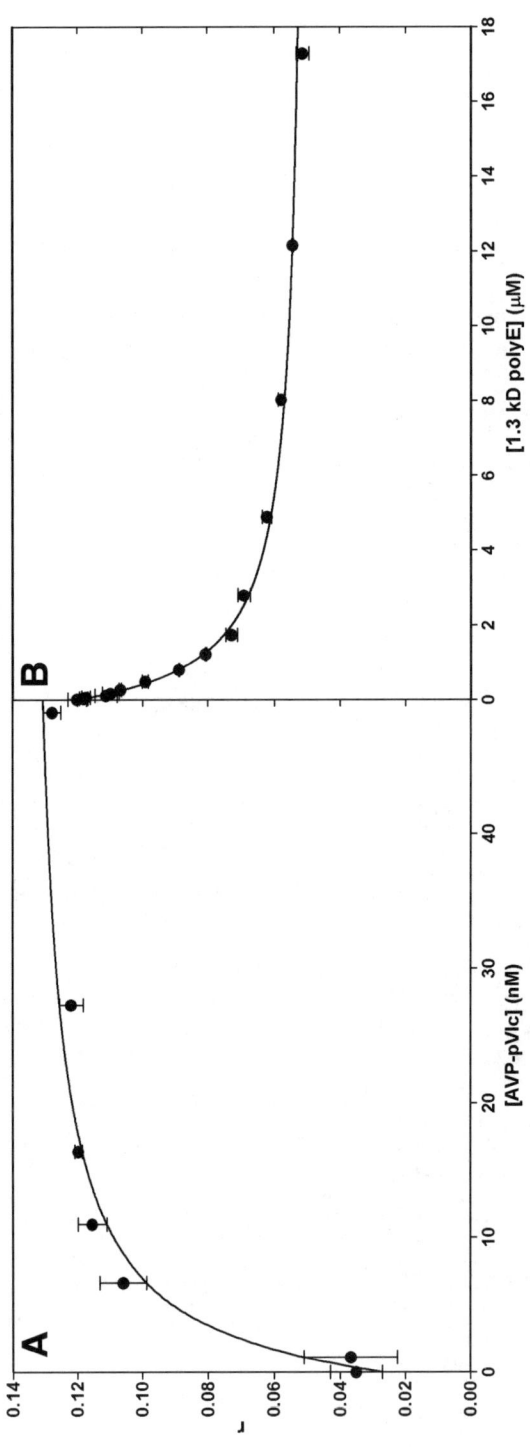

Fig. 2. Equilibrium dissociation constant for the binding of 12-mer ssDNA to adenovirus proteinase (AVP)–pVIc (A) and for the binding of 1.3-kDa polyE to AVP–pVIc (B). (A) The anisotropy (r) of a 1.8-mL solution of 1 nM 12-mer ssDNA with fluorescein attached to the 5'-end was measured at 25°C. Then, an aliquot of AVP–pVIc was added, and after 90 s the anisotropy was measured. This was repeated until there was no further increase in anisotropy. (B) Next, an aliquot of 1.3-kDa polyE was added, and after 90 s the anisotropy was measured. This was repeated until no further decrease in anisotropy was observed. The equilibrium dissociation constant from the data in (A) was calculated as described in **Subheading 3.3.1.8.** The equilibrium dissociation constant from the data in (B) was calculated as described in **Subheading 3.3.3.5.**

3. All measurements are made at a setting of 18 MHz in solutions of 25 mM *N*-2-hydroxyethylpiperazine-*N*'-2-ethanesulfonate (HEPES) (pH 8.0) containing 12.5 nM 18-mer dsDNA labeled at one of its 5'-ends with fluorescein in the absence or presence of 235 nM AVP–pVIc (*see* **Note 7**).

3.3.3. Stoichiometry of Binding by Fluorescence Polarization

Fluorescence anisotropy under tight-binding conditions can be used to ascertain the stoichiometry of binding of 5'-fluorescein-labeled dsDNA to AVP.

1. The equipment and solutions used to measure the stoichiometry of binding by fluorescence anisotropy are described in **Subheading 3.3.1.**, with some exceptions listed below.
2. The anisotropy of a 100 nM solution of 12-mer dsDNA is measured in which 1 nM of the DNA is labeled at its 5'-end with fluorescein. This is a concentration 20-fold higher than the K_d for the binding of 5'-fluorescein-labeled 12-mer dsDNA to AVP–pVIc complexes.
3. An aliquot of AVP–pVIc is added, and, after binding equilibrium is reached, the change in anisotropy is measured.
4. Additional aliquots of AVP–pVIc are added until saturation is reached.
5. The data are plotted as the change in anisotropy (r_{obs}-r_o) vs the AVP–pVIc concentration, where r_{obs} is the anisotropy and r_o is the anisotropy in the absence of any added ligand. The graph is interpreted as described for a similar graph in **Subheading 3.2.7.**

3.3.4. Indirect Binding Method

In the indirect method, a fluorophore is attached to a ligand, and its displacement by another ligand is monitored by changes in anisotropy. This method can be used to obtain indirectly the K_d of the competing, unlabeled ligand. For example, the K_d for the binding of 1.3-kDa polyE to AVP–pVIc is determined indirectly by measuring the displacement of 5'-fluorescein-labeled 12-mer ssDNA from AVP–pVIc–(12-mer ssDNA) complexes by increasing concentrations of 1.3-kDa polyE.

1. The experimental setup and solutions are as described in **Subheadings 3.3.1.1–3.3.1.3.**, except that AVP–pVIc at a concentration fivefold greater than the K_d for 12-mer ssDNA and 1 nM 5'-fluorescein-labeled 12-mer ssDNA are incubated for 90 s.
2. The anisotropy is measured repeatedly for a period of 30 s and the numbers averaged.
3. An aliquot of polyE is added. After 90 s the anisotropy is measured.
4. The process in **step 3** is repeated until no further decrease in anisotropy is observed.
5. To get the equilibrium dissociation constant of the nonfluorescent ligand–receptor complex, the observed fluorescence anisotropy vs total concentration of unlabeled ligand data (**Fig. 2B**), are fitted to the following recursive equation (*18*):

$$r_{i+1} = r_o + \frac{(r_s - r_i) \cdot R_0}{L_0 \cdot \dfrac{r_s - r_i}{r_s - r_0} + \dfrac{U_0}{\dfrac{r_i - r_0}{r_s - r_i} + \dfrac{K_U}{K_L}} + K_L}$$

where r_s is the initial anisotropy value when the receptor is preloaded with the fluorescent ligand, r_0 is the limiting anisotropy value at complete saturation with unlabeled ligand, and r_i is the observed anisotropy; K_L and K_U are the dissociation constants of the labeled and unlabeled ligand–receptor complexes, respectively; R_0, L_0, and U_0 are the total molar concentrations of receptor, labeled, and unlabeled ligands, respectively.

4. Notes

1. Expensive quartz cuvets are not required because Rhodamine 110 compounds absorb maximally in the 490-nm range, not in the ultraviolet range.
2. The NaCl is present to prevent nonspecific absorption of protein to the walls of the cuvette.
3. NaCl is absent, because the binding of DNA or polyE to AVP is sensitive to ionic strength *(19)*.
4. The K_m of AVP–pVIc and (Leu-Arg-Gly-Gly-NH)$_2$-Rhodamine is 2.1 μ*M*, and the K_m of AVP–DNA and (Leu-Arg-Gly-Gly-NH)$_2$-Rhodamine is 9.2 μ*M* *(9)*. These are the concentrations that will give half-maximal velocity.
5. G is the grating factor, $G = I_{HV}/I_{HH}$, where I_{HV} and I_{HH} are the measured emission intensities with the excitation beam (first subscript) horizontally polarized ($_H$) and the emission beam (second subscript) vertically ($_V$) or horizontally ($_H$) polarized. The grating factor is used to correct for optical artifacts introduced by the emission monochromator.
6. The lifetime of the excited state of the fluorescein moiety attached to one of the 5'-ends of 18-mer dsDNA is 2.3 ± 0.4 ns, and it did not change by more than 10% when the DNA is saturated with ligand *(4)*. As a control, the lifetime of the excited state of free fluorescein is measured using as a standard reference scattering solution 0.1% (w/v) magnesium oxide. Free fluorescein consistently gave a lifetime of 3.7 ± 0.9 ns. Each lifetime measurement is the average of 10 readings.

Acknowledgments

Research was supported by the Office of Biological and Environmental Research of the U.S. Department of Energy under Prime Contract No. DE-AC0298CH10886 with Brookhaven National Laboratory and by National Institute of Health Grant AI41599.

References

1. Mangel, W. F., McGrath, W. J., Toledo, D. L., and Anderson, C. W. (1993) Viral DNA and a viral peptide can act as cofactors of adenovirus virion proteinase activity. *Nature* **361,** 274–275.
2. Baniecki, M. L., McGrath, W. J., McWhirter, S. M., et al. (2001) Interaction of the human adenovirus proteinase with its eleven amino-acid cofactor pVIc. *Biochemistry* **40,** 12,349–12,356.
3. McGrath, W. J., Baniecki, M. L., Peters, E., Green, D. T., and Mangel, W. F. (2001) Roles of two conserved cysteine residues in the activation of human adenovirus proteinase. *Biochemistry* **40,** 14,468–14,474.
4. McGrath, W. J., Baniecki, M. L., Li, C., et al. (2001) Human adenovirus proteinase: DNA binding and stimulation of proteinase activity by DNA. *Biochemistry* **40,** 13,237–13,245.
5. Bajpayee, N. S., McGrath, W. J., and Mangel, W. F. (2005) Interaction of the adenovirus proteinase with protein cofactors of high negative charge density. *Biochemistry* **44,** 8721–8729.
6. Brown, M. T. and Mangel, W. F. (2004) Interaction of actin and its 11-amino acid C-terminal peptide as cofactors with the adenovirus proteinase. *FEBS* **563,** 213–218.
7. Brown, M. T., McBride, K. M., Baniecki, M. L., Reich, N. C., Marriott, G., and Mangel, W. F. (2002) Actin can act as a cofactor for a viral proteinase, in the cleavage of the cytoskeleton. *J. Biol. Chem.* **277,** 46,298–46,303.
8. Joshua-Tor, L., Xu, H. E., Johnston, S. A., and Rees, D. C. (1995) Crystal structure of a conserved protease that binds DNA: the bleomycin hydrolase, Gal6. *Science* **269,** 945–950.
9. Mangel, W. F., Baniecki, M. L., and McGrath, W. J. (2003) Specific interactions of the adenovirus proteinase with the viral DNA, an 11-amino-acid viral peptide, and the cellular protein actin. *Cell. Mol. Life Sci.* **60,** 2347–2355.
10. Mangel, W. F., Toledo, D. L., Ding, J., Sweet, R. M., and McGrath, W. J. (1997) Temporal and spatial control of the adenovirus proteinase by both a peptide and the viral DNA. *Trends Biochem. Sci.* **22,** 393–398.
11. Rancourt, C., Keyvani-Amineh, H., Sircar, S., Labrecque, P., and Weber, J. M. (1995) Proline 137 is critical for adenovirus protease encapsidation and activation but not enzyme activity. *Virology* **209,** 167–173.
12. Ruzindana-Umunyana, A., Sircar, S., and Weber, J. M. (2000) The effect of mutant peptide cofactors on adenovirus protease activity and virus infection. *Virology* **270,** 173–179.
13. Russell, W. C. and Precious, B. (1982) Nucleic acid-binding properties of adenovirus structural polypeptides. *J. Gen. Virol.* **63,** 69–79.
14. Ding, J., McGrath, W. J., Sweet, R. M., and Mangel, W. F. (1996) Crystal structure of the human adenovirus proteinase with its 11 amino-acid cofactor. *EMBO J.* **15,** 1778–1783.

15. Brown, M. T., McGrath, W. J., Toledo, D. L., and Mangel, W. F. (1996) Different modes of inhibition of human adenovirus proteinase, probably a cysteine proteinase, by bovine pancreatic trypsin inhibitor. *FEBS Lett.* **388**, 233–237.
16. Hinkle, D. and Chamberlin, M. (1972) Studies of the binding of *E. coli* RNA polymerase to DNA. II. The kinetics of the binding reaction. *J. Mol. Biol.* **70**, 187–196.
17. Spencer, R. D. and Weber, G. (1969) Measurement of subnanosecond fluorescence lifetimes with a cross-correlation phase fluorometer. *Ann. NY Acad. Sci.* **158**, 361–376.
18. Kuzmic, P., Moss, M. L., Kofron, J. L., and Rich, D. H. (1992) Fluorescence displacement method for the determination of receptor-ligand binding constants. *Anal. Biochem.* **205**, 65–69.
19. McGrath, W. J., Ding, J., Sweet, R. M., and Mangel, W. F. (1996) Preparation and crystallization of a complex between human adenovirus serotype 2 proteinase and its 11-amino-acid cofactor pVIc. *J. Struct. Biol.* **117**, 77–79.

21

Characterization of the Adenovirus Fiber Protein

Jeffrey A. Engler and Jeong Shin Hong

Summary

Entry of the adenovirus (Ad) capsids during the early stages of infection is a multistep process that includes initial attachment of the virus capsid to the cell surface followed by internalization of the virus into early endosomes. The Ad fiber protein, a complex of three apparently identical subunits, mediates the initial attachment step. In this chapter, methods for the purification and characterization of the Ad fiber protein are presented. Chromatographic methods for the isolation of the protein from infected cells can yield substantial quantities of protein for biochemical analysis. Protocols for characterization of the protein by Western blot and by indirect immunofluorescence of infected cells are also presented. The specificity of different monoclonal and polyclonal antibodies that recognize Ad fiber is also discussed. Ad fiber from a number of serotypes also contains a posttranslational modification, O-linked N-acetyl-glucosamine; methods for detection and characterization of this modification are also provided. With these tools and protocols, one can address important questions about this protein, which helps direct the tissue tropism of Ad.

Key Words: Adenovirus; fiber; virus entry; posttranslational modification; monoclonal antibodies; polyclonal antibodies; indirect immunofluorescence; O-GlcNAc; Western blot; purification; tissue culture; virus infection.

1. Introduction

Entry of the adenovirus (Ad) capsids during the early stages of infection is a multistep process that includes (1) initial attachment of the virus capsid via a high affinity interaction with a cell surface receptor *(1–4)* (the Coxsackie-Ad receptor protein, in the case of Ad serotypes 2 and 5; **ref. 5**); and (2) a second interaction of the penton base protein in the capsid with $\alpha_V\beta3$ and/or $\alpha_V\beta5$ integrins *(6–8)*, which mediates internalization of the virus capsid into early endosomes *(3)*.

From: *Methods in Molecular Medicine, Vol. 131:*
Adenovirus Methods and Protocols, Second Edition, vol. 2:
Ad Proteins, RNA, Lifecycle, Host Interactions, and Phylogenetics
Edited by: W. S. M. Wold and A. E. Tollefson © Humana Press Inc., Totowa, NJ

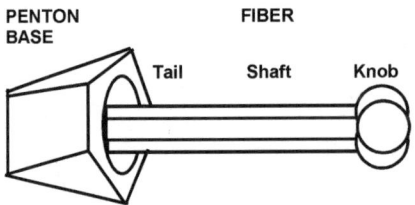

Fig. 1. Proposed structure of the fiber trimer.

With the use of Ad for transfer of genetic information into cells and tissues for therapeutic gene therapy and for basic research, studies of the structure and function of the fiber protein have enjoyed a renewed popularity, because this protein mediates the initial attachment of the virus capsid to a cell surface receptor *(1–5)*. The fiber protein is a complex of three apparently identical subunits and is found at each of the vertices in the icosahedral virus particle *(9)*; by sodium dodecyl sulfate–polyacrylamide gel electrophoresis (SDS-PAGE), the molecular weights of the monomeric and the trimeric form of Ad type 2 (Ad2) and Ad5 fiber are 62,000 and approx 180,000–210,000, respectively *(9,10)*. This trimeric structure has been proposed to consist of *(2,11)* (**Fig. 1**) an N-terminal domain that interacts with the virus penton base and contains signals for transport of the protein to the nucleus, a shaft that contains a series of repeating motifs, and a C-terminal knob that mediates the attachment of the protein complex to the cell surface receptor. This C-terminal knob also contains determinants for type-specific antigens. X-ray crystallographic studies of the structure of Ad2 and Ad5 fiber confirm this structural organization for the fiber trimer *(12–15)*.

\In other serotypes of fiber, the apparent lengths of the trimers are different, based on electron microscopy measurements and comparison of the sequences of the fiber genes from different serotypes *(16–19)*; this difference is generally attributed to the number of repeating motifs found in the shaft domain. In Ad40 and Ad41, the virus encodes two fiber proteins, one of which protrudes from each vertex of the virus capsids; the role played by these additional fibers is not clear *(20,21)*. The Ad fiber protein from serotypes 2, 5, and 12 is also known to contain a posttranslational modification, *O*-linked *N*-acetyl-glucosamine (O-GlcNAc) *(22–25)*; one site of addition of O-GlcNAc on Ad5 fiber has recently been mapped by mass spectroscopy *(26)*.

In order to study the structure and function of fiber and of fiber mutants, a number of strategies have been employed:

1. A number of laboratories have studied temperature-sensitive mutants in the Ad2 and Ad5 fiber gene (for example, *see* **refs. 23** and **27**). Temperature-sensitive mutations (e.g., H5<u>ts</u>142) that affected trimer formation and that affected the

ability to be labeled by ^{3}H-glucosamine (and presumably deficient in the addition of O-GlcNAc) were identified by this approach.

2. Expression of mutant fiber protein has also been achieved in heterologous expression systems, such as *Escherichia coli (28)*, baculovirus *(29)*, and vaccinia virus systems *(30,31)*. Using site-directed mutagenesis, specific mutations in the fiber gene can be constructed and the resulting gene product studied for alteration of its structure and/or function.

An advantage of these systems is that production of the fiber protein does not depend on the necessity of forming an infectious Ad capsid for propagation of the mutant protein, although adenovirions lacking fiber can be produced and appear to be somewhat infectious *(32)*.

2. Materials

2.1. Viruses and Cells

Ad2 and Ad5 viruses can be grown on plates of human HeLa (American Type Culture Collection [ATCC] #CCL2) or other permissive cell lines (A549 [ATCC no. CCL185], KB[ATCC no. CCL17], 293[ATCC no. CRL1573]); spinner cultures of human HeLa cells (HeLa S3 [ATCC no. CCL2.2]) can also be used. Generally infections for preparation of fiber protein are performed at a multiplicity of infection (MOI) of approx 20–30 PFU/cell. Several protocols are available for preparation of virus from infection, such as the one used in **Subheading 3.1.**

2.2. Antibodies

Several polyclonal and monoclonal antibodies (MAbs) are available for studies on Ad fiber. Polyclonal antibodies such as R72 (against Ad2 full-length fiber *[3,33]*) or RAF2Nat (against nondenatured Ad2 full-length fiber *[34]*) can be useful for immunoprecipitations or Western blots of mutant fiber polypeptides; antibody RaF2Nat is also useful for immunofluorescence of Ad2 fiber truncation mutants. Polyclonal antibodies against virion components (available from the ATCC) may also be useful in some cases, but generally will recognize all three major capsid proteins (hexon, penton base, and fiber). Alternatively, a large number of MAbs are available. Several MAbs made in our laboratory have proven very useful for structural studies of fiber by Western blot and by indirect immunofluorescence:

1. 4D2 recognizes both monomer and trimer fiber proteins of Ad2, Ad5, and Ad7 *(30)*; the epitope lies between residues 10 and 17 *(35)*.
2. 2A6 recognizes only the trimeric form of the polypeptide and recognizes an epitope in the N-terminal half of the shaft of the trimer between residues 60 and 260; deletions or alterations in the shaft may remove this epitope (unpublished observation).

Other antibodies that are type specific and that recognize other domains in the protein have also been described. Chroboczek and colleagues have described a series of MAbs made against Ad2 fiber knob *(36)*. Hong et al. *(37)* have described two antibodies (1D6.3 and 7A2.7) that block attachment of Ad5 to the Coxsackie-Ad receptor.

2.3. Purification of Fiber From Virus-Infected Cells

1. S-minimum essential medium (GIBCO), supplemented with 7% bovine calf serum.
2. Phosphate-buffered saline (PBS).
3. TE: 10 m*M* Tris-HCl, pH 8.1, 1 m*M* ethylene diamine tetraacetic acid (EDTA).
4. Freon 113.
5. Solution of CsCl density 1.4 g/mL: weigh 54.8 g CsCl powder and dissolve to a final volume of 100 mL in TE (10 m*M* Tris-HCl [pH 7.9], 1 m*M* EDTA).
6. Solution of CsCl density 1.2 g/mL: weigh 27.7 g CsCl powder and dissolve to a final volume of 100 mL in TE.
7. Saturated solution of ammonium sulfate: to make a saturated ammonium sulfate solution, add 750 g of ammonium sulfate to 1 L of deionized H_2O in a beaker. Stir the solution with a magnetic stirrer at room temperature for a minimum of 20 min. The clear supernatant solution above the undissolved solids that will settle to the bottom of the beaker is the saturated solution.
8. 50 m*M* Sodium phosphate buffer, pH 6.8.
9. Diethylamino ethanol (DEAE)-Sepharose column equilibrated with 50 m*M* sodium phosphate buffer, pH 6.8.
10. 0.01 *M* Potassium phosphate, pH 6.8.
11. Hydroxyapatite column equilibrated with 0.01 *M* potassium phosphate, pH 6.8.

2.4. Indirect Immunofluorescence of Infected Cells

1. Cover slips, grade no. 1.
2. 4% Formaldehyde (Ultrapure, Polysciences, Inc.) in PBS.
3. Nikon Optiphot microscope equipped for fluorescent illumination or similar equipment.
4. Tris-buffered saline (TBS).
5. TBS or PBS containing 1% Triton X-100 (TBS-TX or PBS-TX).
6. TBS-TX or PBS-TX supplemented with 1% bovine serum albumin (BSA; TBS-TX-BSA or PBS-TX-BSA).
7. Primary antibodies: mouse monoclonal 4D2 or 2A6, and control IgG.
8. Fluoroscein isothiocyanate-labeled goat anti-mouse IgG (Jackson Immunoresearch, West Grove, PA) or other fluorescent secondary antibody.
9. Hoechst 33258 (20 µg/mL in PBS).
10. Mounting medium: Vectashield with 4',6-diamidino-2-phenylindole (DAPI; Vector Laboratories, Burlingame, CA).
11. Alternative mounting medium (0.1% *p*-phenylenediamine in 9:1 glycerol/PBS): dissolve 0.05 g *p*-phenylenediamine in 5 mL of PBS in a 50-mL conical tube.

Vortexing the solution in a tube wrapped with foil is preferred. Then, add 45 mL of glycerol.

2.5. Western Blot

1. Standard Laemmli sample buffer.
2. Laemmli sample buffer with 0–0.1% SDS (instead of 2% SDS) to detect trimers.
3. Polyacrylamide gels: between 6 and 8% polyacrylamide.
4. 4–20% Gradient gels.
5. Nitrocellulose or polyvinylidene fluoride membrane or any other appropriate membrane.
6. Coomassie blue.
7. Ponceau S stain for Western blots: dissolve 0.5 g of Ponceau S (Sigma Chemical, cat. no. P 3504) in 1 mL of glacial acetic acid. Bring to 100 mL with H_2O.
8. 5% Nonfat dry milk in PBS (BLOTTO hybridization solution).
9. Primary antibody (anti-fiber): mouse MAb 4D2 for Western blots (available from Neomarkers, Inc., Fremont, CA, www.labvision.com; antibody AB-4).
10. Goat anti-mouse IgG antibody conjugated to alkaline phosphatase (Fisher Biotech). For chemiluminescence detection, use an antibody conjugated to horseradish peroxidase (HRP) instead.
9. Appropriate substrate for alkaline phosphatase or HRP: use according to manufacturer's instructions.

2.6. Detection of O-GlcNAc on Fiber

1. Monoclonal antibody, RL-2 (available from Alexis Biochemicals, www.alexis-corp.com, cat. no. ALX-804-111).
2. MAb that recognizes only O-GlcNAc on proteins (CTD110.6; available from Covance Research Products, store.crpinc.com, cat. no. MMS248R).
3. 10% Polyacrylamide gels.
4. 5% BSA in PBS (BSA hybridization solution).
5. UDP-(U-^{14}C)-galactose (approx 250 mCi/mmol; PerkinElmer Life Sciences; cat. no. NEC429050UC).
6. UDP-galactose (approx 1.5 mg; disodium salt; available from Sigma Chemical, cat. no. U 4500).
7. Cacodylate buffer: 200 mM sodium cacodylate, pH 6.8, and 5 mM $MnCl_2$.
8. β-Mercaptoethanol.
9. Protease inhibitor, such as complete miniprotease inhibitor (Roche).
10. 85% Ammonium sulfate: dissolve 122 g of ammonium sulfate in a final volume of 200 mL H_2O at 25°C.
11. Resuspension buffer: 25 mM N-2-hydroxyethylpiperazine-N'-2-ethanesulfonic acid (HEPES), pH 7.4, 2.5 mM $MnCl_2$, and 50% glycerol.
12. 50 mM HEPES.
13. Bovine milk 4β-galactosyltransferase (4.7 U/mg; Sigma Chemical, cat. no. G-5507).
14. ^{14}C-Glucosamine (50–300 mCi/mmol).

3. Methods

3.1. Purification of Fiber From Virus-Infected Cells

Infection of human HeLa or other permissive human cell types by Ad2 and Ad5 leads to production of a large fraction of capsid proteins (including fiber) that are not incorporated into virions. As far as can be determined experimentally, there are no significant biochemical differences between these unincorporated virus proteins and those found in the virion.

This protocol is modified from that of Boulanger and Puvion *(38)* and is a modification of earlier protocols from Philipson *(39)* and Maizel et al. *(40)*. This protocol continues to be the most popular and reliable method for purifying fiber protein. Although this protocol uses HeLa spinner cells for infection, fiber protein can be isolated by a similar procedure from many other cell lines that can support Ad infection (*see also* **Notes 1** and **2**).

1. HeLa S3 cells are grown in suspension to 5×10^5 cells/mL S-minimum essential medium (GIBCO), supplemented with 7% bovine calf serum, then infected with an MOI of 20–50 PFU/cell. For cells grown on plates, infect 20 15-cm dishes with cells at 90% confluence.
2. After 30–60 h of infection, harvest the cells, wash cells three times with 10 pellet volumes of PBS, and resuspend in 10 volumes of 10 mM Tris, pH 8.1, 1 mM EDTA.
3. Subject the cell pellet to five cycles of quick freezing and thawing. Vortex after each thaw.
4. Extract the solution with an equal volume of Freon 113 with vortexing.
5. Separate the upper aqueous layer containing virus and cellular proteins away from the lower Freon layer by low-speed centrifugation (1500–2000g) for 10 min at room temperature.
6. Layer the upper aqueous phase after low-speed centrifugation on top of a discontinuous cesium chloride step gradient (CsCl density 1.4 g/mL layered below a layer of CsCl density 1.2 g/mL; *see* **Subheading 2.3.**, **items 5** and **6**).
7. Separate the virus by ultracentrifugation using an SW 28 rotor at 24,000 rpm (76,000g) for at least 16 h.
8. After centrifugation (**Fig. 2**) there is a flocculent band (containing unincorporated virus proteins) above a thinner tight band (virions). The supernatant above the virion band is the source of Ad-soluble fiber. This band can be removed from the tube with a syringe and needle or with careful pipetting. (For isolating banded virus, two CsCl separations are recommended. The supernatant above the virus band from each separation may be pooled and used in the next steps.)
9. Slowly (dropwise with stirring) add a saturated solution of ammonium sulfate (*see* **Subheading 2.3.**, **item 7**) at 4°C to a final concentration of 55% ammonium sulfate. To calculate the amount of saturated ammonium sulfate to add, measure the volume of the fiber fraction and multiply it by 1.22 (i.e., if the fiber solution has a volume of 10 mL, add 12.2 mL of saturated ammonium sulfate). Allow the ammonium sulfate precipitate to form for 15 h at 4°C and pH 6.5; do not stir the solution during this incubation.

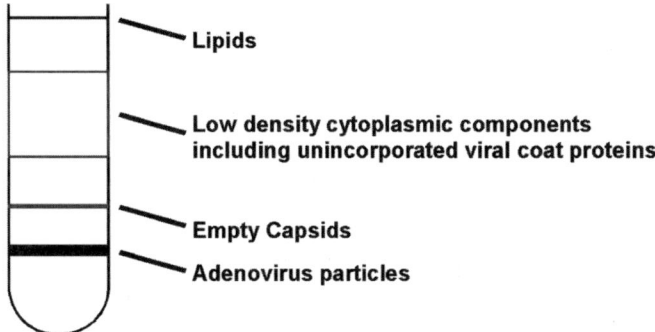

Fig. 2. Diagram of expected bands obtained after centrifugation of virus lysates to purify virions and unincorporated virion proteins, including fiber. The dense virion particle band is generally observed at the boundary between the 1.2 and 1.4 g/mL CsCl layers. The diffuse area of virus and cellular proteins is generally observed above the 1.2 g/mL CsCl layer.

10. Spin the precipitate at 5000*g* for 20 min at 4°C.
11. Dissolve the pellet in a minimal volume of 50 m*M* sodium phosphate buffer, pH 6.8, at 4°C.
12. Dialyze the sample at 4°C against 10 vol of 50 m*M* sodium phosphate buffer, pH 6.8, over 72 h with at least five changes of buffer.
13. After dialysis, centrifuge the solution at 110,000*g* for 2 h. This supernatant is ready for column purification or for storage at –20°C.
14. Load the sample onto a DEAE-Sepharose column equilibrated with 50 m*M* sodium phosphate buffer, pH 6.8, at 4°C.
15. Wash the column with at least 10 column volumes of 50 m*M* sodium phosphate buffer, pH 6.8, at 4°C.
16. Elute the column with a linear gradient of sodium chloride (from 0.0 to 0.5 *M*), in 50 m*M* sodium phosphate buffer, pH 6.8. Collect fractions and assay for the fractions containing protein by absorbance at 278 nm.
17. Fractions containing fiber are identified by SDS-PAGE of aliquots of fractions on a 10% gel. Boiled Ad2 or Ad5 fiber protein should run with an apparent molecular weight of 62,000. Fractions containing fiber can also be identified by Western blot of the gel, using an antibody specific for fiber (*see* **Subheading 3.3.**).
18. Pool the fractions containing fiber protein and dialyze the pooled fractions against 0.01 *M* potassium phosphate, pH 6.8 (*see* **Note 3**).
19. Load this dialyzed solution onto a hydroxyapatite column equilibrated with 0.01 *M* potassium phosphate buffer, pH 6.8.
20. Wash the column with at least 10 column volumes of buffer.
21. Elute the fiber protein with a linear gradient of 0.01 to 0.50 *M* potassium phosphate, pH 6.8.
22. Collect fractions and test for protein concentration and the presence of fiber as described in **steps 16–17**.

3.2. Indirect Immunofluorescence of Infected Cells

Indirect immunofluorescence of infected cells provides one of the most direct means for assessing the expression of fiber and fiber mutants within infected cells. With the use of appropriate conformation specific antibodies, information about the ability to form appropriate trimers inside the infected cell can also be determined.

1. Grow the cells to be infected on cover slips (grade no. 1) that can be viewed in an appropriate light microscope. This growth is most easily accomplished by placing sterile cover slips (*see* **Note 4**) in the bottom of a 35-mm tissue culture dish, into which the cells will be seeded in appropriate medium (Dulbecco's modified Eagle's medium + 5% fetal bovine serum). Alternatively, chamber slides can be used. The cells should be seeded at least 24–36 h prior to infection to ensure that the infected cells remain attached to the cover slip (*see* **Note 5**).
2. When 60–80% confluent, infect the cells with Ad (MOI approx 10 PFU/cell) or other recombinant virus that expresses fiber.
3. After 24–60 h of infection, wash the cells on the cover slip once with PBS and fix them with 4% formaldehyde in PBS for 10–15 min at room temperature.
4. Wash the cells with TBS or with PBS. Do not allow the cells to dry during the procedure.
5. Incubate the cells on the coverslip with TBS or PBS containing 1% Triton X-100 (TBS-TX or PBS-TX) to permeabilize the cells for 10 min at room temperature. As an alternative, the cells can be permeabilized by treating them with ice-cold methanol for 2 min.
6. Block nonspecific binding with TBS-TX or PBS-TX supplemented with 1% BSA (TBS-TX-BSA or PBS-TX-BSA) in the same buffer for 30 min at room temperature.
7. Incubate the cells on the cover slip with a 1:100 to 1:500 dilution of antibody (ascites fluid of either 4D2 or 2A6; the appropriate dilution should be determined experimentally) for 1 h at room temperature or 37°C in a humidified chamber. Control IgG should be used as a negative control. If you use tissue culture supernatant taken from hybridoma cells, use it without further dilution (*see* **Note 6**).
8. Wash the cells on the cover slip with TBS-TX-BSA or PBS-TX-BSA for 15–30 min with three changes of buffer.
9. Add an appropriate dilution (approx 1:100, but the appropriate dilution should be determined experimentally) of fluorescein isothyocyanate-labeled goat anti-mouse Ig as a secondary antibody for 1 h at room temperature.
10. The cells on the cover slip are then washed with three changes of TBS-TX-BSA or PBS-TX-BSA (5 min per wash).
11. Optional: if you need to specifically visualize the cell nuclei, treat the cells with Hoechst 33258 (20 µg/mL in PBS) for 4 min at room temperature to stain the nuclei. Rinse the cells on the cover slips briefly with TBS or PBS. Alternatively, one can use a mounting medium containing DAPI to stain the nuclei, such as Vectashield with DAPI.

12. Mount the cover slips on slides using a mounting medium such as Vectashield. Alternatively, one can use 0.1% *p*-phenylenediamine in 9:1 glycerol/PBS (*see* **Subheading 2.4., item 12**). Remove any excess mounting medium by absorbing with a filter paper (the bigger the pore size, the better).

13. Seal the edges of the cover slips with nail polish. Allow the polish to dry thoroughly before proceeding further. Protect the cover slips from light by covering with aluminum foil while drying.

14. View the infected cells using a Nikon Optiphot microscope equipped for fluorescent illumination or similar equipment.

15. The slides can be stored at –20°C for a limited period, but must be kept in an opaque slide box or wrapped in foil to minimize exposure to light.

3.3. Detection of Fiber Monomers and Trimers by Western Blot

Another means to detect the presence of trimeric fiber complexes is by Western blotting of boiled and unboiled samples of fiber protein *(31,41,42)*. Fiber trimers are stable to electrophoresis in the standard polyacrylamide gel containing SDS (SDS-PAGE), provided that the samples are not boiled prior to loading.

1. Resuspend samples of fiber protein (either from crude lysates of infected cells or from purified protein) in standard Laemmli sample buffer *(43)* to detect monomers or in Laemmli sample buffer with 0–0.1% SDS (instead of 2% SDS) to detect trimers.

2. To detect trimers from Ad2 and Ad5 fiber, use a 6–8% polyacrylamide gel to allow the trimer to enter the gel matrix. When boiled and nonboiled samples are to be resolved on the same gel, it is preferable to use a 4–20% gradient gel.

3. Prior to loading the samples on the polyacrylamide gel, heat those samples in which the monomeric size of the protein is to be determined. For samples used to detect trimers, leave the samples on ice or at room temperature.

4. Load the samples into the wells of the polyacrylamide gel and subject to SDS-PAGE. Avoid excess heating of the gel during electrophoresis.

5. Transfer the gel onto nitrocellulose or polyvinyl difluoride membrane or any other appropriate membrane, following the standard protocol of the manufacturer.

6. Use a monoclonal or polyclonal antibody to detect the apparent molecular mass of the fiber on the blot. After transfer, stain the gel with Coomassie blue to assure complete transfer (there should be few or no proteins left in the gel to be stained). Alternatively, stain the blot with Ponceau S stain (*see* **Subheading 2.5., item 7**), to visualize the proteins on the blot.

7. Prehybridize the blot for 30 min at room temperature with 5% nonfat dry milk in PBS (BLOTTO hybridization solution).

8. Dilute the anti-fiber antibody (1:1000–1:2000 for a typical ascites fluid preparation of a MAb) into the BLOTTO hybridization solution. Incubate at room temperature for 1–2 h. We prefer using MAb 4D2 for Western blots (Neomarkers, Inc., antibody AB-4).

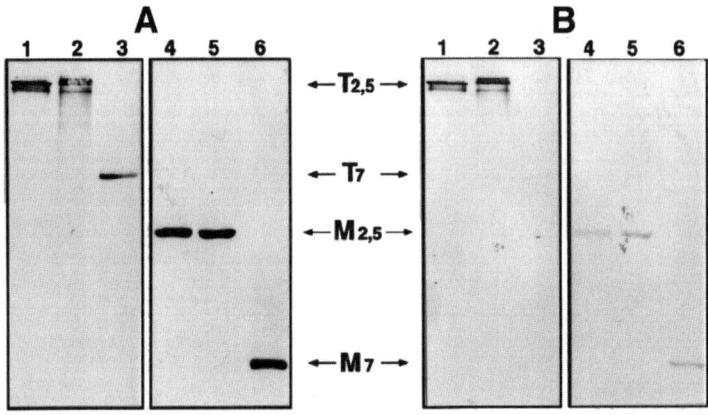

Fig. 3. Western blot of fiber monomers and trimers. Aliquots of cell lysates from Ad2-, Ad5-, and Ad7-infected A549 cells were separated by sodium dodecyl sulfate–polyacrylamide gel electrophoresis on a 10% gel and blotted onto Immobilon membrane prior to developing with monoclonal antibody 4D2-5 (**A**) or 2A6-36 (**B**). The sizes expected for monomers (M) and trimers (T) of the various molecules are indicated by the arrows. In each panel, the order of the lanes is: lane 1, Ad2 fiber unboiled; lane 2, Ad5 fiber unboiled; lane 3, Ad7 fiber unboiled; lane 4, Ad2 fiber boiled; lane 5, Ad5 fiber boiled; lane 6, Ad7 boiled.

9. Remove the hybridization solution and wash three times with fresh BLOTTO hybridization solution.
10. Add an appropriate dilution goat anti-mouse IgG antibody conjugated to alkaline phosphatase (Fisher Biotech) if a MAb is used. For chemiluminescence detection, use an antibody conjugated to HRP instead. The required dilution should be determined experimentally. Dilute with BLOTTO hybridization solution. Incubate for 60 min at room temperature (*see* **Note 7**).
11. Wash three times with PBS (5 min each wash).
12. Develop the color with an appropriate substrate for alkaline phosphatase, according to the manufacturer's instructions. Alternatively, chemiluminescent detection can be used. The intensity of the color depends on many factors, including the length of incubation with the color reagent.
13. Trimers generally run at a molecular mass approximately three times larger than that of monomers (*see* **Fig. 3**; *see also* **Notes 8** and **9**).

3.4. Detection of O-GlcNAc Addition on Fiber

As discussed in the introduction, fibers from some serotypes of Ad contain O-GlcNAc modifications. There are three primary means by which the presence of this O-GlcNAc can be detected on fiber. Use of O-GlcNAc-specific antibodies is the easiest method but may suffer from lack of sensitivity (**Subheading 3.4.1.**). More quantitative approaches are the use of bovine milk galactosyltransferase to transfer radioactive galactose onto the O-GlcNAc (**Subheading 3.4.2.**) or to label protein with radioactive glucosamine during infection (**Subheading 3.4.3.**; *see also* **Note 10**).

3.4.1. Detection by RL-2 Antibody Reactivity on Western Blot

The easiest means for detection of O-GlcNAc on the fiber protein is the use of the MAb, RL-2 *(44)*, as a probe of fiber immobilized on a Western blot. The antibody RL-2 was originally isolated because of its reactivity to nuclear pore proteins that contain O-GlcNAc; this antibody also recognizes the O-GlcNAc on fiber *(25)*. However, RL-2 also recognizes a protein component in addition to O-GlcNAc, so it may or may not detect all potential O-GlcNAc sites on fiber. More recently, a MAb that recognizes only O-GlcNAc on proteins (CTD110.6) has been described *(45)*; whether this antibody will detect O-GlcNAc on fiber has not been tested.

1. Separate a minimum of 1–2 µg of fiber (per lane) by SDS-PAGE on a 10% gel. In crude lysates from infected cells, more protein may have to be loaded per lane in order to detect the fiber present in the extract. If the infection is efficient, the major band detected by RL-2 will be fiber.
2. Electroblot onto nitrocellulose membrane, Immobilon, or any other appropriate membrane for Western blotting. After transfer, stain the gel with Coomassie blue to assure complete transfer (there should be few proteins left in the gel to be stained). Alternatively, stain the blot with Ponceau S stain (**Subheading 2.5., item 7**) to visualize the proteins on the blot.
3. Prehybridize the blot for 30 min at room temperature with 5% BSA in PBS (BSA hybridization solution). For RL-2, one cannot use a hybridization solution containing nonfat dry milk because of the high reactivity of the proteins in the blocking buffer to RL-2; this potential pitfall will reduce the signal obtained on the blot.
4. Dilute the RL-2 antibody (1:1000–1:2000 for a typical ascites fluid preparation of enzyme) into BSA hybridization solution. Incubate at room temperature for 1–2 h.
5. Remove the hybridization solution and wash three times with fresh BSA hybridization solution.
6. Add goat anti-mouse IgG antibody conjugated to alkaline phosphatase (Fisher Biotech) or to HRP and incubate for 1 h at room temperature (*see* **Note 7**). In our experience, RL-2 blots developed by chemiluminescence methods show multiple minor bands in crude cell lysates because of the high sensitivity of the procedure.
7. Wash three times with PBS (5 min each wash).
8. Develop the color with an appropriate substrate for alkaline phosphatase, according to the manufacturer's instructions.

3.4.2. Detection by Labeling of Fiber With Radioactive UDP-Galactose by Bovine Milk Galactosyltransferase

Either ^{3}H- or ^{14}C-UDP-galactose (PerkinElmer Life Sciences) can be used for this assay; we have used ^{14}C-UDP-galactose in this protocol, but either compound can be used. For the best results with this assay, one needs purified fiber protein. Previous work has shown that the O-GlcNAc moieties on fiber are relatively inaccessible to the enzyme *(25)*. If impure preparations are used, the amount of label added to fiber could be significantly lower because many of the possible contaminating proteins may also contain O-GlcNAc that is more accessible to galactosyltransferase than are the O-GlcNAc moieties on fiber (*see also* **Note 11**).

3.4.2.1. PREGALACTOSYLATION OF THE GALACTOSYLTRANSFERASE ENZYME

1. Prior to the reaction, the enzyme itself must be pretreated with unlabeled UDP-galactose, because the enzyme itself contains significant quantities of terminal O-GlcNAc *(46–49)*. 1.5 mg of bovine milk 4β-galactosyltransferase is added to a 1 mL solution that contains 200 mM sodium cacodylate, pH 6.8, and 5 mM MnCl$_2$. As an alternative, 50 mM HEPES, pH 7.3–7.5, may be substituted for the cacodylate buffer.
2. Add 2 mM unlabeled UDP-galactose (approx 1.5 mg), 3.5 µL of a 1:50 dilution β-mercaptoethanol in H$_2$O, and 10 µL aprotinin. The reaction is carried out for 30 min on ice, then for 60 min at 37°C.
3. Add 5.66 mL of saturated ammonium sulfate solution at 4°C (85% final concentration after dilution) and incubate on ice for 30 min.
4. Spin at 10,000g for 15 min in a refrigerated centrifuge. Typically, this centrifugation is performed in an SS-34 rotor (Sorvall).
5. Remove and save the supernatant.
6. Resuspend the pellet in 5 mL of 85% ammonium sulfate (**Subheading 2.6., item 10**) prechilled to 4°C. Incubate for 30 min on ice.
7. Spin at 10,000g for 15 min in a refrigerated centrifuge. Typically, this centrifugation is performed in an SS-34 rotor (Sorvall).
8. Remove and save the supernatant.
9. Resuspend the pellet in 1 mL of 25 mM HEPES, pH 7.4, 2.5 mM MnCl$_2$, and 50% glycerol. The approximate enzyme activity is 25 mU/µL.

3.4.2.2. GALACTOSYLATION OF FIBER PROTEIN WITH GALACTOSYLTRANSFERASE

1. In a final volume of 100 µL, resuspend purified fiber (generally 15 µg) in a solution that contains 200 mM sodium cacodylate, pH 6.8, 5 mM MnCl$_2$, and 0.4 µCi UDP-(U-^{14}C)-galactose (approx 250 mCi/mmol; PerkinElmer Life Sciences). 50 mM HEPES may be substituted for the cacodylate in the buffer.
2. Add 50 mU of the pre-treated 4β-galactosyltransferase to the reaction and incubate for 30–120 min at 37°C. The time depends on how completely the reaction must proceed.

3. After reaction, use immunoprecipitation with an excess of antibody to remove the labeled fiber from the solution. Alternatively, fiber protein can be immunoprecipitated from the chide cell lysate prior to galactosylation (*see* **Note 11**).

4. Analyze a portion of the immunoprecipitate by SDS-PAGE on a 10% gel. The fiber protein can be visualized by staining with Coomassie blue. The band can then be excised from the gel and the amount of ^{14}C-label determined by scintillation counting. Alternatively, labeled bands in the gel can be visualized by fluorography, by impregnating the gel with EN[3]HANCE or other similar product, and exposing the gel at −70°C to preflashed X-ray film. Another alternative is to develop the image on a Phosphorimager and quantitate the intensity of the bands by densitometry. If tritiated UDP-galactose is used, do not treat the gel with EN[3]HANCE if you intend to use a Phosphorimager to detect the incorporated radioactivity; EN[3]HANCE will ruin the tritium-sensitive Phosphorimager screen.

5. For quantitative results, two controls must be performed.
   a. After 1 h of reaction, add an additional 15-μg aliquot of purified fiber protein to the reaction and incubate for an additional 60 min. The amount of ^{14}C-cpm incorporated into the fiber band should be twice as high, indicating that the reaction is linear with time.
   b. Take an aliquot of the original reaction at 60 and at 120 min. If the reaction has gone to completion, the amount of ^{14}C-cpm incorporated should be nearly identical.

3.4.3. Detection by ^{14}C-Glucosamine Labeling of Ad2- or Ad5-Infected Cells (22,24,50–52)

Another method for detecting O-GlcNAc on fiber is in vivo labeling of Ad-infected cells with ^{14}C-glucosamine. This approach suffers from the expense of the radioactive reagent and that many of the other abundant cellular proteins also contain O-GlcNAc, so that much of the label incorporated is found in proteins not related to fiber.

1. Grow HeLa cells to 50–70% confluence (on plates) or 5×10^5 cells/mL (spinner culture) in Dulbecco's modified Eagle's medium supplemented with serum.
2. Infect HeLa cells with Ad2 or Ad5 virus (MOI of 10–20 PFU/cell).
3. After 12 h of infection, add 100 μCi ^{14}C-glucosamine (50–300 mCi/mmol) and label cells for an additional 24 h.
4. Remove radioactive medium and dispose of properly according to your local regulations.
5. Collect infected cells and purify by one of the purification methods listed above. Alternatively, the labeled fiber can be immunoprecipitated if only a qualitative answer is required.

4. Notes

1. An alternate protocol for purification of fiber has been developed by M. S. Horwitz (personal communication). This procedure uses MonoQ and MonoS column chromatography as an alternative to the procedure provided in **Subheading 3.1.**

2. It is also possible to express fiber mutants and fiber truncations in *E. coli*, in baculovirus, and in vaccinia virus. Procedures for purification of the fibers produced in these systems may vary somewhat from the protocol given. Similarly, procedures for purification of fibers from other serotypes may also differ.

3. The protein concentration in the pooled fractions can be increased by ammonium sulfate precipitation (**steps 9** and **10** in **Subheading 3.1.**), followed by dialysis against 0.01 *M* potassium phosphate, pH 6.8.

4. Cover slips may be sterilized by dipping in ethanol and flaming or by baking in an oven after putting in a glass Petri dish. Alternatively, cover slips can be sterilized by UV irradiation in a UV light chamber (Bio-Rad); be sure to irradiate both sides of the cover slip. We do not recommend autoclaving for sterilization; the cover slips will stick together.

5. For indirect immunofluorescence using vaccinia virus to express fiber, HeLa T4 cells may be preferable because they keep their morphology intact for a longer period of time after virus infection.

6. The polyclonal anti-virion antibodies from ATCC do not give good immunofluorescence results because of high background reactivity. Use of MAbs is generally preferable.

7. Alternatively, incubate with goat anti-mouse Ig antibody conjugated to biotin and then develop with alkaline phosphatase-conjugated streptavidin; this strategy is more sensitive but subject to higher background. Chemiluminescence using HRP-conjugated antibodies also provides a sensitive method for detection.

8. Even when the sample is not boiled, a fraction of the fiber may run at a size consistent with monomers. The fraction of monomers and trimers may also be affected by whether mutations or insertions (such as retargeting ligands for gene therapy) have been incorporated into the protein. In our experience, many insertions in the protein result in a less stable trimer protein and may reduce the yield of infectious virus particles.

9. In the Western blotting protocol, occasionally antibodies that recognize only trimers (e.g., 2A6) will react with monomer size bands, probably because of the reassembly of the fiber trimer within the band on the blot. This artifact is often more prevalent if the blots are used days or weeks after the initial protein transfer.

10. Lectin affinity chromatography is a popular method for detection and identification of carbohydrate modifications on protein *(53)*. In the case of Ad fiber from serotypes 2 and 5, wheat germ agglutinin-Sepharose can be used as a matrix for detection of these proteins. The efficiency of binding of protein to wheat germ agglutinin-Sepharose is affected by the concentration, accessibility, and clustering of the moieties on the protein; in the case of the fiber protein, these considerations limit the utility of this approach, especially as a method for purification or analysis of fiber. Ad2 and Ad5 fiber proteins have no affinity to other lectins, such as *ricinus communis* agglutinin or concanavalin A.

11. For the galactosylation reaction with radioactive UDP-galactose, the fiber can be immunoprecipitated prior to galactosylation. However, antibodies contain a significant amount of terminal O-GlcNAc, so that the resulting products will contain label on both fiber and the precipitating antibodies. To minimize this

extraneous labeling, one needs to pregalactosylate the antibody with unlabeled UDP-galactose prior to immunoprecipitation using the protocol described in **Subheading 3.4.2.2., steps 1** and **2**.

References

1. Svensson, U., Oersson, R., and Everitt, E. (1982) Virus-receptor interaction in the Ad system. I. Identification of virion attachment proteins of the HeLa cell plasma membrane. *J. Virol.* **38**, 70–81.
2. Devaux, C., Caillet-Boudin, M.-L., Jacrot, B., and Boulanger, P. (1987) Crystallization, enzymatic cleavage, and the polarity of the Ad type 2 fiber. *Virology* **161**, 121–128.
3. Greber, U. F., Willetts, M., Webster, P., and Helenius, A. (1993) Stepwise dismantling of Ad 2 during entry into cells. *Cell* **75**, 477–486.
4. Louis, N., Fender, P., Barge, A., Kitts, P., and Chroboczek, J. (1994) Cell-binding domain of Ad serotype 2 fiber. *J. Virol.* **68**, 4104–4106.
5. Bergelson, J. M., Cunningham, J. A., Droguett, G., et al. (1997) Isolation of a common receptor for Coxsackie B viruses and Ades 2 and 5. *Science* **275**, 1320–1323.
6. Bai, M., Harfe, B., and Freimuth, P. (1993) Mutations that alter an Arg-Gly-Asp (RGD) sequence in the Ad type 2 penton base protein abolish its cell-rounding activity and delay virus reproduction in flat cells. *J. Virol.* **67**, 5198–5205.
7. Wickham, T. J., Mathias, P., Cheresh, D. A., and Nemerow, G. R. (1993) Integrin αvβ3 and αvβ5 promote Ad internalization but not virus attachment. *Cell* **73**, 309–319.
8. Wickham, T. J., Filardo, E. J., Cheresh, D. A., and Nemerow, G.R. (1994) Integrin αvβ5 selectively promotes Ad mediated cell membrane permeabilization. *J. Cell Biol.* **127**, 257–264.
9. van Oostrum, J. and Burnett, R. M. (1985) Molecular composition of the Ad type 2 virion. *J. Virol.* **56**, 439–448.
10. Sunquist, B., Pettersson, U., Thelander, L., and Philipson, L. (1973) Structural proteins of Ades. IX. Molecular weight and subunit composition of Ad type 2 fiber. *Virology* **51**, 252–256.
11. Green, N. M., Wrigley, N. G., Russell, W. C., Martin, S. R., and McLachlan, A. D. (1983) Evidence for a repeating cross-β sheet structure in the Ad fibre. *EMBO J.* **2**, 1357–1365.
12. Henry, L. J., Xia, D., Wilke, M. E., Deisenhofer, J., and Gerard R. D. (1994) Characterization of the knob domain of the Ad type 5 fiber protein expressed in Escherichia coli. *J. Virol.* **68**, 5239–5246.
13. Xia, D., Henry, L. J., Gerard, R. D, and Deisenhofer, J. (1994) Crystal structure of the receptor-binding domain of Ad type 5 fiber protein at 1.7 A resolution. *Structure* **2**, 1259–1270.
14. van Raaij, M. J., Louis, N., Chroboczek, J., and Cusack, S. (1999) Structure of the human Ad serotype 2 fiber head domain at 1.5 A resolution. *Virology* **262**, 333–343.

15. van Raaij, M. J., Mitraki, A., Lavigne, G. , and Cusack, S. (1999) A triple beta-spiral in the Ad fibre shaft reveals a new structural motif for a fibrous protein. *Nature* **401,** 935.

16. Signäs, C., Akusjärvi, G., and Petterson, U. (1985) Ad 3 fiber polypeptide gene: implications for the structure of the fiber protein. *J. Virol.* **53,** 672–678.

17. Hong, J. S., Mullis, K. G., and Engler, J.A. (1988) Characterization of the early region 3 and fiber genes of Ad7. *Virology* **167,** 545–553.

18. Kidd, A. H., Erasmus, M. J., and Tiemessen, C. T. (1990) Fiber sequence hetero-geneity in subgroup F Ades. *Virology* **179,** 139–150.

19. Ruigrok, R. W. H., Barge, A., Albiges-Rizo, C., and Dayan, S. (1990) Structure of Ad fibre II. Morphology of single fibres. *J. Mol. Biol.* **215,** 589–596.

20. Pieniazek, N. J., Slemenda, S. B., Pieniazek, D., Velarde, J., Jr., and Luftig, R. B. (1990) Human enteric Ad type 41 (Tak) contains a second fiber protein gene. *Nucleic Acids Res.* **18,** 1901.

21. Kidd, A. H., Chroboczek, J., Cusack, S., and Ruigrok, R. W. (1993) Ad type 40 virions contain two distinct fibers. *Virology* **192,** 73–84.

22. Ishibashi, M. and Maizel, J. V. (1974) The polypeptides of Ad VI. Early and late glycoproteins. *Virology* **58,** 345–361.

23. Chee-Sheung, C. C. and Ginsberg, H. S. (1982) Characterization of a tempera-ture-sensitive fiber mutant of type 5 Ad and effect of the mutation on virion assembly. *J. Virol.* **42,** 932–950.

24. Caillet-Boudin, M.-L., Strecker, G., and Michalski J.-C. (1989) O-linked GlcNAc in serotype-2 Ad fibre. *Eur. J. Biochem.* **184,** 205–211.

25. Mullis, K. G., Haltiwanter, R. S., Hart, G. W., Marchase, R. B., and Engler, J. A. (1990) Relative accessibility of N-acetylglucosamine in trimers of the Ad types 2 and 5 fiber proteins. *J. Virol.* **64,** 5317–5323.

26. Cauet, G., Strub, J.-M., Leize, E., Wagner, E., Van Dorsselaer, A. and Lusky, M. (2005) Identification of the glycosylation site of Ad type 5 fiber protein. *Bio-chemistry* **44,** 5453–5460.

27. Caillet-Boudin, M.-L., Lemay, P., and Boulanger, P. (1991) Functional and struc-tural effects of an Ala to Val mutation in the Ad serotype 2 fibre. *J. Mol. Biol.* **217,** 477–486.

28. Albiges-Rizo, C. and Chroboczek, J. (1990) Ad serotype 3 fibre protein is expressed as a trimer in *E. coli*. *J. Mol. Biol.* **212,** 247–252.

29. Novelli, A., and Boulanger. P. (1991) Deletion analysis of functional domains in baculovirus-expressed Ad type 2 fiber. *Virology* **185,** 365–376.

30. Hong, J. S., and Engler, J. A. (1991) The amino terminus of the Ad fiber protein encodes the nuclear localization signal. *Virology* **185,** 758–767.

31. Hong, J. S. and Engler, J. A. (1996) Domains required for assembly of Ad type 2 fiber trimers. *J. Virol.* **70,** 7071–7078.

32. Falgout, B. and Ketner, G. (1988) Characterization of Ad particles made by dele-tion mutants lacking the fiber gene. *J. Virol.* **62,** 622–625.

33. Baum, S. G., Horwitz, M. S., and Maizel, J. V., Jr. (1972) Studies of the mecha-nism of enhancement of human adenovirus infection in monkey cells by simian virus 40. *J. Virol.* **10,** 211–219.

34. Caillet-Boudin, M. L., Novelli, A., Gesquiere, J. C., and Lemay, P. (1988) Structural study of adenovirus type 2 fibre using anti-fibre and anti-peptide zero. *Annales de l'Institut Pasteur Virol.* **139**, 141–156.

35. Hong, S.-S., and Boulanger, P. (1995) Protein ligands of the human Ad type 2 outer capsid identified by biopanning of a phage-displayed peptide library on separate domains of wild-type and mutant penton capsomers. *EMBO J.* **14**, 4714–4727.

36. Fender, P., Kidd, A. H., Brebant, R., Öberg, M., Drouet, E., and Chroboczek, J. (1995) Antigenic sites on the receptor-binding domain of human Ad type 2 fiber. *Virology* **214**, 110–117.

37. Hong, S. S., Karayan, L., Tournier, J., Curiel, D. T., and Boulanger, P. A. (1997) Ad type 5 fiber knob binds to MHC class I alpha2 domain at the surface of human epithelial and B lymphoblastoid cells. *EMBO J.* **16**, 2294–2306.

38. Boulanger, P. and Puvion, F. (1973) Large-scale preparation of soluble Ad hexon, penton and fiber antigens in highly purified form. *Eur. J. Biochem.* **39**, 37–42.

39. Philipson, L. (1960) Separation on DEAE cellulose of components associated with Ad reproduction. *Virology* **10**, 459–465.

40. Maizel, J. V., Jr., White, D. O., and Scharff, M. D. (1968) The polypeptides of Ad. II. Soluble proteins, cores, top components and the structure of the virion. *Virology* **36**, 126–136.

41. Smith, J. A. (1992) Analysis of proteins, in *Current Protocols in Molecular Biology* (Ausubel, F. M., Brent, R., Kingston, R. E., et al., eds), John Wiley & Sons, Inc., New York, Section 10-7.

42. Michael, S. I., Hong, J. S., Curiel, D. T., and Engler, J. A. (1995) Addition of a short peptide ligand to the Ad fiber protein. *Gene Ther.* **2**, 660–668.

43. Laemmli, U. (1970) Cleavage of structural proteins during the assembly of the head of bacteriophage T4. *Nature* **227**, 680–685.

44. Snow, C. M., Senior, A., and Gerace, L. (1987) Monoclonal antibodies identify a group of nuclear pore complex glycoproteins. *J. Cell Biol.* **104**, 1143–1156.

45. Comer, F. I., Vosseller, K., Wells, L., Accavitti, M. A., and Hart, G. W. (2001) Characterization of a mouse monoclonal antibody specific for O-linked N-acetylglucosamine. *Anal. Biochem.* **293**, 169–177.

46. Torres, C.-R. and Hart, G. W. (1984) Topography and polypeptide distribution of terminal N-acetylglucosamine residues on the surfaces of intact lymphocytes. *J. Biol. Chem.* **259**, 3308–3317.

47. Whiteheart, S. W., Passaniti, A., Reichner, J. S., Holt, G. D., Haltiwanger, R. S., and Hart, G. W. (1989) Glycosyltransferase probes. *Methods Enzymol.* **179**, 82–95.

48. Roquemore, E. P., Chou, T.-Y., and Hart, G. W. (1994) Detection of O-linked N-acetylglucosamine (O-GlcNAc) on cytoplasmic and nuclear proteins. *Methods Enzymol.* **230**, 443–460.

49. Hayes, B. K., and Hart, G. W. (1996) Analysis of saccharide structure and function using glycosyltransferases glycoconjugates, in *Current Protocols in Molecular Biology* (Ausubel, F. M., Brent, R., Kingston, R. E., Moore, D. D., Seidman, J. B., and Struhl, K., eds.), John Wiley & Sons, Inc., New York, Section 17.6.

50. Yurchenco, P. D., Ceccarini, C., and Atkinson, P. H. (1978) Labeling complex carbohydrates of animal cells with monosaccharides. *Methods Enzymol.* **50,** 175–204.

51. Varki, A. (1991) Radioactive tracer techniques in the sequencing of glycoprotein oligosaccharides. *FASEB J.* **5,** 226–235.

52. Diaz, S. and Varki, A. (1996) Metabolic radiolabeling of animal cell glycoconjugates, in *Current Protocols in Molecular Biology* (Ausubel, F. M., Brent, R., Kingston, R. E., et al., eds), John Wiley & Sons, Inc., New York, Section 17.4.

53. Merkle, R. K. and Cummings, R. D. (1987) Lectin affinity chromatography of glycopeptides. *Meth. Enzymol.* **138,** 232–259.

22

Phylogenetic Analysis of Adenovirus Sequences

Balázs Harrach and Mária Benkő

Summary

Members of the family *Adenoviridae* have been isolated from a large variety of hosts, including representatives from every major vertebrate class from fish to mammals. The high prevalence, together with the fairly conserved organization of the central part of their genomes, make the adenoviruses one of (if not *the*) best models for studying viral evolution on a larger time scale. Phylogenetic calculation can infer the evolutionary distance among adenovirus strains on serotype, species, and genus levels, thus helping the establishment of a correct taxonomy on the one hand, and speeding up the process of typing new isolates on the other. Initially, four major lineages corresponding to four genera were recognized. Later, the demarcation criteria of lower taxon levels, such as species or types, could also be defined with phylogenetic calculations. A limited number of possible host switches have been hypothesized and convincingly supported. Application of the web-based BLAST and MultAlin programs and the freely available PHYLIP package, along with the TreeView program, enables everyone to make correct calculations. In addition to step-by-step instruction on how to perform phylogenetic analysis, critical points where typical mistakes or misinterpretation of the results might occur will be identified and hints for their avoidance will be provided.

Key Words: Phylogeny; phylogenetic analysis; atadenovirus; siadenovirus; aviadenovirus; PHYLIP; TreeView; distance matrix; bioinformatics; fish; reptiles; wild animals.

1. Introduction

During the 8 yr that have elapsed since the first edition of this book *(1)*, DNA sequencing and the use of computer-aided phylogenetic analyses on partial or full viral sequences have enormously escalated and have become routine methods in both research and diagnostic laboratories. The large-scale ground-gaining of polymerase chain reaction (PCR) and sequencing resulted in the fast accumulation of a great wealth of viral sequence information available in public databases, which, in turn, has significantly ameliorated reliability of the

From: *Methods in Molecular Medicine, Vol. 131:*
Adenovirus Methods and Protocols, Second Edition, vol. 2:
Ad Proteins, RNA, Lifecycle, Host Interactions, and Phylogenetics
Edited by: W. S. M. Wold and A. E. Tollefson © Humana Press Inc., Totowa, NJ

results of phylogenetic calculations. The initial phylogenetic analyses of adenoviruses (Ads) have facilitated the recognition of two novel genera, which are nowadays officially approved *(2)*. With the continuously growing number of partially or fully sequenced adenoviral genes and genomes, more delicate distinctions can be made among Ads of the same host animal, and the relatedness of certain Ad lineages within a genus can also be discerned. More recently, the results of phylogenetic analyses led to the hypothesis of a close co-evolutionary history of Ads with their vertebrate hosts *(3)*.

For several decades, the family *Adenoviridae* contained only two genera, *Mastadenovirus* and *Aviadenovirus,* for the allocation of mammalian or avian Ad types, respectively. Although the occurrence of Ads in lower vertebrates, including snake, frog, and fish, has been described, no such Ad has been taxonomically classified. However, each of the two valid genera contained a few exceptional members. In the genus *Mastadenovirus*, certain bovine Ad types (BAdV-4 to 8) were separated as "Subgroup 2" vs "Subgroup 1" BAdVs that did not show divergence from mammalian Ads *(4)*. The separation was originally based on biological properties, such as, among others, the requirement of primary cells for propagation, characteristic morphology of inclusion bodies, and lack of antigenic cross-reaction with "conventional" mastadenoviruses. The distinctness was later strengthened by DNA studies revealing special restriction enzyme pattern, smaller genome size *(5)*, and lack of cross-DNA hybridization with subgroup 1 BAdVs *(6)*. Similarly, the genus *Aviadenovirus* also contained two exceptions classified as Group II and III avian Ads *(7)*. Group III has been established for the egg drop syndrome virus *(8)*, officially named duck Ad 1 (DAdV-1) *(9,10)*, which has even been proposed as a, or the, candidate member of a new genus *(11)*. Later, a new ovine Ad isolate (OAV287) that differed from the officially accepted OAdV serotypes was characterized *(12)*. The genome of OAV287, by now approved as OAdV type 7 (OAdV-7), has been completely sequenced and found to have an organization strikingly different from that of human Ad type 2 (HAdV-2) and other mastadenoviruses *(13,14)*. A third genus, *Atadenovirus (15)*, has been proposed for the correct classification of these unconventional BAdVs, together with DAdV-1 and OAdV-7 *(16)*. Common ancestry of the atadenoviruses has been supported by their shared unique genomic organization, characterized by a novel structural protein p32K, close to the left genome terminus *(17,18)*, as well as supposedly early regions on the two genome ends occupying the place of the mastadenoviral E1B and E4 regions *(14,17)*. The role and function of these putative novel genes, named left- and right-hand (LH and RH) open reading frames, have been partially elucidated and are subjects of intensive scrutiny *(19,20)*. The genus name *Atadenovirus* refers to another shared characteristic of its first members, namely, the biased base composition of the

DNA toward A+T. The need for the establishment of a fourth genus emerged when the first full genome sequence of an Ad from a lower vertebrate host, more precisely from a frog, was determined *(21)*. In phylogenetic analyses, frog Ad (FrAdV-1) clustered together with the other former exception from the *aviadenovirus* genus. The so-called turkey hemorrhagic enteritis virus, officially named turkey Ad type 3, had been classified as Group II. Turkey hemorrhagic enteritis virus and FrAdV-1 have not only been shown to be in close phylogenetic relation, but their short genomes (26 kb) have also been found to be co-linear and unique among Ads *(21)*. They have five putative characteristic genes in common, one of which seems to be related to the genes of certain bacterial sialidases; therefore, this genus is named *Siadenovirus (22)*.

With the increase in numbers of full genomic sequences from every Ad genus, four distinct types of gene-arrangement patterns, each characteristic for one genetic virus lineage, have been recognized. It became clear that only the inner part of the adenoviral genome is fully conserved; the two extremities, however, show considerable diversity in size and compartment according to the genera *(23)*. Several genes and transcription units that had been thought to be indispensable/essential in mastadenoviruses have no counterparts in all or any members of the other three genera. The consistency found between the phylogenetic clusters and their genomic organization has been considered strong supportive evidence for the division of the virus family into four genera. The two traditional genera have homogeneous (mammalian or avian) host origin and presently do not contain any exceptions. Members of the two novel genera, however, originate from miscellaneous (avian, mammalian, amphibian) host animals. Comparison of phylogenetic trees of the AdVs and that of their hosts suggested a co-evolution, and a hypothesis as to the reptilian origin of atadenoviruses was born *(24)*. To challenge this hypothesis, we have sequenced the full genome of an Ad strain isolated from a corn snake (SnAdV-1). Both the genome organization and the phylogenetic analysis prove that SnAdV-1 is indeed the most ancient atadenovirus known to date *(25;* manuscript in preparation). Interestingly, the presence of a putative atadenovirus in a marsupial animal, the brushtail possum, has also been confirmed by DNA sequencing *(26)*. To provide further evidence for the reptilian origin of atadenoviruses, a collection of reptilian samples, originating from animals with suspected adenoviral infection, was tested with an exceptionally sensitive, novel, nested PCR method targeting the most conserved part of the DNA polymerase gene of Ads *(27)*. All the six partial gene sequences obtained from different lizard species have clustered within the atadenovirus clad on a distance matrix tree *(27)*. Preliminary results of the genome-sequencing project on a fish Ad isolate from white sturgeon, ongoing in our lab, forecast the likely existence of a fifth genus *(28,29)*. Because the sequence of the two ends of the fish Ad genome has not yet been determined, the proposal of a new genus (with a tentative name *Ichtadenovirus*) has been based solely on the appearance of a fifth great branch on the phylogenetic trees.

A new challenge that modern virus taxonomy is facing is the establishment and definition of virus species as novel taxons. For Ads, phylogenetically close serotypes are grouped into species *(30)*. Obviously, correct phylogenetic calculations are needed to do this grouping properly. Once the species demarcation is established, new adenoviral isolates need to be analyzed (and constantly reanalyzed as further sequences become available) to determine their phylogeny and thus the correct classification. Host switches between closely related host animals are less evident, and their recognition requires cautious and thorough scrutiny. Human Ad species have been created along with the former "groups" or "subgenera," consequently the 51 HAdV serotypes known to date are now classified into six HAdV species from HAdVs A to F. The accumulation of novel genome sequences from different ape and monkey Ads *(31–35)* suggests that HAdV species might represent the Ad lineages of different primate hosts *(3,33,36)*.

The use of computers to perform systematic phylogenetic analysis of viral sequences (for summary, *see* **ref. 37**) opened the possibility of giving quantitative answers to such questions as how different certain Ads are, and whether or not the genetic distance merits the establishment of a new genus or species. We would like to demonstrate that computerized phylogenetic analysis is an indispensable tool in modern virus taxonomy and should be applied to complement the results of traditional virological methods. Ultimately, the different methods together should form the final picture. In the case of Ads, the results of phylogenetic analyses usually support and sometimes can explain old observations that have been difficult to interpret at the time of their original description.

2. Materials

2.1. Hardware

Today's usual personal computers, either IBM-compatible, Macintosh, or UNIX-based, will suffice. However, having more power will yield speed. This may be especially important in bootstrap calculation (*see* **Subheading 3.3.3.1.**), when distance analysis with a normal data set of 100 samplings, for example, from an appropriate, approx 700-residue-long amino acid (aa) sequence of the hexon available from 60–70 Ad serotypes may require an overnight run.

2.2. Software

The Basic Local Alignment Search Tool (BLAST) homology search program *(38,39)* and database of the National Center for Biotechnology Information (NCBI) was used to obtain all the available sequences for a certain comparison (*see* **Subheading 3.2.**). For multiple alignments, the MultAlin program was used *(40)* (available online). For phylogenetic calculations, we used the PHYLIP (the PHYLogeny Inference Package) Version 3.63, a package of programs for inferring phylogenies (evolutionary trees) written and freely distributed by Joseph Felsenstein *(41,42)*. (However, version 3.65 was released just at the time of submission of this manuscript, and constant improvements can be anticipated). (How to obtain, install, and use PHYLIP is described in detail in **Subheading 3.**) Finally, the program TreeView version 1.6.6 was applied to visualize the phylogenetic tree *(43)*.

3. Method

3.1. Setup of the PHYLIP and TreeView Programs

3.1.1. How to Get the PHYLIP Package

You can get the necessary files (including the documentation) through the Internet by any browser program. Submit `phylip` (anything to be typed into the computer will be shown in `Courier`) to a web-searching program such as Google, or go directly to the following address:

`http://evolution.genetics.washington.edu/phylip.html`

Here, one can find interesting information about these and similar phylogenetic programs. Clicking on the "Get me PHYLIP" title will take you to a page where the necessary files can be obtained. For Intel-compatible processors with Windows95, Windows98, WindowsNT, Windows 2000, Windows ME, or Windows XP, download the three files (the compiled executables) under the title "Windows:"

> Documentation and C source code
> Windows95/98/NT/2000/me/xp executables, part 1
> Windows95/98/NT/2000/me/xp executables, part 2

Save these files (phylip.exe, phylipwx.exe, and phylipwy.exe) in your main directory (C:\).

3.1.2. How to Install the PHYLIP Programs

1. By clicking on them, start all the three programs that actually are archives, i.e., collections of several programs and documents. The programs with their built-in self-extracting program will create a folder "C:\phylip3.65" and three folders within it and will extract and place the different files into them. The documents describing the use of the different programs will be in the "doc" subdirectory (in html format). You can read them by clicking on their name. The programs and fonts used by the programs will be in the "exe" subdirectory, and special icons help their identification. The folder "src" contains the programs (even some fur-

ther programs) in nonexecutable forms (e.g., in computer language C) and icons, but practically you do not need those files.

2. Make a link to the subdirectory "exe" (containing all the programs) by clicking on the table with the right mouse button, choosing "new," then "shortcut," and finally browsing to "C:\phylip3.65\exe\." After establishing this shortcut, you can start it by clicking on it. You will see 35 icons for different programs (and also 6 font files to select from in case you would not be satisfied with the automatically chosen default font1).

3.1.3. How to Get and Install the TreeView Program

1. Submit `treeview` into the search window of Google or write the following address into your browser: `http://taxonomy.zoology.gla.ac.uk/rod/treeview.html`
2. Click on the title "Win32 (Windows 95 or Windows NT) (version 1.6.6)" and save the file "treev32.zip" into the "C:\phylip3.65\" folder.
3. Extract (unzip) the TreeView files by any extracting program (WinZip, etc.), or in Windows XP just click on the treev32.zip file with the right button of the mouse and choose "Unzip all." Accept the offered "C:\phylip-3.65\treev32\" as target folder for the setup files of TreeView.
4. Run "setup.exe" by clicking on it. Accept the target folder offered "C:\Program_Files\Rod Page\TreeView\."
5. Copy the treev32.exe file, i.e., TreeView (Win32) beside the other PHYLIP programs: "C:\phylip3.65\exe\" to make your life easier in the future (*see* **Subheading 3.3.4.**).

3.2. Selection and Compilation of Sequences for Analysis

Both, aa and nucleotide (nt) sequences can be analyzed, the only condition being that they are homologous (all sequences originate from a single ancestor). These programs analyze the occurrence and number of (supposedly) independent mutations of originally identical sequences. One can expect to find homologous sequences in the genes (because they code proteins with special conserved structures required for specific biological role). To the contrary, the noncoding sequences may evolve faster with large deletions or insertions instead of point mutations. To gain trustable results, the length of the sequences should be over a certain limit. The required minimal length varies according to the heterogeneity of the studied sequences. Dealing with Ads from different hosts, we found that aa sequence alignments should preferably be longer than 160–190 residues (after removing the nonhomologous regions; *see* **Subheadings 3.3.2.** and **3.3.4.**). This means that shorter sequences may give only tentative data. (However, for preliminary or pilot studies, short fragments determined in a single sequencing reaction can also be very useful [*see* **Subheadings 4.5.2.** and **4.5.3.**].) Certain Ad proteins are simply too small for phylogenetic analysis. For example, the aa sequence of pX protein of different Ads is not feasible for good phylogenetic calculations. (In pX, "p" stands for "precursor" and is not an abbreviation for "protein"!) Similarly, pVII seems to be useless for this purpose because its aa sequences are not only quite short all together, but the clearly homologous parts (that would be appropriate for the analysis) are even much shorter. However, alignments of these precursor proteins that are substrates for the adenoviral protease have been used for demonstrating different conserved recognition and cleavage sites in members of different Ad genera *(23,25)*.

For an easy start, the comparison of aa sequences may be suggested. It is easier to retrieve protein sequences because they are stored individually in the databanks. Genes are often parts of full genomes or longer nt sequences, and one needs to identify manually where a gene starts and ends. With aa sequences, it is also easier to make a correct alignment because they give more explicit patterns than DNA sequences. nt sequences are composed of only four different nts and are therefore prone to wrong alignment.

To find all the available homologous sequences, you may query the GenBank (NCBI). Submit the words genbank or ncbi into Google or use the address:

> http://www.ncbi.nlm.nih.gov/

Click on "BLAST." You will see a page where all the available versions of the heuristic homology search program BLAST are shown *(38,39)*. You can submit an aa sequence to search the database for homologous proteins (program blastp). By submitting a nt sequence you can search for similar nt sequences (blastn). You can even request the server to translate your nt sequence in all theoretically possible ways (in three reading frames on both strands) and make a homology search with the proteins (blastx). The program blastx is recommended when you wish to make the first check of a newly determined sequence because it can identify most efficiently any Ad sequence already in the GenBank. From the homologous protein sequences retrieved or chosen, take one sequence (copy it with the help of the mouse) to the "Search" window of blastp and start the program by clicking on the "BLAST!" button. You will get all the similar proteins available in GenBank (and probably also a few "nonsense" hits). You have to use common sense to decide which proteins are indeed homologous ones. They need to have a high enough score (more than approx 32, depending on the length); they need to be from Ad; they will preferably have the same protein name (or open reading frames in case they have not been identified yet); and the pairwise comparison should show several aa long identities (not only accidental sparse identities). Actually, the search can be limited. For Ads, type Adenoviridae in the box of "Limit by entrez query." The identified homologous proteins can be retrieved by a click on their (blue) names. Copy all of them into a word processor file (e.g., Microsoft Word) in FASTA format.

FASTA format means a title in the first line after the sign ">." It is best is to give names of the Ad types or isolates (not just accession numbers). The length of names should not exceed 10 characters to be compatible with the PHYLIP format (*see* **Subheading 3.3.2.2.**). Then the sequence is copied in its simplest format after this line. The next title comes immediately without any line space.

```
>HAdV-1
SALTHPMPWGPPLNPYERALAARAWQQALDLQGCKIDYFDARLLPGIFTVDAD
PPDETQLDPLPPFCSRKGGRLCWTNERLRGEVATSVDL
>HAdV-2
SALTHPMPWGPPLNPYERALAARAWQQALDLQGCKIDYFDARLLPGVFTVDAD
PPDETQLDPLPPFCSRKGGRLCWTNERLRGEVATSVDL
>HAdV-5
SALTHPM...
```

Sequences could also be searched by key words, but the naming of viruses/ genes/proteins is not consistent in the databanks, and therefore the coverage might be limited. For example, the word `protease` will not identify "protein- ase." You can find all Ad sequences in GenBank if you choose "Taxonomy" in the first window ("Search") and then write `Adenoviridae` in the second win- dow ("for"). Once on the page that shows all the Ads grouped (more or less correctly) by their taxonomy, you can click the "Nucleotide," "Protein," and "Genome Sequences" boxes to see the number of nt, protein, and full genome sequences for a certain Ad. "Genome Sequences" actually means the separate Reference Sequences database (Entrez Genomes) of NCBI *(44)*. It may not contain all the available genomes, but only those that have been checked and included by the curators. The idea is to show here only one (the "most repre- sentative") genome sequence from each species. In the case of Ads, this means that only one serotype is shown from the several types that make up that spe- cies. Other serotypes or isolates (or independently determined sequences) belonging to the same adenoviral species are shown as "neighbors" represented by their number. This number is linked to a page showing an informative graphical comparison of the available sequences of a certain Ad species:

`http://www.ncbi.nlm.nih.gov/genomes/VIRUSES/10508.html`

One may find it disturbing that these RefSeq genomes are just duplicates (with somewhat different/better annotation) of the full genomes submitted originally to GenBank. Although RefSeq makes it easy to overview the present situation, the BLAST program is still the best to compile homologous proteins, because its result shows the identical duplicates in a common group (but listing all their names with all the links). For inclusion into an alignment, any of the identical sequences can be chosen. In your alignment you can use all of them under different names, or just one, but preferably under a common name that shows all the sources (e.g., "HAdV-2,5,6"). The later solution can make a crowded tree more understandable.

Alternatively, our special web page can also be used as a quick reference. Here a graphical presentation shows continuously all the available Ad sequences, including the partial ones. (These are real nonredundant data, since each serotype is represented only once by its longest available sequence.)

```
http://www.vmri.hu/~harrach/ADENOSEQ.HTM
```

3.3. Phylogenetic Analysis

The phylogenetic calculation consists of a certain (arbitrarily chosen) set of steps. You can choose what kind of programs you wish to use, whether you want to apply bootstrapping (*see* **Subheading 3.3.3.1.**), and the form in which you want the results to be displayed. This means that there are different routes to perform the complete analysis. Actually, the best is to make several different calculations both for comparison and for gaining bootstrap values (*see* **Subheading 3.3.3.1.**). Recommended sequences of calculations are depicted in **Fig. 1**.

3.3.1. Multiple Sequence Alignment

To define the homologous regions, the sequences must first be aligned. For this purpose, many multiple alignment programs are available. Clustal V *(45,46)* or its improved version, Clustal W *(47)*, are freely distributed or can be used on-line at

```
http://www.ebi.ac.uk/clustalw/
```

(For some unknown reason, when the Clustal alignment files are copied to a Microsoft Word 2002 document in my computer and I start to edit it, they freeze the editor program.) You may have access to some general sequence analysis programs in your institute, e.g., on-line from a university server. In the Wisconsin program package GCG, you may use the PILEUP for alignment. The Lasergene package of DNASTAR offers MegAlign. The freely distributed BioEdit Sequence Alignment Editor uses Clustal W and offers a lot of very useful options to edit (manually correct) and color the gained alignment *(48)*. A huge advantage of BioEdit is that you can make your own database and run the BLAST search against your private sequences. It can be very handy when confidential sequences are analyzed. However, comparison of too many and too long sequences by the Clustal W of BioEdit can take quite a long time on a personal computer. This is why our favorite is an alignment program available on-line. The MultAlin *(40)* is very quick at today's Internet speed. Submit the word MultAlin into Google or use the address:

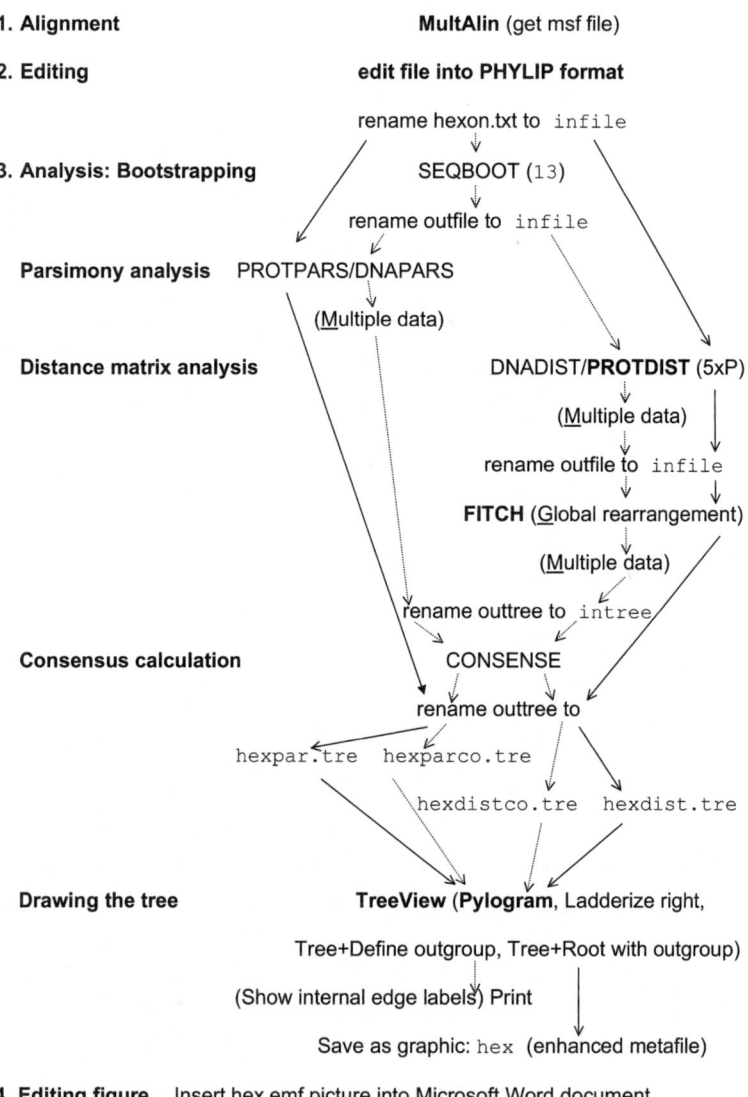

Fig. 1. Application flowchart of the PHYLIP programs. Continuous line shows single alignment analyses; dotted line depicts analyses combined with bootstrapping. The use of programs typed in **bold** is highly recommended; other programs are optional. (For bootstrap values for a tree of different branch lengths, both the PROTDIST+FITCH, and the SEQBOOT+ PROTDIST+FITCH+CONSENSE programs must be performed.) The necessary renamings (anything to be typed) are shown

http://prodes.toulouse.inra.fr/multalin/multalin.html

You may wish to analyze all the available sequences (to see the overall situation) or just select some interesting or informative sequences to get more demonstrative trees. Cut and paste the selected sequences in FASTA format from your document to the window of the program and click the "Start Multalin !" button. The ready alignment will appear in colored format, but it should be saved as a simple text file: "Results as a text page (msf)." This file can be opened subsequently in word processor to do the necessary editing (*see* **Subheading 3.3.2.**).

Of course, there are differences in the calculation and the accuracy of the different alignment programs, not least because of the default parameters and symbol comparison table. These can be changed, but most of us would use the default parameters. For example, if the terminal gaps are penalized (in the default of MultAlin they are not), the sequences of very different lengths (in case not all of them has been sequenced in full length) will not align well because the program will try to create identical lengths by introducing a lot of gaps. For nt sequences, the "Symbol comparison Table" of MultAlin should be changed for "DNA" or "alternate DNA" ("AltDNA").

It is advisable to start the alignment with the HAdVs (these should be first in the document containing the sequences compiled in FASTA format). Because of the high number and close similarity of HAdV sequences, the program will likely create a correct alignment. We prefer to start with members of species HAdV-C (HAdV-1, -2, -5, and -6) that are very similar, and the hexon is available from all of them. In the order of sequences to be aligned, the most "exotic" or questionable ones should be at the end. We usually place the single known fish Ad (originating from white sturgeon) there. In this way we can ensure that the sequence most likely to be divergent will be inserted last to a fairly good consensus alignment. In some alignment programs or when working with variable proteins, the order of sequence in the input file may significantly influence the processing and result in different alignments.

If coding DNA sequences are aligned, one should first make a protein alignment to ensure that triplets coding for identical aa are aligned correctly. Practically, this means that the size of the gaps in a DNA sequence alignment can only be numbers dividable by three. In the case of proteins, when the homology is less obvious (more gaps occur), one may experience many such mis-

---

Fig. 1. (*continued from opposite page*) in `Courier`. Options, different from default, are shown in parenthesis. (5xP) means to type `P` then push <Enter> consecutively five times to select "Categories model" that sometimes provides more consistent results than the default (Jones–Taylor–Thorton matrix). The hexon-derived file names are just examples.

takes in the alignment made by the programs. Manual correction and constant comparison to the aa alignment make this work much more time-consuming than analyzing protein alignments. However, if the sequences are very similar (isolates of the same serotype are studied) or the sequences are extremely short, the use of nt sequences might provide enough variability to ensure meaningful results. Because the majority of aa have multiple codons (different triplets) and "synonymous mutations" (i.e., nt changes that have no effect on the aa sequence) are more frequent, short DNA fragments that code for identical aas may vary at the nt level. Analysis of nt sequences is also useful for the confirmation of the results obtained by the analysis of proteins.

3.3.2. Creating Infile

Theoretically, this step consists of two separate tasks. First, the homologous regions applicable in the analysis must be selected. Second, a certain format has to be given to the file to make it understandable for the PHYLIP programs. When using word processors, single line spacing and proportional font (i.e., characters of uniform width, e.g., Courier New) should be applied.

3.3.2.1. SELECTING THE HOMOLOGOUS REGIONS

This is the artistic step of the analysis, where one can use intuition, because no exact rules can be given as to what to include in an analysis. However, do not worry. If the sequences are long enough and the differences you want to demonstrate are significant, then even considerable changes in the selection of the sequence will result in similar evolutionary trees. Actually, this is the basis of the bootstrap analysis (*see* **Subheading 3.3.3.1.**) when different parts of the alignment are randomly sampled to test if they would yield the same tree or not.

Our usual policy is the removal of any region suspected to be nonhomologous based on the following facts:

1. Deletions (gaps) of more than four to seven residues are present. A deletion of more residues or nt may be the result of a single event and should not be counted as independent mutations. However, smaller deletions may be very conclusive, appearing in every closely related virus, and can be considered as shared derived characteristics of some viral lineages. (Some programs can automatically remove every column in which gaps occur.)
2. Alignment regions do not have homologous residues for a length of more than five to eight residues. There is no resemblance, no common aa in (at least some of) the sequences.
3. Some sequences are longer than others. The analysis should be ended at the stop codon of the shortest sequence. The stop codon may be marked by an asterisk. (Theoretically there is an option to put question marks after the last residue, implying that anything could be there if it did not end. However, we found in our

practice that this may lead to miscalculations.) A different case occurs when, in obtaining preliminary results, a short sequence is compared to full-length sequences. In this case some question marks may be introduced (but not gaps, because gaps mean existing deletions) (**Fig. 2**), yet the sequences must be abbreviated almost to the size of the shortest sequence. Analyzing both the short alignment (with the added new sequence) and the long alignment will show the reliability of the results gained with the short alignment.

The computer alignment may be corrected manually if some homology is obvious. Most often the deletions tend to occur at incorrect positions. If they can be found in the same number just shifted by one column left or right, one can presume that they are the results of the same ancient deletion (especially if they occur in closely related serotypes). In such cases the alignment can be safely edited accordingly. There are occasions when reading frame shifts several residues long can easily be seen in an alignment. In such cases, even the most remote or exotic Ad has the conserved residues except for the one with the frame shift. Obviously, one cannot know (if it is not your own sequence) if the frame shift indeed occurs in nature (and the virus can tolerate it) or if it is just a sequencing error. Because it is caused by a single step, and not by point mutations, it should not be counted that way. Our policy is to replace such stretches (6–10 residues) with question marks.

3.3.2.2. PHYLIP FORMAT

The file containing the data prepared for analysis must start by the number of aligned sequences (after several spaces) and the number giving the total length of the sequence alignment (46 and 91 in **Fig. 2**). The so-called "interleaved" (aligned) format, which we recommend, is as follows. Each sequence starts on a new line, has a 10-character species name (if shorter it must be blank-filled up to 10; do not use parentheses or other special symbols), followed immediately by the aa sequence in one-letter code (**Fig. 2**). The blocks of sequences can be separated by an empty line. This format is similar to that produced by most of the multiple alignment programs. The names can be present only in the first block, and no sequence numbering and signs showing the homologous residues are allowed. The blocks do not have to have the same length, so the nonhomologous regions can be deleted anywhere. Blanks are allowed, for example, after every tenth character. However, we prefer to remove them; thus, a text editor can be used to find out how many letters are in a line. Just move to the place before the last letter, and most editors will show the number of characters in that line. But do not forget to deduct 10 from this number of the first line (title of the sequence). The blank line separating the blocks should not contain any extra character, not even a space. Non-IUPAC

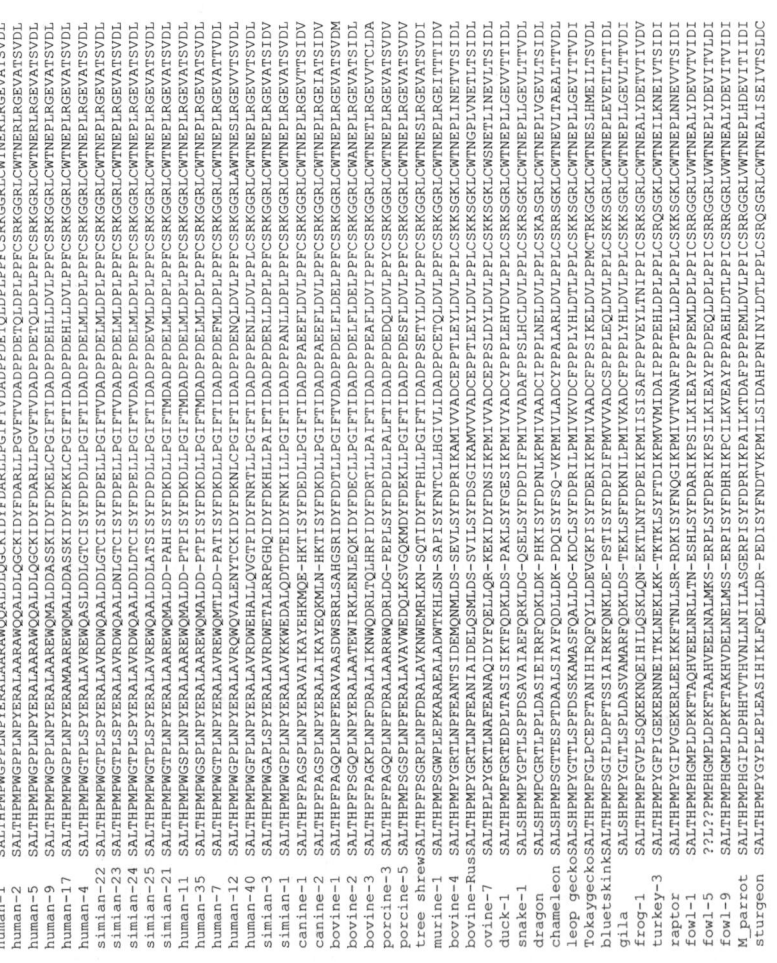

Fig. 2. Multiple alignments edited in PHYLIP format. Partial amino acid sequences derived from a fragment of the adenoviral DNA polymerase gene that can be amplified by PCR (*27*). Official virus names or the host animal's name are used (abbreviated as not to exceed the maximal 10 spaces). (Full names are used in **Fig. 3**.) The sequences are deposited in GenBank except fowl Ad 5 and white sturgeon Ad (unpublished). The fowl Ad 5 sequence is not complete (the PCR product has been sequenced from one end only); the as-yet-unknown residues are replaced by question marks (not gaps).

abbreviations (e.g., "." instead of gap: "-" or end of file symbol: "//") in the sequences or blanks at the ends of the lines are not allowed.

In an alternative format called "sequential," data are entered continuously. This format is obligatory for certain PHYLIP programs (gene frequencies, discrete and quantitative characters programs) not described herein.

A simple text editor might be efficient when working with short sequences or introducing minor changes (corrections). For longer sequences or extensive editing, however, more sophisticated word processors should be used. It is easier to remove every name after the first block and all the numbers after the sequences by deleting columns. To do this, the <control+Shift+F8> can be used in Microsoft Word. Alternatively, just push the <alt> button and then start to select any column by pushing continuously the left button of the mouse while pushing the <alt>. This latter is especially useful when you wish to delete very long columns (numbers, blanks, or names) through many pages. Just draw the mouse down, and you will find yourself at the end of the text no matter how long it is. The results must be saved as "text only."

3.3.3. Performing the Analysis

The use of these programs requires copying and constant renaming of the text files. Life can be made easier by using Norton Commander or similar programs (Total Commander, Salamander). By opening the "exe" folder with the icon of the programs and the commander program in a smaller window, you can easily perform the sequential steps of the analysis.

A flowchart of the necessary steps to perform different calculations and to get the final drawing is depicted in **Fig. 1**. The best solution is to keep all your alignments ready for calculation in a separate folder (subdirectory) under informative names that will help to go back to any of them if needed later. The content of this folder should be shown in the left panel, and the content of the "exe" folder in the right. The latter should show the file according to date, so that the newest file will always be on the top and you will not have to search for it. As a first step, copy the requested file (in PHYLIP format) into the exe folder where the programs are.

Rename your edited file to `infile`.

(The program will automatically look for such a file.) In the folder "exe," click on the icon of the program you wish to use. After asking some questions, the program will do the calculation and will usually create an "outfile" and often the final result in the form of an "outtree." If you have to perform several calculations between two steps, you must create the new "infile" for the next program—usually the outfile (or the outtree) must be renamed for the next infile. The initial alignment file will be lost during these overwriting steps, which is why we suggest that the original alignments remain in a separate folder with a unique name.

3.3.3.1. Bootstrapping (Optional)

Bootstrapping *(42)* is used to test statistical validity of the results after the first round of phylogenetic calculations (**Fig. 1**). In the first round you should skip this program and go to **Subheading 3.3.3.2.** However, in the second round the calculations should be started with bootstrapping. The bootstrapping program will use certain columns of the alignment while deleting others. (The selected columns are used several times to preserve the original length of the alignment.) Generally, 100 such random samplings of the alignment are analyzed to study whether the results would be the same if only certain (different) parts of the alignment were analyzed. Ultimately, the CONSENSE program will count how many of the 100 samplings yielded the consensus result (the most frequent tree topology).

Click on the icon of Seqboot. The program will prompt you to choose an odd number (more exactly: $4n + 1$, e.g., $4 \times 3 + 1 = 13$), and type in the requested number of samplings (usually: 100). Rename the obtained outfile to infile.

3.3.3.2. Phylogenetic Calculations

Different programs, mathematical approaches, and comparison tables can be used in an attempt to describe evolutionary happenings in the past that can be deduced from the differences existing presently. Our favorite program is the distance matrix analysis. The maximum parsimony analysis can be used in parallel to confirm the results of the distance matrix analysis. The mathematical formulas and different options can be found in the .doc files. One may, however, start with the default values of the programs. Any necessary changes will be suggested below.

3.3.3.2.1. Parsimony Analysis. For the analysis of aa sequences, the program PROTPARS (for DNA analysis the DNAPARS) should be used. The options do not need to be changed, except after bootstrapping, when "Multiple data set analysis" must be chosen, by typing M, followed by the requested number of bootstrap sampling, e.g., 100. In the tree construction programs based on parsimony and distance matrix analyses, the details of the search for different trees will depend on the order of input of species. The Jumble (J) option ("Randomize the input order of sequences") can be selected to ensure that the program uses a random number generator for choosing the input order of species. You will be prompted for a "seed" for the random number generator ($4n + 1$, e.g., 25). Different seeds lead to different sequences, so changing them and rerunning the programs can yield other/better trees. When using the J option, you will be asked how many times you want to restart the process (e.g.,

10). The results will reflect this entire search process (the best trees found among all runs, not the best trees from each individual run). Rerunning the calculations will therefore take more time. With our Ad sequences, we could not see much improvement by using it. In the case of using multiple data sets, the J option will start automatically, and the two requested numbers should be given.

If a single alignment is studied, the results can be visualized right after the calculation by the program TreeView (*see* **Subheading 3.3.4.**). To do it,

> rename outtree to `whatever.tre.`

("Tre" is the extension for the TreeView programs.) Now you can proceed to drawing the tree (*see* **Subheading 3.3.4.**).

In case of multiple data sets, the CONSENSE program must be used before visualization of the tree (*see* **Subheading 3.3.3.3.**).

3.3.3.2.2. Distance Analysis. This is our preferred PHYLIP program because it gives not only the order but also the extent of the similarity of the studied genes/proteins. The extent is proportional with branch length, and such trees are very informative.

The parsimony analysis leaves the original infile unchanged so that it can be used again in the second analysis. The two steps of the distance matrix analysis, however, lead to the loss of the original infile. Therefore, it is advisable to use the programs in this order. If the distance matrix analysis is used first; then copying and renaming of the original alignment file will be necessary.

1. Distance matrix: Click on PROTDIST (or for DNA analysis on DNADIST). You can accept the default "Jones-Taylor-Thornton" matrix (*49*), but we prefer Felsenstein's "Categories model." To get it, type letter P and push <Enter> consecutively five times, and the programs will rotate the applicable matrices until the Categories model appears as the one selected. If bootstrapping has been applied, choose multiple analyses (M). For the next questions you have to clarify that you wish to use multiple data sets (D) and write the number of the data sets (100). The multiple data set analysis can take a very long time when the sequences are long and numerous. You may start it and see the progress, namely, how much time it takes on your computer to analyze the "first set of data." You may stop it with the <Esc> button. We leave such calculations running overnight (monitor turned off).

   > Rename outfile to `infile`.

2. Tree building (FITCH): Click on FITCH (for the Fitch-Margoliash calculation), preferred, or, for experimenting, try KITSCH. (KITSCH presumes the validity of an exact molecular clock, which is actually not really correct. In the case of vertebrate Ads reflecting several hundreds of million years of evolution, we are getting unbelievable results not consistent with other results obtained by different methods.) Select multiple analysis if bootstrapping was applied (M then 100). In

this case the program will change the input order after each cycle of calculations, so it seeks again an odd number (e.g., 29). We strongly recommend the application of the global rearrangements option (G), which results in removing and re-adding each possible group after the last species has been added to the three. This method increases the reliability of the results because the position of every sample is reconsidered. The only disadvantage could be that it triples the run time. (On today's computers, even that is significant only if bootstrapping is applied.) If you choose the J option, provide the seed number and the number of repetitions desired. In the case of multiple analyses, calculate the consensus (*see* **Subheading 3.3.3.3.**); otherwise

rename outtree to `whatever.tre.`

3.3.3.3. Consensus Calculation

If bootstrapping has been applied, the consense program can calculate the most probable tree (based on all samplings) and give the bootstrap values.

Rename outtree to `intree.`

Click on consense and accept the default "Extended majority rule." The resulting consensus tree will have numbers on the branches indicating the number of times the partition of the species into the two sets, which are separated by that branch occurring among the trees out of 100 trees. More simply, the number at each fork indicates at what percentage the group consisting of the species to the right of (descended from) the fork occurred. To see details (usually unnecessary), the output file (outfile) shows the actual number of all the most frequent topologies, i.e., which groups occurred together in all 100 trees, which of them 99, etc. However, you should visualize the final tree (*see* **Subheading 3.3.5.**). If the sequence is long enough and the evolutionary distance is considerable, then the same clustering of the viruses would occur 95 of 100 times from the 100 samplings. These data can ensure that the selected sequence was adequate (representative over all or long enough) for such analysis. Where the bootstrap value is much lower (<70), the tree topology is uncertain. In that case the genetic distance is too small (or there are no enough data, e.g., the sequences are too short) for definite conclusions. Rename outtree to whatevercons.tre. (Using some hints in the name like "cons" after the protein name will help you finding a consensus file with the bootstrap values later when you need those data again; *see* **Subheading 3.3.5.**)

3.3.4. Visualization of the Result

The PHYLIP programs make a so-called outfile that shows a very primitive presentation of the tree just for quick viewing. For a nicer presentation, the program package offers the DRAWGRAM and DRAWTREE programs, and to change the appearance of the tree it offers RETREE. However, there is a much handier program for displaying the tree, Rod Page's excellent supplementary program, the TreeView *(43)*, which we recommend. It can even display the bootstrap values and supports the standard Macintosh PICT and Windows metafile or Enhanced metafile formats for output, allowing tree pictures to be used and further edited in other applications. You can easily compare several trees by TreeView or change their root or print them with any of your available printers.

Click on the icon of TreeView (we placed in the "exe" folder with all other programs). Open (one or more) treefile(s). Alternatively, clicking on the name of any file having an extension of "tre" will automatically launch the program and open the file. However, if you have another freely available program, Mega 3 (Molecular Evolutionary Genetics Analysis) *(50)*, installed on your computer, clicking on any files with "tre" extension might invoke Mega 3 instead of Treeview, unless your commander program is instructed differently. Choose the desired format of the tree: Radial, Slanted cladogram, Rectangular cladogram, or Phylogram. (You can select also by clicking on the appropriate icon.) The Phylogram view (and button) is available only if the tree has branch lengths, but including consensus trees (without branch lengths). The Radial and Phylogram presentations can visualize the calculated branch length differences, which is very important. The other two presentations will only show which sequences were grouped together by the calculations. The Radial presentation gives a topology similar to that of a real tree viewed from above. No evolutionary directions are shown, just the differences among the sequences. It is very easily understood by nonspecialists. However, if there are many sequences, some part of the tree can be overcrowded. In such trees, the virus names may overlap, and eventually one needs to arrange them manually (by dragging and displacing) so that they all can be seen clearly.

In the Phylogram format, the names appear at exact distance from each other (**Figs. 3** and **4**), so that they can be read easily. It presents the genetic distance by variable branch lengths, and thus this is our favorite tree presentation form. Evidence for different proposals and conclusions can be clearly demonstrated, such as whether the distance between the clusters of the proposed genera or species is indeed sufficient for the demarcation of those taxons. Similarly, the sufficiently small genetic distance among Ad types makes clear the need for grouping them into common Ad species (**Fig. 4**).

The Phylogram format yields a tree, which seems to be a so-called rooted tree (with known direction of evolution). But the calculations are generally performed so as to result in unrooted trees (unless an appropriate root can be specified at the start of the calculation). However, during the visualization of the tree, one can choose a so-called outgroup (the sequence where the visual-

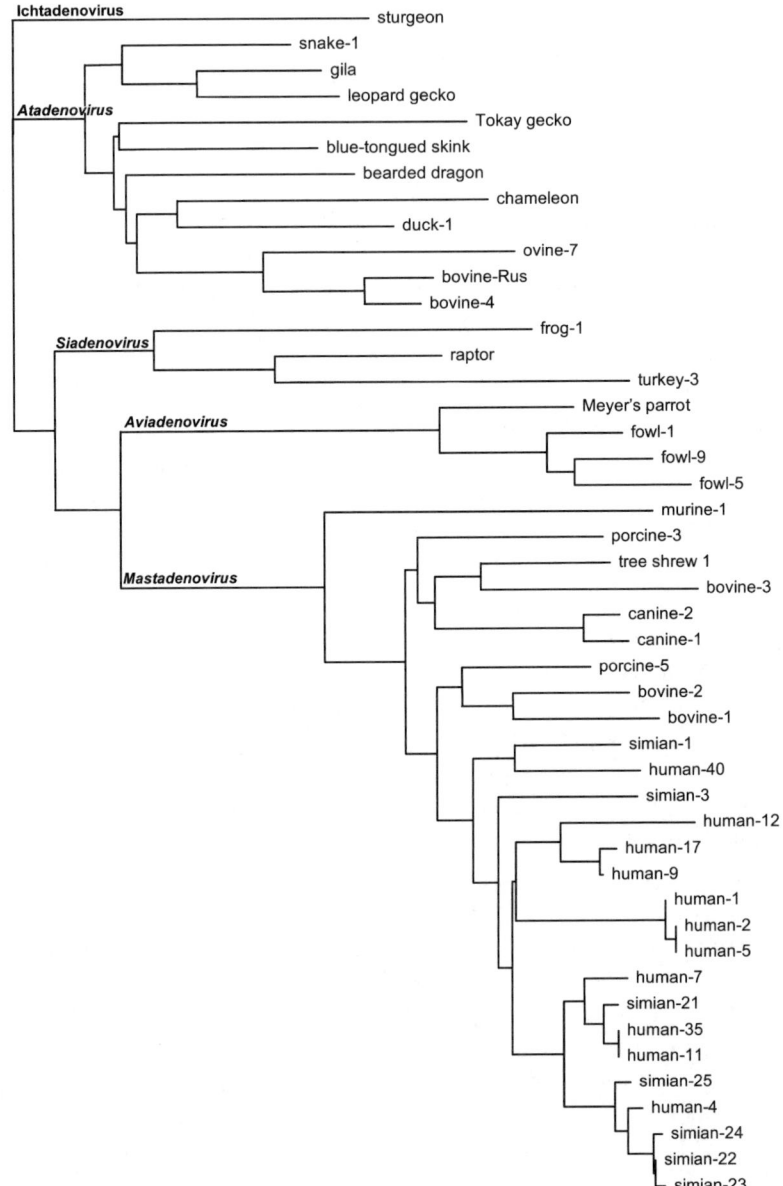

Fig. 3. Phylogenetic tree obtained from the analysis (PROTDIST with Categories scoring matrix, followed by FITCH with Global rearrangements) of the alignment in **Fig. 2**. Separation of the four official (in *Italics*) and fifth proposed genera is clear. A novel siAd (from dead raptors) clusters with turkey Ad type 3 *(51)*. All reptilian Ads are placed into the genus *Atadenovirus*. The length of the branches is proportional to the phylogenetic distance between the different viruses. Unrooted tree, white sturgeon Ad used as outgroup. Simian Ad types 1–20 originate from Old World monkeys, whereas types 21–25 originate from chimpanzees. Bootstrap values were not calculated because the alignment is too short to allow firm conclusions.

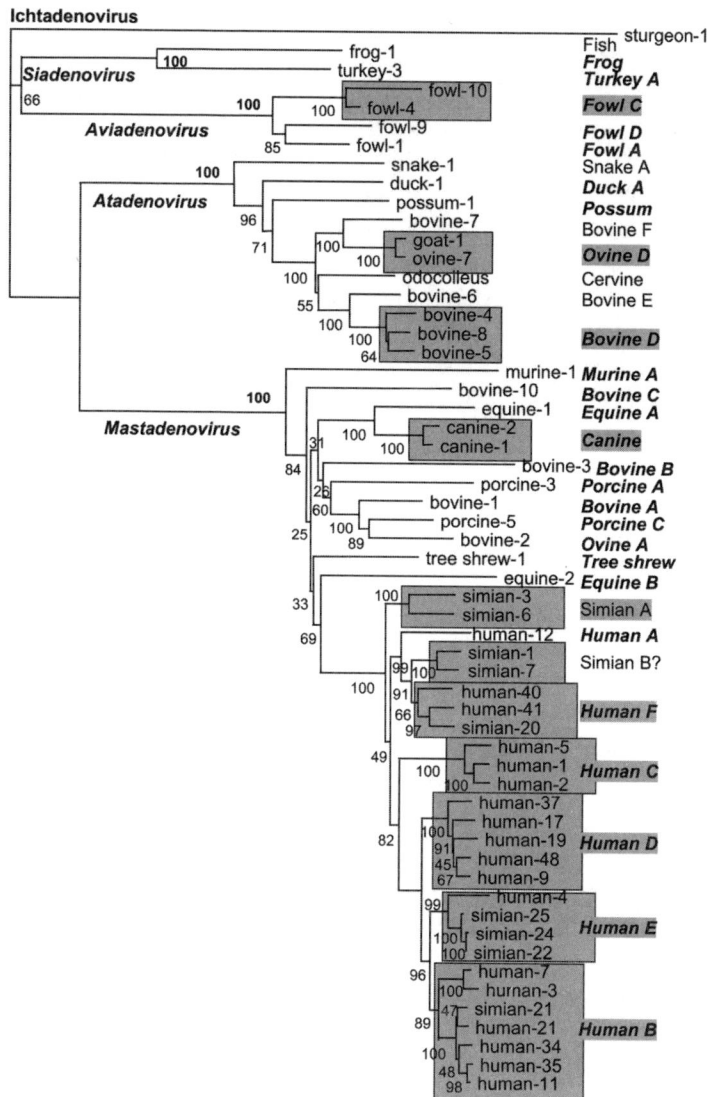

Fig. 4. Phylogenetic tree obtained from distance matrix analysis (with bootstrapping) of hexon amino acid sequences. Adenovirus types grouped into Ad species are boxed and labeled with the *accepted* or proposed names. The length of the sequence alignment (after removal of variable parts) was 811 residues. HAdV-16 and SAdV-23, although available in GenBank, were excluded from the analysis because their hexons are recombinants. Two hexon sequences (HAdV-19 and -37) have been assembled from partial sequences from GenBank. Missing residues have been replaced by question marks in the alignment. The sequence of BAdV strain Rus is unpublished. The amino acid sequence of BAdV-2 retrieved from GenBank has been corrected (by eliminating an obvious reading frame shift close to its 3'-end). Only the high bootstrap values are shown (those for the official genera in **bold**). Unrooted tree; sturgeon Ad used as outgroup.

ized tree should start). Define outgroup ("Tree" then "Define outgroup") by double-clicking on the requested species of the list shown in a window (then click "OK"). Select "Root with outgroup," or the change regarding the outgroup will not take place. Selection of an outgroup during the visualization should not cause any problems as long as we remember that the tree, in reality, is unrooted. (This fact should be mentioned in figure legends [e.g., **Fig. 3**], because it is not obvious to the readers just from the form of the tree whether the whole calculation was rooted or not.) In the case of Ads, we generally select the white sturgeon Ad as the outgroup supposing that the fish Ad must be the most ancient Ad lineage known to date, because fish evolved from a more ancient branch of vertebrates than amphibians, reptiles, birds, or mammals.

The outlook of the tree can be changed to "grow" from up-to-down or the opposite. Click on "Tree," then on "Order," and finally choose either "Ladderize right" or "Ladderize left." The font type and size can be changed, and they can be different on the internal labels. Your preference for Tree style (including if internal labels should be shown), Tree order, and Fonts (both in the "Document font" and the "Internal labels") can be fixed ("Edit" then "Preferences...").

After performing a bootstrap calculation, the outtree file of CONSENSE should be visualized (best to be printed) to see and use the bootstrap values and to compare the form of the consensus tree to that resulting from the calculation performed on a single alignment. To see the bootstrap values, click on "Tree," then on "Show internal edge labels."

By using "Window" and then "Tile," all the opened trees will be arranged on the screen, thus making the comparison of the results of different calculations easy. Finally, "Save as graphic" will create an Enhanced metafile (*.emf) or Windows metafile (*.wmf).

For cosmetic reasons the tree can be rearranged, the order of the species on a branch can be inverted, subtrees can be flipped (rotated) at a node, and the outgroup species can be changed (rerooted) in an unrooted tree. None of these changes is fraudulent because a certain tree topology can be shown in different ways. Only the length of the branches and their order (clustering/grouping) must stay fixed, but flipping of certain branches does not influence the result. PHYLIP offers RETREE to do it, but it is a bit complicated. Some other programs offer easier (graphical/clicking) methods to do it, but you have to convert the tree file to formats that are recognized by those programs. TreeView has an "Edit tree" subprogram, but you lose branch lengths if you open any file with it. Actually, you can also manually flip/rotate the branches if you feel that it would look more understandable in that format. However, there is no need for these manipulations. Errors in the names can also be corrected at this stage (no need for complete recalculations). When publishing a tree, longer names might be needed to replace the 10-character-long names used during the calculations.

3.3.5. Editing the Picture

Insert the picture, saved in .emf (or .wmf) format, into a Word document as a picture, and edit it by clicking on it with the right button of the mouse, then choosing "Edit picture." For publications it is usually necessary to show the bootstrap values on the figures. The tree produced by CONSENSE does not have branch lengths. This is why we do not show the consensus tree in the publication, although it has the bootstrap values right away. Instead, we take the tree that was obtained from a single analysis of the complete alignment (not bootstrapped). You can insert a text box to add the bootstrap values to the correct branches (forks) of the tree. Print out the tree that had been produced by the CONSENSE program, and write the values one by one. Sometimes, you may find that the consensus tree differs from the one you are working on. This means that not all branching in the topology had been confirmed. You can skip these branches (not writing anything there), and it can be referred to in the figure legend with the wording: "bootstrap values shown when bootstrapping confirmed the calculations" or "values indicated where above 75%" or "bootstrap values shown for adenoviral species and genera" (but not for the very similar isolates with questionable order). The most important values can be colored/enlarged/boxed, as are the numbers of 100 in **Fig. 4** showing that members of every (old or recently approved) genus always clustered together. The final (colored) picture can be printed, copied to PowerPoint presentations, sent by e-mail, or displayed on your web page. (You may have to group objects and text together to be able to select them in a single step.)

3.4. Interpretation of the Results

We found that different applications of the PHYLIP package provide very useful methods for the comparison of Ads for taxonomic or diagnostic/identification purposes. Initially, the results of phylogenetic analyses were used as a basis for the demarcation of novel genera *(1,21)*, but phylogeny also proved to be feasible for the species and type determination of Ads *(30,33)*.

Because only available Ads can be studied, our calculations are necessarily unrooted. Unrooted trees show merely the relations that presently exist among the different virus species. To determine the direction of evolution, an outgroup virus from a different but closely related virus family would be needed for rooted calculation. As no viral order exists above Adenoviridae, no such virus is available at the moment. However, the distance between the five clusters corresponding to the genera is convincing, and it is reasonable to presume that the fish Ad can be considered as the most ancient Ad type known to date because its host, the white sturgeon, belongs to Chondrostei, one of the most ancient lineages of vertebrates *(29)*.

If each of the Ad genera did indeed co-evolve with one major class of vertebrates, as we suppose *(3)*, they must have separated from each other hundreds of millions of years ago. It seems that this very long time caused masking of the exact evolutionary distance of these genera. The result is the uncertain tree topology involving the deepest branches (the "trunk") of the tree. It is possible that Ads in warm-blooded hosts, and especially in birds that have a relatively short life cycle, might have evolved much faster than in poikilotherm reptiles, so that the eventual monophyletic relation between avi- and atadenoviruses can no longer be traced. However, if we use as the outgroup a virus from a different genus, then the direction of evolution can be discerned within a genus. With such an outgroup, we can root our tree even if the calculation was "unrooted." Alternatively, rooted phylogenetic calculation can also be performed with members of a single genus with the use of an outgroup. For example, we have used bovine Ad 1 or 3 as the root in the comparison of primate (monkey, ape, and human) Ads *(33,34)*.

Evaluation of the results may be started with the identification of Ad types that cluster together in distance matrix analysis and preferably also in parsimony analysis. These Ad types probably evolved together. We found and proved the surprising fact that certain Ads cluster together independently of their host origin (BAdV-4 to 8, OAdV-7, DAdV-1, possum Ad, SnAdV-1, etc.; **Figs. 3** and **4**), whereas other types clustered into two additional groups according to their mammalian or avian host. Altogether five clusters appeared at a considerable distance from each other (**Figs. 3** and **4**). It can be concluded that phylogenetic analyses demonstrated great distance between these five clusters with high confidence. The bootstrap values characterizing the separation of the four officially recognized genera proved to be 100 both in distance matrix (**Fig. 4**) and in parsimony analysis (not shown) of the hexon aa alignment. The five independent lineages of Ads infect members of five vertebrate classes (fish, amphibians, reptiles, birds, and mammals).

The genus Siadenovirus and the fifth cluster (proposed Ichtadenovirus genus) cannot be further studied at the moment until newer candidate members are recognized. We can, however, pinpoint a number of Atadenovirus types that are assumed to have undergone host switches *(3)*. These include several closely related viruses from a variety of ruminant hosts such as cattle, sheep, goat, mule deer (*Odocoileus* spp.), and moose *(52,53)*. The Atadenoviruses of ruminant origin that have been characterized at sequence level so far seem to have diverged from a common ancestor: they are all monophyletic on **Figs. 3** and **4**. Therefore, we concluded that a rather ancient, single host switch must have taken place. On the other hand, Atadenovirus, the only Atadenovirus infecting birds, clustered closer to reptilian Atadenoviruses (**Figs. 3** and **4**) and is assumed to have changed host more recently. No significant variation among the different isolates and strains of DAdV-1 could be demonstrated, and this fact also supports a relatively newer host switch. The branching pattern of the different reptilian Atadenoviruses is fascinating in that it implies that each reptilian host species could have specific Ads *(27)*. However, because the tree in **Fig. 3** is based on sequences from short PCR products, the co-evolution of reptiles and Atadenoviruses needs further confirmation, preferably by analyses with longer (hexon or protease) gene sequences. Nonetheless, the reptilian origin of this genus is convincingly reinforced by the fact that every Ad identified in reptiles proved to be an Atadenovirus.

Additional challenging observations could be made within the AviAd and MastAd genera concerning coevolution of Ads with their hosts. Among these, perhaps the most fascinating is the relationship found among simian and human Ad species. The presently recognized 51 HAdV types are classified into six clusters, called species according to the new taxonomy *(2,30)*. Although this grouping (from A to F) had been introduced long before any sequence data were available and had been based on different biological properties, the results of phylogenetic analyses are in perfect agreement with it. Members of the different species vary in their prevalence, pathogenicity, and tissue or organ tropism. The variable organization of the E3 and E4 regions, having slightly different gene compartments according to the species, is very likely related to the tropism *(33,54)*. It was reasonable to hypothesize that Ads have these six major lineages and the Ads from different animal hosts will ultimately belong to one or another of them. However, thorough examination of the mastadeno-virus cluster on the phylogenetic trees might imply another scenario. Interestingly, a large number of simian Ads are scattered among the HAdV species, while others form independent SAdV species (**Fig. 4**). Species HAdV-E includes a single human type, namely HAdV-4, but oddly enough, four chimpanzee Ads occur here. On the other hand, many human Ad serotypes belong to species HAdV-B, and only a single chimpanzee serotype (SAdV-21) clusters here. We hypothesize that HAdV-B might represent the real human Ad lineage that always evolved with humans, and the only chimpanzee isolate here is the result of a recent switch from human to chimpanzee (**Fig. 4**). Species HAdV-E is assumed to be the chimpanzee Ad lineage with a single virus, HAdV-4, that has switched to human. The origin of species HAdV-C and -D cannot be discerned, but they have possibly co-evolved either with humans or with other hominoid apes (gibbons, gorillas, or orangutans). Unfortunately, no SAdV isolates are known from these hosts to test this hypothesis. On the other hand, the limited number of human serotypes that belong to HAdV-A and -F species seem to be more similar to Old World monkey Ads. Indeed, HAdV-40 and -41 are rather peculiar in that they cannot be propagated on conventional human cell lines and have two fiber genes that are expressed, and fibers of two different lengths appear alternately at the vertices of the virions. The genomes of five Old World monkey Ads have been fully sequenced *(33–35)*, and the similar genome organization (short E3 region, and duplicated fiber genes in SAdV-1) seems to confirm these host-switch theories *(34)*. The phylogenetic results shown in **Fig. 4** (and other biological data, such as hemagglutination and genome organization characteristics) suggest that SAdV-3 and -6 merit the establishment of a new Ad species *(Simian Ad A)*. However, the exact species designation for the other Old World monkey Ads is more challenging because of their high but not always convincing similarity to HAdV-40 and -41. Notably, SAdV-1 and -7 could either form a separate species or be part of HAdV-F. To decide these taxonomical questions, we would need more sequences from both SAdVs and HAdV-41. Furthermore, the full genome sequence of the two other HAdV-A serotypes (HAdV-18 and -31) would also be a tremendous help in identifying their exact phylogeny (and in demarcation of the species barrier that is actually just an arbitrary, man-made selection of the level of evolutionary distance).

Obviously, we have to accept the fact that very close relationships cannot be resolved with satisfactory confidence, and in such cases, different genes or different programs might give contradictory results (*see also* **Heading 4.**). Although all methods have limitations, the validity of consistent data should be accepted. High bootstrap values suggest reliable results, whereas low values (around or below 50) show the inadequacy of the analysis to resolve that particular evolutionary relation. Analyses of longer or/and other sequences may ameliorate the estimation. The constantly growing number of sequences available for analysis will result in increasing reliability.

4. Notes

4.1. Alternative Versions

The use of the PHYLIP program on computers with older disk operating systems (such as DOS) is also possible. It was described in the earlier version of this chapter *(1)*, and the programs can be found at the PHYLIP home page (together with instructions on how to use the programs with Linux/Unix, Macintosh, etc.).

4.2. Alternative Programs

Several alternative phylogenetic programs are also available. A very good collection, descriptions, and direct Internet connections to many of them can be found on the PHYLIP www pages (http://evolution.genetics.washington.edu/phylip.html). The PAUP program package is described as the most sophisticated parsimony program that also includes other programs. It costs a small fee. It is also available for both Macintosh and Windows systems. However, the use of the Windows version is not easy for beginners because the commands to perform the different steps need to be typed. On the other hand, a very pleasant option is that one may request all the subsequent steps to be performed at the start of a calculation.

Mega 3 is a free program package that can perform some of the most important calculations, such as neighbor-joining and parsimony *(50)*. It can do also the bootstrap analysis—and conveniently in one step, unlike PHYLIP. (The bootstrap values appear on a tree of different branch lengths.) A further advantage of Mega 3 is its very high speed. However, some of the more sophisticated methods are not included, and we have not tested the reliability of the very quick calculations (no possibility for global rearrangements).

Finally, we would like to emphasize that the hypothetical phylogenetic tree ("dendrogram") obtained from the pairwise similarity scores of Clustal should not be considered as an alternative to the trees resulting from the sophisticated phylogenetic programs. The dendrogram produced in Clustal is a preliminary step to help the final multiple alignment. Based on this tree, the program will start to align first the sequences that were found to be most similar. Thus, it is just a guide tree—the second step during the three steps of the calculations of the alignment. This is why the topology of Clustal trees may change considerably depending on the order of entering the studied species.

4.3. Illogical Results May Point to Homologous Recombination

In some cases we found that an Ad is placed in an unexpected cluster together with very different viruses. For example, HAdV-16 is known to belong to species HAdV-B, yet in distance analysis of the hexon it would be found in the cluster of HAdV-E. Another example is a member of species HAdV-E, the chimpanzee Ad SAdV-23 that occurs in the cluster of HAdV-D. This seeming anomaly can be the sign of homologous recombination of the gene. Meticulous study of the aa alignment can even reveal the points of the recombination. The place of such recombinant proteins on the calculated tree will depend on how long a sequence has been captured from the different hexon of the other serotype. Inclusion of such recombinants should be avoided in the calculation, as they would not reflect the phylogeny of the whole virus. Preferably another gene or protein of such viruses should be used to infer their correct phylogeny. An alternative solution is to diminish the length of multiple alignments to contain the nonrecombinant region only. Thus, one could make two trees: one showing the phylogeny of the real virus species (as shown by other proteins from it) and one that shows the phylogeny of the other virus that provided the insert.

4.4. Troubleshooting

4.4.1. Typical "Fatal" Errors

Error messages or symptoms	Solution of the problem
a. "can't find input file "infile" Please enter a new file name>"	Rename whatever-name to `infile` or type in the valid name.
b. "ERROR: SEQUENCES OUT OF ALIGNMENT"	Check the number and the total length of sequences. No blanks in the lines separating the inter leaved sequences! Invalid letters (e.g., /, U, number, etc.)?
c. "BAD CHARACTER STATE"	Illegal character or illegal numerical value.
d. "Error allocating memories" or program freezes after starting	Check the numerical values and whether infile is correct/intact (it may have been overwritten). Check if the titles (including blanks) are 10 spaces long.
e. Unexpected tree is drawn after bootstrapping	It shows the *first* tree out of 100 trees; use CONSENSE to build the consensus tree. *Multiple* analysis has not been requested in FITCH, it analyzed only the first sampling.
f. "ERROR: END-OF-LINE OR END-OF-FILE IN THE MIDDLE OF A SPECIES NAME"	Species number is higher than the real number of sequences (the program located an "empty" line, i.e., the separating line).
g. Instead of renaming the outfile to infile, I gave "outfile" as sequence name, but the program froze	As the program creates a new (empty) outfile first of all, the outfile containing data (e.g., matrices from PROTDIST) have been destroyed before the program could load them.
h. "ERROR: INCONSISTENT NUMBER OF SPECIES IN DATA SET 2"	Analysis of multiple data sets has been requested, but the infile contains only one set (rename outfile [of SEQBOOT] to `infile`).
i. "ERROR: CANNOT FIND SPECIES: <second species>"	CONSENSE was started, but the multiple outtree has not been renamed `intree`.
j. "INFINITE DISTANCE..."	The sequences in the alignment are too different from each other. (Try to use a smaller number of sequences, or other program.)

k. "ERROR: a function asked for an inappropriate amount of memory: 0 bytes" — Incorrect input file, e.g., you are trying to use PROTDIST second time but the generated file (matrix) is correct only for FITCH, etc.

Generally most of the errors are caused by a mistake in the infile; check and count it carefully.

4.4.2. I Forgot Which Program Was Used to Create a Certain Picture

1. The analyzed viruses themselves will suggest a gene, e.g., (presently) if there are lizard Ads on the tree, it can be only the DNA polymerase; guinea pig Ad sequence (partial) is available only from hexon; porcine Ad 2 may show pVIII; etc.
2. The form of the tree can help to identify the program applied to obtain it:
   a. Different branch lengths, it was distance analysis; if the names are aligned to the right, KITSCH was applied (the lengths are shown backward from present time); if the names are not in one column, FITCH was used.
   b. Phylogram format, all the names are in one column—parsimony analysis.
   c. Phylogram format but the names are not in one column (no different branch lengths)—CONSENSE tree after distance or parsimony analysis.

4.5. Considerations

4.5.1. Selecting the Right Programs

Parsimony analysis calculates how many steps are needed to change a sequence into the second one, then into the third one, and so on, supposing that the tree with the shortest overall length, i.e., with the smallest number of required changes, most likely represents the past events. Evolution, however, may proceed through a more complicated rather than the simplest way. Convergence (back change) also exists, yet it is not considered by this sort of calculation.

In Joe Felsenstein's parsimony analysis, change from one aa into another by the change of two nts (both resulting in coding new aas) is calculated as two changes. (However, if one of the two necessary changes was only a synonymous change not yielding a new aa, it is not counted because such synonymous changes occur relatively frequently.)

Distance matrix analysis consists of two steps. First, every sequence is compared to all others producing a distance matrix (a table of pairwise similarities). In case of aa sequences, you can give different weight to certain aa changes. Dayhoff's PAM 001 matrix calculates with experimentally identified weights. The kind of changes that occur most easily have been calculated from actual sequences. A later version of such a comparison table was made from much more numerous sequences by Jones et al. *(49)*, so it is preferred over the original PAM model. It is the default model in the described version of the PROTDIST. On the other hand, Kimura's distance simply measures the fraction of aas that differ in two sequences. The PMB (Probability Matrix from Blocks) model is derived from the Blocks database of conserved protein motifs. Felsenstein's Categories distance gives more weight to change from a basic aa into an acidic one than to (a conservative) change to another basic aa. According to our results, from the available possibilities the Categories model seems to give the most consistent results even in case of short sequence alignments, so we tend to prefer that model. In the case of DNA sequences, different weight can be given if a purine changes to pyrimidine or to another purine base. The first, second, and third characters of the codons can also be weighted differently.

In the second step, different methods produce a tree from these pairwise comparisons. We prefer the Fitch-Margoliash calculation method to that of the Neighbor-joining or the KITSCH programs (Fitch-Margoliash and least squares methods with evolutionary clock). In the case of large genetic distances, however, the choice of method does not seem to significantly influence the results. FITCH starts with the two nearest neighbors; then the sequence closest to them will be selected, and so on. We did not find the use of the Jumble option necessary because it did not seem to make a great improvement but took very much time.

Distance analysis is sometimes criticized for being a too simplistic approach: Why would two sequences be close relatives just because they look similar. Of course both parsimony analysis and distance matrix analysis may have their theoretical limitations; in practice they usually produce fairly similar results on Ad proteins. The reason could be that these sequences are variable and usually long enough to diminish the problems of convergence, and the incidental similarities do not receive too much weight compared with informative similarities and differences. However, after many years of experience with these programs and having analyzed sequences from many closely related Ad types, we concluded that the distance matrix analysis (with FITCH) gives more consistent and thus more reliable data than the parsimony analysis program of the PHYLIP package. By consistency, we mean that similar tree topology is obtained with different genes (except in the case of homologous recombination). And this is indeed what we expect, presuming that the 16 genes preserved in every Ad known today have been inherited from a common ancestor. Because the distance matrix analysis also gives more expressive trees by showing branch lengths, we prefer it to other calculation methods.

4.5.2. Which Genes to Analyze?

Preferably, long nonvariable genes that are sequenced from many virus types from different hosts should be selected. Unfortunately, many sequences are available only as unannotated sequences from patents. Moreover, there are sequences that contain smaller or larger mistakes. It may be not a popular step, but we tried to identify the presumed mistakes, reinterpret, and annotate or reannotate most of the adenoviral genomes in the GenBank *(22,29)*. We submitted these sequences as third-party annotations *(22,29)*. GenBank is still reluctant to accept them as third-party annotations because there are no new experimental data. However, we believe that "pure" bioinformatic methods are also useful because they can inexpensively and quickly increase the number of proteins retrievable from the databases for homology search and phylogenetic analysis by other users. (Actually, GenBank has just started a special new numbering system for our reinterpreted sequences. These sequences are labeled by lettering AC, e.g., AC_000001 to 20.) Our laboratory and many colleagues have also determined many sequences from Ads of fish, reptiles, domestic and wild birds, as well as different mammals such as possum, cattle, sheep, and squirrel. Some of these sequences are still preliminary and/or partial. A further problem is that the policy of releasing unpublished sequences is being seriously breached by authors who use such sequences for analysis and publish them but fail to cite the submitting authors. We try to show the available Ad sequences in an up-to-date graphical presentation (together with links to the sequences):

```
http://www.vmri.hu/~harrach/ADENOSEQ.HTM
```

A gene that is least appropriate for analysis is the fiber gene. The middle part of the fiber protein, the shaft, has repetitions of variable number. The first problem is making the alignment, because a fiber fragment will be similar to different parts of the other fibers. Practically, only the two extremities of this protein—the tail and the knob—could be used for correct alignments and meaningful calculations. However, they are too short to give reliable results.

It is interesting to mention (more or less for historical reasons) that the inverted terminal repeat was the first Ad sequence analyzed by phylogenetic program, because it was relatively easy to "find," clone, and sequence *(55)*. However, by now it is clear that the inverted terminal repeat is too short and too variable (ranging in length from 36 to 371 nt) to be used for such purposes.

Some practical problems of choosing the objects of analysis are listed below.

4.5.2.1. NUMBER OF SPECIES

Certain genes that have been studied more extensively, such as the protease or hexon genes of Ads, are more popular and consequently better represented in databases. Sequences from a larger number of viruses will result in more detailed (more valuable) phylogenetic trees. A very sparse tree cannot give further new results, but it can be used to confirm other results.

A nested PCR method that has recently been elaborated targets the adenoviral DNA polymerase gene *(27)*. The method seems to be very sensitive and has efficiently detected every Ad type from different genera tested so far. It can be considered as a "pan-Ad" PCR method. Although the size of the amplification product is too small to provide reliable results in phylogenetic analysis, the tree seems to be very informative concerning the genus classification (**Fig. 3**). This PCR should be the first choice for the detection of yet-unknown Ads in clinical samples or exotic animals. Apparently birds can be infected with very different Ads that belong to three separate genera. It is not difficult to predict that homologous sequences obtained by this PCR will soon appear in the GenBank from Ads from a large variety of hosts.

4.5.2.2. "INTERESTING" SPECIES

The different Ad types are represented very unevenly regarding DNA sequences; therefore, the selection of one gene or another might be influenced by the questions that a particular analysis is intended to answer.

1. Our original interest was in BAdVs because they have been described as exhibiting significant diversity in biological properties. Initially, from the group of BAdVs belonging to the *Atadenovirus* genus), only the protease sequence of BAdV-7 was available. However, because of the extensive functional studies on the adenoviral protease, this gene's sequences were also available from a large number of conventional mastadenoviruses and also from a number of avian Ads. We have aimed at sequencing the protease gene from the exceptional Ads that we wanted to compare *(16)*.

2. Some host species are represented only by "rare" sequences that are not available from many Ads, and therefore direct comparison cannot be done. For example, the virus-associated RNA gene has been sequenced from most simian Ads *(56)*. Unfortunately, this gene is not only too short for trustworthy phylogenetic results, but is present only in primate Ads, and apparently nonprimate mastadenoviruses lack it. There are a number of other genes that, being genus-specific, cannot be used for general comparisons. In mastadenoviruses protein IX and V are genus-specific, whereas in Atadenoviruses p32K is. For correct taxonomy and comparative analyses, there is an obvious need for further sequences, especially from the underrepresented host species.

4.5.3. Can We Believe the Results of a Certain Phylogenetic Analysis?

There will never be absolute proof of what evolution was like hundreds of millions of years ago, so it is hard to say how reliable our mathematical approaches and calculations are. Yet the results of a phylogenetic analysis that has been performed with sufficient care should be accepted if:

1. The results obtained by different methods (parsimony and distance matrix analysis) on the nt or aa sequences of different genes are consistent, the resulting tree topologies are comparable, and these are in agreement with the biological data.
2. The bootstrap values indicate high significance.
3. A special topology is in accordance with earlier established data.
   - The accepted genera and species are monophyletic (they cluster together).
   - Murine Ad 1 is the first branch among the mastadenoviruses. HAdV-B, -D, and -E are monophyletic.

Yet, if you find it disturbing that some genes or some sequence lengths give different results, you are kindly reminded that the statistics are based on the assumption that biological systems are variable. As with any other method, one is expected to evaluate cautiously the significance of new results and test them with other methods if possible. The solution is to perform several calculations to see their overall tendency. The more sequences become available, the more exact the results that are obtained. Finally, we would like to emphasize that the main advantage of computerized phylogenetic analysis (if performed properly) is that differences among Ads can be characterized by quantifiable data, whereas classical biological methods can usually give yes-or-no answers only.

Acknowledgments

We are obliged to Joseph Felsenstein, Florence Corpet, and Rod Page for providing their free programs. We would like to express our sincere gratitude to Krisztina Ursu, Orsolya Angyal, Gabriella Bartók, Ádám Dán, Cyril Barbezange, Tibor Papp, Gyözö Kaján, and Endre R. Kovács for the long hours of laboratory work to produce sequences from Ad isolates or field samples generously provided by Vilmos Palya, Éva Ivanics, Róbert Glávits, Petra Zsivanovits, Silvia Blahak, Sandra Essbauer, Rachel Marschang, Winfried Ahne, Elliott R. Jacobson, Brian Adair, David Graham, Tony Sainsbury, and Richard Gough. Special thanks are also owed to Gábor M. Kovács, Szilvia L. Farkas, Péter Élö, James F. X. Wellehan, Darelle Thomson, and Andrew J. Davison for providing aa sequences before publications. Part of this work was supported by National Research Fund of Hungary grant OTKA T043422.

References

1. Harrach, B. and Benkö, M. (1998) Phylogenetic analysis of Ad sequences. Proof of the necessity of establishing a third genus in the *Adenoviridae* family, in *Adenovirus Methods and Protocols. Methods in Molecular Medicine*, vol. 21 (Wold, W. S. M. ed.), Humana Press Inc., Totowa, NJ, pp. 309–339.
2. Benkö, M., Harrach, B., Both, G. W., et al. (2005) Family *Adenoviridae*, in *Virus Taxonomy. Eighth Report of the International Committee on Taxonomy of Viruses* (Fauquet, C. M., Mayo, M. A., Maniloff, J., Desselberger, U., Ball, L. A., eds.), Elsevier, New York, pp. 213–228.

3. Benkö, M. and Harrach, B. (2003) Molecular evolution of Ads, in *Adenoviruses: Model and Vectors in Virus–Host Interactions. Current Topics in Microbiology and Immunology* (Doerfler, W. and Böhm, P., eds.), Springer, Berlin, pp. 3–35.

4. Bartha, A. (1969) Proposal for subgrouping of bovine Ads. *Acta Vet. Hung.* **19**, 319–321.

5. Benkö, M., Bartha, A., and Wadell, G. (1988) DNA restriction enzyme analysis of bovine Ads. *Intervirology* **29**, 346–350.

6. Benkö, M., Harrach, B., and D'Halluin, J. C. (1990) Molecular cloning and physical mapping of the DNA of bovine Ad serotype 4: study of the DNA homology among bovine, human and porcine Ads. *J. Gen. Virol.* **71**, 465–469.

7. McFerran, J. B. and Smyth, J. A. (2000) Avian Ads. *Rev. Sci. Tech.* **19**, 589–601.

8. McFerran, J. B., McCracken, R. M., McKillop, E. R., McNulty, M. S., and Collins, D. S. (1977) Studies on a depressed egg production syndrome in Northern Ireland. *Avian Pathol.* **7**, 35–47.

9. Adair, B. M., McFerran, J. B., Connor, T. J., McNulty, M. S., and McKillop, E. R. (1979) Biological and physical properties of a virus (strain 127) associated with the Egg Drop Syndrome 1976. *Avian Pathol.* **8**, 249–264.

10. Gelderblom, H. and Maichle-Lauppe, I. (1982) The fibres of fowl Ads. *Arch. Virol.* **72**, 289–298.

11. Wigand, R., Bartha, A., Dreizin, R. S., et al. (1982) Adenoviridae: second report. *Intervirology* **18**, 169–176.

12. Adair, B. M., McKillop, E. R., and Coackley, B. H. (1986) Serological identification of an Australian Ad isolate from sheep. *Aust. Vet. J.* **63**, 162.

13. Vrati, S., Boyle, D., Kocherhans, R., and Both, G. W. (1995) Sequence of ovine Ad homologs for 100K hexon assembly, 33K, pVIII, and fiber genes: early region E3 is not in the expected location. *Virology* **209**, 400–408.

14. Vrati, S., Brookes, D. E., Strike, P., Khatri, A., Boyle, D. B., and Both, G. W. (1996) Unique genome arrangement of an ovine Ad: identification of new proteins and proteinase cleavage sites. *Virology* **220**, 186–199.

15. Benkö, M. and Harrach B. (1998) A proposal for a new (third) genus within the family *Adenoviridae. Arch. Virol.* **143**, 829–837.

16. Harrach, B., Meehan, B. M., Benkö, M., Adair, B. M., and Todd, D. (1997) Close phylogenetic relationship between egg drop syndrome virus, bovine Ad serotype 7, and ovine Ad strain 287. *Virology* **229**, 302–306.

17. Both, G. W. (2002) AtAd. *Adenoviridae, in The Springer Index of Viruses* (Tidona, C. A. and Darai, G., eds.), Springer, Berlin, pp. 2–8.

18. Elö, P., Farkas, L. S., Dán, A., and Kovács, G. (2003) The p32K structural protein of the atadenovirus might have bacterial relatives. *J. Mol. Evol.* **56**, 175–180.

19. Both, G. W. (2002) Identification of a unique family of F-box proteins in atadenoviruses. *Virology* **304**, 425–433.

20. Gorman, J. J., Wallis, T. P., Whelan, D. A., Shaw, J., and Both, G. W. (2005) LH3, a "homologue" of the mastadenoviral E1B 55-kDa protein is a structural protein of atadenoviruses. *Virology* **342**, 159–166.

21. Davison, A. J., Wright, K. M., and Harrach, B. (2000) DNA sequence of frog Ad. *J. Gen. Virol.* **81**, 2431–2439.

22. Davison, A. J. and Harrach, B. (2002) SiAd. *Adenoviridae,* in *The Springer Index of Viruses* (Tidona, C. A. and Darai, G., eds.) Springer, Berlin, pp. 29–33.
23. Davison, A. J., Benkö, M., and Harrach, B. (2003) Genetic content and evolution of Ads. *J. Gen. Virol.* **84** , 2895–2908.
24. Harrach, B. (2000) Reptile Ads in cattle? *Acta Vet. Hung.* **48,** 485–490.
25. Farkas, S. L., Benkö, M., Élö, P., et al. (2002) Genomic and phylogenetic analyses of an Ad isolated from a corn snake (*Elaphe guttata*) imply common origin with the members of the proposed new genus *atadenovirus. J. Gen. Virol.* **83,** 2403–2410.
26. Thomson, D., Meers, J., and Harrach, B. (2002) Molecular confirmation of an Ad in brushtail possums (*Trichosurus vulpecula*). *Virus Res.* **83,** 189–195.
27. Wellehan, J. F. X., Johnson, A. J., Harrach, B., et al. (2004) Detection and analysis of six lizard Ads by consensus primer PCR provides further evidence of a reptilian origin for the atadenoviruses. *J. Virol.* **78,** 13,366–13,369.
28. Benkö, M., Élö, P., Ursu, K., et al. (2002) First molecular evidence for the existence of distinct fish and snake Ads. *J. Virol.* **76,** 10,056–10,059.
29. Kovács, G. M., LaPatra, S. E., D'Halluin, J. C., and Benkö, M. (2003) Phylogenetic analysis of the hexon and protease genes of a fish Ad isolated from white sturgeon (*Acipenser transmontanus*) supports the proposal for a new Ad genus. *Virus Res.* **98,** 27–34.
30. Benkö, M., Harrach, B., and Russell, W. C. (2000) Family *Adenoviridae,* in *Virus Taxonomy. Classification and Nomenclature of Viruses. Seventh Report of the International Committee on Taxonomy of Viruses* (van Regenmortel, M. H. V., Fauquet, C. M., Bishop, D. H. L., et al., eds.), Academic Press, San Diego, pp. 227–238.
31. Farina, S. F., Gao, G. P., Xiang, Z. Q., et al. (2001) Replication-defective vector based on a chimpanzee Ad. *J. Virol.* **75,** 11,603–11,613.
32. Roy, S., Gao, G., Clawson, D. S., Vandenberghe, L. H., Farina, S. F., and Wilson, J. M. (2004) Complete nucleotide sequences and genome organization of four chimpanzee Ads. *Virology* **324,** 361–372.
33. Kovács, G. M., Davison, A. J., Zakhartchouk, A. N., and Harrach, B. (2004) Analysis of the first complete genome sequence of an Old World monkey Ad reveals a lineage distinct from the six human Ad species. *J. Gen. Virol.* **85,** 2799–2807.
34. Kovács, G. M., Harrach, B., Zakhartchouk, A. N., and Davison, A. J. (2005) The complete genome sequence of simian Ad 1—an Old World monkey Ad with two fiber genes. *J. Gen. Virol.* **86,** 1681–1686.
35. Roy, S. and Wilson, J. M. (2005) Methods of generating chimeric Ads and uses for such chimeric Ads. Patent: WO 2005001103-A 4; *GenBank* Accession No. CQ982399 to 401.
36. Purkayastha, A., Ditty, S. E., Su, J., et al. (2005) Genomic and bioinformatics analysis of HAdV-4, a human Ad causing acute respiratory disease: implications for gene therapy and vaccine vector development. *J. Virol.* **79,** 2559–2572.
37. Leigh Brown, A. J. (1994) Methods of evolutionary analysis of viral sequences, in *The Evolutionary Biology of Viruses* (Morse, S. S., ed), Raven, New York, pp. 75–84.

38. Altschul, S. F., Gish, W., Miller, W., Myers, E. W., and Lipman D. J. (1990) Basic local alignment search tool. *J. Mol. Biol.* **215**, 1403–1410.

39. Gish, W. and States, D. J. (1993) Identification of protein coding regions by database similarity search. *Nat. Genetics* **3**, 266–272.

40. Corpet, F. (1998) Multiple sequence alignment with hierarchical clustering. *Nucl. Acids Res.* **16**, 10,881–10,890.

41. Felsenstein, J. (1989) PHYLIP–Phylogeny inference package (version 3.2). *Cladistics* **5**, 164–166.

42. Felsenstein, J. (1985) Confidence limits on phylogenies: an approach using the bootstrap. *Evolution* **39**, 783–791.

43. Page, R. D. (1996) TreeView: an application to display phylogenetic trees on personal computers. *Comput. Appl. Biosci.* **12**, 357–358.

44. Bao, Y., Federhen, S., Leipe, D., et al. (2004) National Center for Biotechnology Information Viral Genomes Project. *J. Virol.* **78**, 7291–7298.

45. Higgins, D. G. and Sharp, P. M. (1989) Fast and sensitive multiple sequence alignments on a microcomputer. *Comput. Appl. Biosci.* **5**, 151–153.

46. Higgins, D. G., Bleasby, A. J., and Fuchs, R. (1992) CLUSTAL V: improved software for multiple sequence alignment. *Comput. Appl. Biosci.* **8**, 189–191.

47. Thompson, J. D., Higgins, D. G., and Gibson, T. J. (1994) CLUSTAL W: improving the sensitivity of progressive multiple sequence alignment through sequence weighting, position specific gap penalties and weight matrix choice. *Nucl. Acids Res.* **22**, 4673–4680.

48. Hall, T. A. (1999) BioEdit: a user-friendly biological sequence alignment editor and analysis program for Windows 95/98/NT. *Nucl. Acids Symp. Ser.* **41**, 95–98.

49. Jones, D. T., Taylor, W. R., and Thornton, J. M. (1992) The rapid generation of mutation data matrices from protein sequences. *Comput. Appl. Biosci.* **8**, 275–282.

50. Kumar, S., Tamura, K., and Nei, M. (2004) MEGA3: Integrated software for Molecular Evolutionary Genetics Analysis and sequence alignment. *Briefings Bioinformatics* **5**, 150–163.

51. Zsivanovits, P., Monks, D. J., Forbes, N. A., Ursu, K., Raue, R., and Benkö, M. (2005) Presumptive identification of a novel Ad in a Harris hawk (*Parabuteo unicinctus*), a Bengal eagle owl (*Bubo bengalensis*), and a Verreaux's eagle owl (*Bubo lacteus*). *J. Avian Med. Surgery* **20**, 105–112.

52. Lehmkuhl, H. D., Hobbs, L. A., and Woods, L. W. (2001) Characterization of a new Ad isolated from black-tailed deer in California. *Arch. Virol.* **146**, 1187–1196.

53. Lehmkuhl, H. D., DeBey, B. M., and Cutlip, R. C. (2001) A new serotype Ad isolated from a goat in the United States. *J. Vet. Diagn. Invest.* **13**, 195–200.

54. Ursu, K., Harrach, B., Matiz, K., and Benkö, M. (2004) DNA sequencing and analysis of the right-hand part of the genome of the unique bovine Ad type 10. *J. Gen. Virol.* **85**, 593–601.

55. Shinagawa, M., Iida, Y., Matsuda, A., Tsukiyama, T., and Sato, G. (1987) Phylogenetic relationships between Ads as inferred from nucleotide sequences of inverted terminal repeats. *Gene* **55**, 85–93.

56. Kidd, A. H., Garwicz, D., and Öberg, M. (1995) Human and simian Ads: phylogenetic inferences from analysis of VA RNA genes. *Virology* **207**, 32–45.

23

Assessment of Genetic Variability Among Subspecies B1 Human Adenoviruses for Molecular Epidemiology Studies

Adriana E. Kajon and Dean D. Erdman

Summary

Adenoviruses exhibit considerable intraserotypic genetic variability. Restriction enzyme analysis of the adenoviral genome is currently the most widely used procedure for the characterization of adenovirus isolates and has been extensively used for molecular epidemiological studies of subspecies B1 adenovirus infections. Comparison of restriction site maps between viral genomes is qualitatively consistent with DNA sequence homology providing that a sufficient number of sites are known. This technique is simple, sensitive, and can be adapted for screening numerous isolates and is therefore particularly useful for analysis of closely related genomes. Restriction enzyme analysis is still the only molecular approach that, at a reasonable cost, can give a "genome-wide" characterization of an adenovirus strain.

Polymerase chain reaction (PCR) amplification followed by sequencing of the generated amplicon is the approach of choice for the detailed analysis of specific regions of the viral genome. Several laboratories have recently adopted PCR amplification of the hexon and/or fiber genes for the determination of adenovirus serotype identity, replacing identification by seroneutralization and hemmaglutination-inhibition. This approach permits rapid and objective type-specific identification of human adenoviruses and is especially useful for the characterization of serologically intermediate strains frequently identified among field strains of subspecies B1 adenoviruses.

Key Words: Adenovirus; genetic variability; genome typing; polymorphism; polymerase chain reaction.

1. Introduction

1.1. The Species B Adenoviruses in Human Disease

The genus *Mastadenovirus*, species B, formerly denominated subgenus B, is comprised of one simian serotype and nine human serotypes that have been further subclassified into two clusters of homology or subspecies (International

From: *Methods in Molecular Medicine, Vol. 131:*
Adenovirus Methods and Protocols, Second Edition, vol. 2:
Ad Proteins, RNA, Lifecycle, Host Interactions, and Phylogenetics
Edited by: W. S. M. Wold and A. E. Tollefson © Humana Press Inc., Totowa, NJ

Committee on Taxonomy of Viruses, http://www.ncbi.nlm.nih.gov/ICTVdb/ICTVdB/). Subspecies B1 includes human adenovirus (hAd) 3, hAd7, hAd16, hAd21 and hAd50 and subspecies B2 includes hAd11, hAd14, hAd34 and hAd35. The subspecies B1 Ads, especially hAd3, hAd7 and hAd21, are well-recognized causative agents of respiratory disease of variable severity *(1–6)*. hAd3 and hAd7 can also cause epidemic outbreaks of conjunctivitis and febrile respiratory illness and are commonly associated with severe central nervous system disease *(7–11)*. hAd16 is rarely detected; its infrequent isolation in clinical specimens has been reported in association with conjunctivitis *(12)*, respiratory disease, and Reye syndrome-like illness *(13)*. hAd50 (prototype strain Wan), one of the two most recently recognized human serotypes *(14)*, was isolated from feces from an AIDS patient in The Netherlands, but its role in the etiology of human disease is unclear.

1.2. Genetic Variability and Molecular Epidemiology of Subspecies B1 Ad Infections

hAds exhibit considerable intraserotypic genetic variability. The use of restriction enzyme fragment analysis in the characterization of adenoviral genomes has revealed a significant intraserotypic genetic variability for all serotypes examined to date, with hAd7 followed by hAd3 being the most thoroughly studied serotypes of subspecies B1. More than 20 DNA variants or genome types of hAd7 have been identified among strains isolated worldwide, and their distinct geographic distribution has been reported *(15–18)*. Over the last two decades, molecular epidemiological studies have shown that certain hAd7 genome types predominate in a geographic area for extended periods of time and then are rapidly replaced by a new genome type. These replacements or shifts have been described in Europe, Australia, North and South America, and Asia *(19–22)*, but the molecular mechanisms underlying these replacements are not completely understood. Recent studies have documented the introduction and dissemination of these viruses from previously geographically restricted areas of circulation *(23–28)*.

A number of publications have provided evidence of extensive genetic variability for hAd3 *(21,29,30)*. However, the occurrence of genome type shifts over time is not as clear as with Ad7. The existence of multiple genome types of hAd21 has been documented *(31)*, but no similar data are available for hAd16 or the recently identified hAd50 *(14)*.

1.3. Ad Evolution

The linear double-stranded hAd DNA genome ranks as medium-sized among the DNA viruses. The genome is characterized by an inverted terminal repeat, and the 5'-ends are covalently linked to a terminal protein *(32)*. In common with other viral DNA genomes, such as the *Herpesviridae (33)* and *Poxviridae (34)*, the genes conserved among all Ads are centrally located in the genome and encode structural proteins and proteins involved in DNA replication and packaging. The genus-specific genes are encoded near the termini in the early regions E1 and E4. A common feature among all three virus families is that many of the genus- and species-specific genes are involved in modulating the host response to infection and, like the telomeric regions of yeast chromosomes *(35)*, appear to exhibit more rapid evolution *(36,37)*. In contrast to E1 and E4, which are present in all genera, E3 is only present in the genomes of the *Mastadenoviruses* (mammalian) and the members of the new genus *Siadenovirus* and shows considerable variability among Ads from different host species and subgroups. Several reports have shown that subtypes of mammalian Ads differing in their E3 regions exhibit different degrees of virulence and differences in host range *(38–40)*.

Whereas the mechanisms producing mutational changes in RNA virus genomes have been extensively studied, far less is known about the evolution of DNA viruses. Viral DNA polymerases have greater fidelity than RNA polymerases as a result of proofreading capability *(41)*, and viral DNA genomes, including adenoviral genomes, have been shown to remain genetically stable under a variety of conditions *(42)*. However, the existence in nature of multiple genome types among Ad serotypes suggests that evolutionary mechanisms are operative. Homologous recombination has been widely acknowledged as an important mechanism of evolution of Ads *(43)*. Recombinants result from one or more crossover events occurring randomly along the viral genome *(44)*. Homologous recombination is possible between closely related serotypes, and crossover sites are confined to regions of high sequence homology *(45,46)*. Serologically intermediate strains exhibiting antigenic determinants of one serotype in the hexon protein and determinants of another serotype in the fiber protein occur in nature and have been reported for the species B Ads *(47–51)*.

Illegitimate recombination is defined as a collection of mutational events resulting in deletions, insertions, duplications, and translocations. Illegitimate recombination results in the joining of two pieces of DNA with limited homology and flanked by short direct repeats at the breakpoint junctions of the joined pieces of DNA *(52)*. The slippage/misalignment model explaining these phenomena *(53,54)* proposes a transient disassociation of the DNA duplex during replication and a forward slipped-strand mispairing between two regions with short microhomologies with subsequent deletion of one of the repeats and the intervening sequence in the daughter strand. Single base substitutions can also arise from localized mispairing of bases within the heteroduplex structure formed from slipped-strand mispairing *(55)*.

Ad serotype evolution among species D Ads and in hAd4 of species E appears to be driven by illegitimate recombination events occurring in the hypervariable regions (HVRs) of the hexon gene, as demonstrated by the work of Crawford-Miksza and colleagues *(56,57)*. Comparison of the sequences of the hexon gene among different hAd7 genome types shows high conservation but suggests that some genetic diversity in this region is also likely to be the result of illegitimate recombination events *(58)*. Our recent characterization of genetic polymorphisms affecting E3 open reading frame (ORF) 7.7K of Ad7 and the analogous ORF of hAd3, hAd16, hAd21, and hAd50 *(59)* shows evidence suggesting that illegitimate recombination events may also be implicated in the generation of sequence diversity in the E3 region of the species B1 Ad genome.

2. Materials
2.1. Viral DNA Extraction From Infected Cells

1. Tissue culture flasks and media.
2. Refrigerated centrifuge, micropipet tips, 1.5- and 2-mL tubes, 15-mL conical tubes.
3. Ultracentrifuge, ultracentrifuge tubes (for extraction protocol B).
4. Cell lines: A549 cells (American Type Culture Collection [ATCC] no. CCL-185) or HEp-2 cells (ATCC no. CCL-23).
5. Growth medium: Eagle's modified essential medium (EMEM) containing 10% newborn calf serum (10% fetal calf serum for HEp-2). Maintenance medium: EMEM containing 2% newborn calf serum (2% fetal calf serum for HEp-2).
6. Absolute ethanol.
7. 75% Ethanol.
8. Isopropanol.
9. Chloroform/isoamyl alcohol (24:1).
10. Distilled autoclaved water (dH_2O).
11. 1X TE buffer: 10 mM Tris-HCl, 1 mM ethylene diamine tetraacetic acid [EDTA], pH 8.0. This solution is prepared by diluting a 10X stock solution (100 mM Tris-HCl, 10 mM EDTA, pH 8.0).
12. TE-saturated phenol: melt phenol by immersing the jar containing the crystallized chemical in a water bath at 60°C. Mix 1 vol of phenol with 1 vol of 10X TE buffer and mix by using a stirring bar for 1 h. To prevent oxidation of phenol, we recommend the addition of 8-hydroxyquinoline (to a final concentration of 0.1%). Allow the phases to separate. Pipet out the aqueous phase and add 1 vol of 1X TE. Stir to mix for 1 h. Allow the phases to separate. Pipet out the aqueous phase and add one volume of 1X TE. Stir to mix for 1 h. Allow the two phases to separate. Pipet out two-thirds of the volume of the aqueous phase.
13. dH_2O-saturated chloroform: mix 1 vol of chloroform with one volume of dH_2O in a glass jar and shaking vigorously. Allow the two phases to separate. Pipet out the aqueous phase and add 1 vol of dH_2O. Shake to mix. Allow the two phases to separate. Pipet out two-thirds of the volume of the aqueous phase.

14. 10% Sodium dodecyl sulfate (SDS): dissolve 100 g of electrophoresis-grade SDS in 900 mL of water. Wear a mask when weighing the chemical. Heat to 68°C to help dissolution. Adjust the pH to 7.2 with concentrated HCl. Adjust the volume to 1 L with deionized water and dispense in aliquots.

15. 5 M NaCl: dissolve 292.2 g of NaCl in 800 mL of deionized water. Adjust the volume to 1 L with deionized water, aliquot, and sterilize by autoclaving.

16. 10 mg/mL Proteinase K.

17. RNase A (DNase-free).

18. Phosphate-buffered saline, pH 7.2.

19. Proteinase K buffer for extraction protocol A: TE 1X, 100 mM NaCl, 0.5% SDS.

20. Proteinase K buffer for extraction protocol B: 100 mM Tris-HCl, pH 7.4, 150 mM NaCl, 12.5 mM EDTA, 1% SDS.

21. 1X TBE buffer for horizontal gel electrophoresis: 0.09 M Tris-borate, 2 mM EDTA, pH 8.0. Prepare a 5X stock solution by dissolving 54 g of Tris base, 27.5 g of boric acid, and 20 mL of 0.5 M EDTA, pH 8.0, in 1 L of deionized water.

22. Agarose.

23. Ethidium bromide.

24. Equipment: horizontal gel electrophoresis apparatus, power supply, and gel imaging system.

2.2. Polymerase Chain Reaction Amplification and Sequence Analysis of Select Regions of the Viral Genome

1. Hexon primers:
   a. Hypervariable regions 1–6:
      Forward: 5'-TIC TTT GAC ATI CGI GGI GTI CTI GA-3'
      Reverse: 5'-CTG TCI ACI GCC TGI TTC CAC AT-3'
   b. Hypervariable region 7:
      Forward: 5'-ATG TGG AAI CAG GCI GTI GAC AG-3'
      Reverse: 5'-CGG TGG TGI TTI AAI GGI TTI ACI TTG TCC AT-3'
   where I = inosine.

2. Species B-specific fiber degenerate primers *(60)* (*see* **Note 1**):
   a. Forward (Tail): 5'-TST ACC CYT ATG AAG ATG AAA GC-3'
   b. Reverse (Knob): 5'-GGA TAA GCT GTA GTR CTK GGC AT-3'
   where S = C or G; Y = C or T; R = A or G; and K = G or T

3. Primers for polymerase chain reaction (PCR)-amplification and sequence analysis of ORF 7.7K *(27)*:
   a. Forward: 5'-CAA AAA GGT GAT GCA TTA C-3'
   b. Reverse: 5'-TGT TCA CCA TAC TGT AAG A-3'

4. Taq DNA polymerase (Roche, cat. no. 1-146-173).

5. 2 mM dNTP mix: 2 mM dATP, 2 mM dCTP, 2 mM dGTP, 2 mM dTTP in dH$_2$O.

6. 15 mM MgCl$_2$ (provided with Taq).

7. 10X reaction buffer (provided with Taq).

8. Montage-PCR (DNA purification columns, Millipore) or QIAQuick kit (QIAGEN).

9. Agarose.
10. TBE electrophoresis running buffer (prepared as described in **Subheading 2.1.**).
11. Equipment: gel electrophoresis apparatus, power supply, thermal cycler, and gel imaging system.

3. Methods

Built on the original early contributions by Wadell and colleagues *(61,62)*, restriction enzyme analysis of the adenoviral genome is currently the most widely used procedure for the characterization of Ad isolates. Comparison of restriction site maps between viral genomes is qualitatively consistent with DNA sequence homology provided that a sufficient number of sites are known. Newly lost or gained restriction sites can be readily mapped to their approximate locations within the Ad genome. This technique is particularly useful for analysis of closely related genomes because it is simple, sensitive, and can be adapted for screening numerous isolates.

Most of the published molecular epidemiological studies involving subspecies B1 Ads have examined genetic variability of Ad3 and Ad7, associated with both respiratory disease and conjunctivitis, in various geographic locations as listed in **Table 1**.

3.1. Viral DNA Extraction From Infected Cells

Prior to infecting the cell monolayer that will be processed for viral DNA extraction, we recommend a passage of the isolate in a small flask (25 cm²) or 60-mm Petri dish to make sure that cytopathic effect (CPE) will develop within 48–72 h postinfection. A 100-µL aliquot coming from a monolayer with 4+ stage of CPE (on a 0–4 scale) is usually adequate to infect a 75-cm² flask. A549 (ATCC no. CCL-185) or HEp-2 (ATCC no. CCL-23) are suitable cell lines for this purpose.

*3.1.1. Extraction Protocol A (see **Note 2**)*

When high DNA yields are required for multiple digestions, we recommend the following adaptation of the protocol originally developed by Shinagawa and colleagues in 1983 *(63)*. This procedure takes advantage of the presence of a viral protein covalently attached to the genome to produce high yields of highly purified viral DNA. For multiple digestions, yields of ≥20 µg of viral DNA are necessary, and therefore we recommend infecting cell monolayers of 75 cm² or larger.

1. Remove growth medium from a confluent 75-cm² cell monolayer.
2. Infect with 100-µL aliquot and adsorb inoculum for 1 h with periodic rocking of the flask.
3. Add maintenance medium and let infection progress until CPE is visible.

Table 1
Genome Type Analysis of Species B Adenovirus Strains and Representative Publications

Serotype	Authors, year of publication	Reference	Published data
3	Adrian et al., 1986	*Arch. Virol.* **91**, 277–290.	Restriction pattern
	Adrian et al., 1989	*J. Clin. Microbiol.* **27**, 1329–1334.	Restriction profiles and restriction site maps
	O'Donnell et al., 1986	*J. Med. Virol.* **42**, 213–227.	Restriction profiles
	Bailey and Richmond, 1986	*J. Clin. Microbiol.* **24**, 30–35.	Restriction profiles
	Li and Wadell, 1988	*J. Clin. Microbiol.* **26**, 1009–1015.	Restriction profiles
	Niel et al., 1989	*J. Med. Virol.* **33**, 123–127.	Restriction patterns and PCRF
	Golovina et al., 1991	*J. Clin. Microbiol.* **29**, 2313–2321.	Restriction profiles and restriction site maps
	Li et al., 1996	*J. Med. Virol.* **49**, 170–177.	Restriction profiles and restriction patterns, PCRF
	Kim et al., 2003	*J. Clin. Microbiol.* **41**, 4594–4599.	Restriction profiles, phylogenetic trees
7	Wadell and Varsanyi, 1978	*Infect. Immun.* **21**, 238–246.	Restriction profiles
	Wadell et al., 1981	*Infect. Immun.* **34**, 368–372.	Restriction profiles and restriction site maps
	Wadell et al., 1985	*J. Clin. Microbiol.* **21**, 403–408.	Restriction profiles and geographic distribution
	Li and Wadell, 1986	*J. Virol.* **60**, 331–335.	Restriction patterns and PCRF, restriction site maps
	Adrian et al., 1986	*Arch. Virol.* **91**, 277–290.	Restriction patterns
	Bailey and Richmond, 1986	*J. Clin. Microbiol.* **24**, 30–35.	Restriction profiles
	Niel et al., 1989	*J. Med. Virol.* **33**, 123–127.	Restriction profiles and restriction site maps
	Golovina et al., 1991	*J. Clin. Microbiol.* **29**, 2313–2321.	Restriction profiles and restriction patterns, PCRF
	Azar et al., 1998	*J. Med. Virol.* **54**, 291–299.	Restriction profiles
	Li et al., 1996	*J. Med. Virol.* **49**, 170–177.	
	Erdman et al., 2001	*Emerg. Infect. Dis.* **8**, 269–277.	Restriction profiles
	Noda et al., 2002	*J. Clin. Microbiol.* **40**, 140–145.	Restriction profiles
	Kim et al., 2003	*J. Clin. Microbiol.* **41**, 4594–4599.	Restriction profiles
	Ikeda et al., 2003	*J. Med. Virol.* **69**, 215–219.	Restriciton profiles
16	No comprehensive studies have been carried out for this stereotype		
21	Van der Avoort et al., 1986	*J. Clin. Microbiol.* **24**, 1084–1088.	Restriction profiles
	Ooi et al., 2003	*Clin. Infect. Dis.* **36**, 550–559.	
50	De Jong et al., 1999	*J. Clin. Microbiol.* **37**, 3940–3945.	Restriction profiles

1

4. When extensive CPE is attained (3+, ideally 48–72 h postinfection), gently scrape the cell monolayer and transfer infected cells and medium to a 15-mL conical tube.

5. Spin the cells for 10 min at 300*g* and discard the supernatant.

6. Resuspend the infected cell pellet in 1.8 mL of 1X TE by gently vortexing the cell suspension, making sure the pellet is completely dissociated.

7. Add 200 μL of 10% SDS (final concentration 1%) and invert the tube a few times. Cell lysis will occur immediately and will be readily visible.

8. Add 500 μL of 5 *M* NaCl dropwise (final NaCL concentration 1 *M*) and observe the instant precipitation of cellular DNA.

9. Refrigerate at 4°C for 4 h to overnight to allow for complete precipitation of cellular DNA.

10. Cut the tip of a 1-mL micropipet tip with a clean pair of scissors (to avoid shearing cellular DNA), and use this tip to transfer the entire content of the tube (approx 2.5 mL) to two 2-mL tubes.

11. Spin at 14,000*g* in a refrigerated microfuge for 20 min at 4°C.

12. A compact white pellet should be formed in the bottom of the tubes. Transfer the supernatant of both tubes into two new 2-mL tubes without disrupting the pellet.

13. Add 700 μL of TE-saturated phenol. Vortex the tubes gently to mix the organic and aqueous phases and spin for 5 min at 14,000*g* in a refrigerated microcentrifuge at 4°C. The tubes coming out of this centrifugation step should show a yellowish organic phase at the bottom (phenol), a clear aqueous phase at the top, and a white interphase of variable thickness (protein-associated adenoviral DNA and cellular and viral capsid proteins; *see* **Fig. 1**).

14. Carefully remove aqueous phase, trying not to disrupt the interphase. Add 1 mL of 1X TE. Vortex gently and spin for 5 min at 4°C.

15. Repeat **step 14** to wash the interphase two more times.

16. After removing the aqueous phase, add 1 mL of absolute ethanol.

17. Incubate for 4 h to overnight at –20°C to precipitate protein-associated viral DNA and protein fraction.

18. Spin at 14,000*g* for 20 min in a refrigerated microcentrifuge to get a compact pellet.

19. Discard supernatant and rinse the pellet twice with dH_2O-saturated chloroform. This step will remove residual phenol from the pellet.

20. Let the pellet air-dry in a fume hood.

21. Resuspend the pellet with 350 μL of Proteinase K buffer. Add 5 μL of 10 mg/mL Proteinase K. Pool both tubes into one 2-mL tube.

22. Incubate overnight at 37°C in a water bath.

23. To remove residual protein, add 500 μL of TE-saturated phenol, vortex gently, and spin for 5 min at 4°C.

24. Transfer the top aqueous phase into a clean tube and extract with 1 vol (500 μL) of dH_2O-saturated chloroform.

25. Transfer the top aqueous phase into a clean tube and precipitate viral DNA with 1 mL of isopropanol for 4 h to overnight at –20°C.

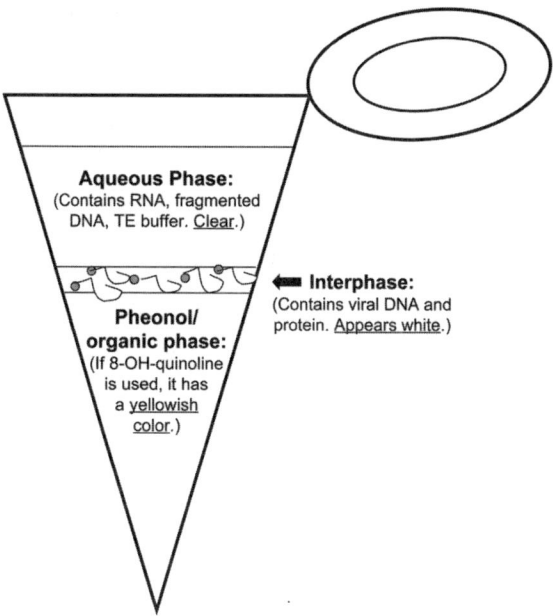

Fig. 1. Initial phase separation in a 2-mL tube after first phenol extraction step of Subheading 3.1.1. protocol A adapted from the original procedure developed by Shinagawa and colleagues *(63)*.

26. Spin tubes at 14,000*g* for 20 min in a refrigerated microcentrifuge. Discard the supernatant and let the pellet air-dry.
27. Resuspend the pellet in 100 µL of dH$_2$O or 1X TE.
28. Determine the concentration and quality of DNA by spectrophotometry. A 1:20 to 1:50 dilution of the DNA suspension should be suitable for this purpose.

3.1.2. Extraction Protocol B (see *Note 2*)

The method of Deryckere and Burgert *(64)* provides a more rapid and convenient means of obtaining purified Ad DNA. Although yields are typically lower (5–20 µg) than for Protocol A (*see* **Subheading 3.1.1.**), sufficient DNA is obtained for restriction with multiple enzymes.

1. Infect a confluent 75-cm^2 flask of A549 cells with Ad at low multiplicity of infection (between 1 and 10 PFU/cell).
2. When CPE reaches 4+, decant supernatant (~20 mL) into a 50-mL centrifuge tube and pellet cells at 300*g* for 10 min.
3. Without dislodging the cell pellet, transfer supernatant to an ultracentrifuge tube (50 Ultra-Clear) and top tubes off with 1X phosphate-buffered saline, pH 7.2. Pellet virus particles at 10,000*g* for 2 h.

4. Carefully decant supernatant and resuspend virus pellet in 400 μL of Proteinase K buffer. Transfer to a 1.5-mL microcentrifuge tube.
5. Treat with 0.05 mg/mL RNase A (DNase-free) for 30 min at 37°C to remove contaminating RNA.
6. Treat with 0.5 mg/mL Proteinase K for 30 min at 50°C to hydrolyze proteins.
7. Extract lysates twice with equal volumes of phenol and chloroform/isoamyl alcohol (24:1) and once with chloroform/isoamyl alcohol alone.
8. Precipitate DNA by adding 20 μL of 5 *M* NaCl and 1 mL of absolute ethanol at –70°C for 30 min (or overnight).
9. Centrifuge for 10 min at 11,000*g*, wash once with 75% ethanol, and resuspend DNA in 50–100 μL of deionized nuclease-free water.

3.2. Genome Typing With a Basic Panel of Restriction Endonucleases

3.2.1. Enzyme Restriction and Gel Electrophoresis

1. Determine DNA concentration using a spectrophotometer. Dilute extracted DNA 1:20 to 1:50 and read absorbance of the dilution at 260 and 280 nm. Calculate concentration by applying the equation:

$$\text{DNA concentration} = \text{Abs}_{260} \times \text{dilution factor} \times 50 \ \mu\text{g/mL}$$

Determine overall quality of the preparation by calculating the ratio $\text{Abs}_{260}:\text{Abs}_{280}$. This value should be around 1.8 or higher for the quality to meet the requirements for most downstream applications.

2. Digest purified DNA by adding approx 1 μg of viral DNA to a 1.5-mL tube containing the appropriate 10X restriction buffer with 1 μL (10 U) of enzyme in a total volume of 25–40 μL and incubate according to manufacturer's instructions. We recommend 4 h to overnight incubation times. Use a panel of at least five enzymes. Endonucleases *Bam*HI, *Bcl*I, *Bgl*II, *Bst*EII, and *Sma*I have been shown to be adequate for the discrimination of genome types and subtypes of subspecies B1 Ad serotypes *(16,29,42,65)*. Ten units of each enzyme should be used for the digestion of 1 μg of viral DNA following the conditions recommended by the manufacturer (*see* **Note 3**).
3. Load digested DNA on a 24-cm, 0.8–1% agarose gel (depending on desired fragment separation and fragment size range) and electorphorese at 100V for 6 h (voltage and time can be adjusted to suit work schedule or desired fragment resolution).
4. Stain the gel for 30 min with 0.5 μg/mL ethidium bromide. A 15-min destain in running water can help resolve faint bands.

3.3. Genome Type Denomination

Following the classification system proposed by Li and Wadell *(16,29)*, Ad "genome types" are denoted by the letter p, for the prototype strain, and a, b, c, etc., based on differences in *Bam*HI restriction profiles obtained with newly identified strains. Examples of hAd3 and hAd7 genome types are shown in

Fig. 2. Using additional restriction enzymes, genetic variants within the genome types are assigned a unique number, e.g., a1, a2, etc. The degree of genetic relatedness of different genome types can then be expressed as the percentage of comigrating restriction fragments. This denomination system has been adopted by most of the published studies.

In 1985 Adrian and colleagues proposed an alternative system to name genome types or DNA variants *(66)*. Restriction endonucleases used are listed in alphabetical order, and restriction patterns found are named 1 (prototype) or 2, 3, 4, etc. for deviating profiles making up an enzyme code. DNA variants are named D1 (prototype), D2, D3, etc. The enzymes recommended by these authors for species B1 serotypes are *Bam*HI, *Bgl*II, *Bst*EII, *Hind*III, *Kpn*I, and *Sma*I, and the prototype strain of any given serotype is designated 111111.

1. Compare the generated profiles with those published in the literature. A list of representative publications is attached in **Table 1**.
2. Assign genome type identity based on the *Bam*HI profile (p, a, b, c, etc.). If a novel profile is identified, then select a letter not previously used.

3.4. Characterization of Genetic Variability by PCR Amplification and Sequencing of Select Regions of the Viral Genome (see Notes 4 and 5)

3.4.1. Characterization of Genetic Variability in Hexon and Fiber Genes That Encode the Major Neutralizing Epitopes

Several laboratories have recently adopted PCR amplification and subsequent sequencing of the DNA fragment generated as a replacement for seroneutralization and hemmaglutination-inhibition for the determination of serotype identity of Ad isolates. This approach permits rapid and objective type-specific identification of hAds and the characterization of serologically intermediate strains.

3.4.1.1. PCR Amplification and Sequence Analysis of the Seven HVRs of the Hexon Gene L2

The HVR of the hexon gene, comprising nucleotides 403–1356 (in the genome of the prototype strain of Ad7 [Gomen]), has been shown to encode the residues that define Ad serotype *(67)*. Genetic variability among Ad7 strains representing different genome types in this portion of the hexon gene has been examined by several laboratories *(24,57,58,67,68)*. Recently, Sarantis and colleagues *(69)* published the use of a single primer pair to amplify HVR7 for Ad serotype determination. We have successfully used and highly recommend the primer pairs described by Crawford-Miksza and colleagues to amplify DNA fragments comprising HVRs 1–7 *(67)*.

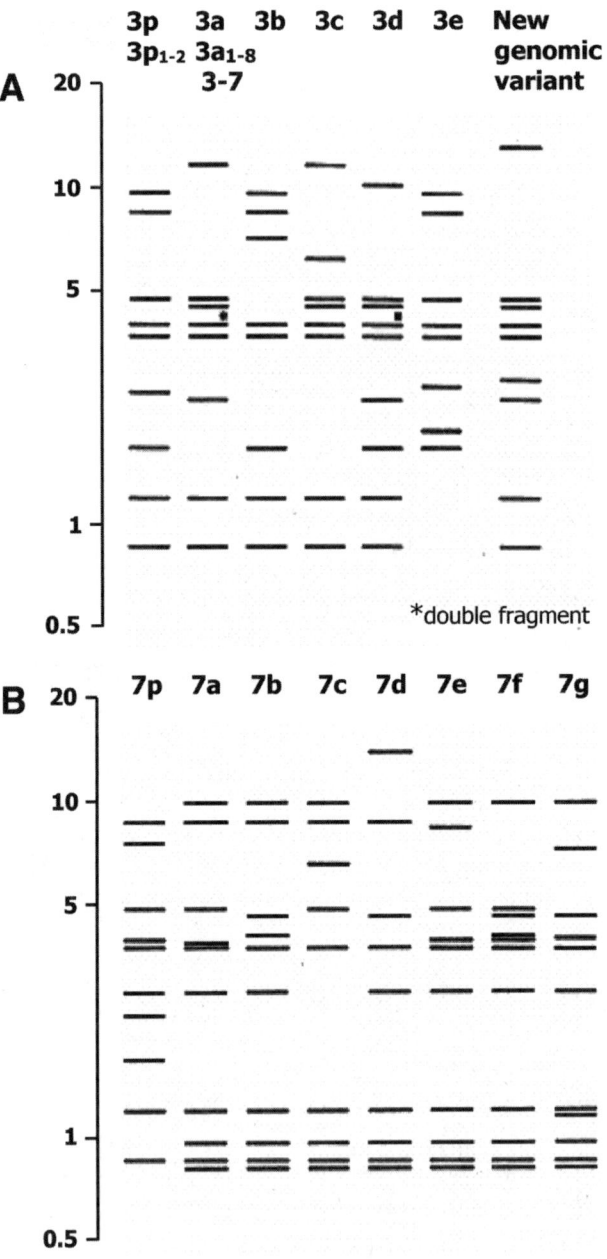

Fig. 2. (A) Genome types of adenovirus (Ad)3 as discriminated by their unique *Bam*HI restriction profiles. Asterisks indicate double fragments. (B) Genome types of Ad7.

1. Perform amplification reactions in 50-µL volumes containing 45 µL of reaction mixture (10 m*M* Tris-HCl, pH 8.3, 1.5 m*M* MgCl$_2$, 50 m*M* KCl, 200 µ*M* of each deoxynucleotide triphosphate, 0.2 µ*M* of each primer, 1 U of Taq DNA polymerase) and 5 µL of DNA extract.
2. Cycling conditions consist of a preliminary denaturation at 94°C for 2 min, followed by 35 cycles of denaturation at 94°C for 1 min, annealing at 45°C for 1 min, and primer extension at 72°C for 1 min, and a final product extension at 72°C for 5 min.
3. Analyze 10 µL of each reaction in a 1% agarose gel to control for size and presence of a single band.
4. Purify PCR fragments for sequencing from the remaining 40 µL (two reactions per sample are recommended for best yields) using Montage-PCR columns (Millipore) or QIAquick PCR purification kit as recommended by the manufacturers.
5. Determine the concentration and purity of the PCR fragment spectrophotometrically as described in **Subheading 3.2.1.**
6. Dilute fragment and primers to meet the needs of the sequencing reaction.

3.4.1.2. PCR Amplification and Sequence Analysis of the Fiber Gene

The fiber gene encodes the Ad structural fiber protein, pIV, which interacts with the cell receptor and is therefore a major determinant of tissue/cell type tropism. Mei and Wadell have shown differences in the fiber sequence between genome types of Ad11 and Ad35 *(70,71)*.

1. Perform PCR in 50-µL volumes containing 5 µL of template DNA and 45 µL of reaction mix (10 m*M* Tris-HCl, pH 8.3, 1.5 m*M* MgCl$_2$, 50 m*M* KCl, 200 µ*M* of each deoxynucleoside triphosphate, 200 m*M* of each primer, 1 U of Taq DNA polymerase).
2. The amplification reaction consists of one step of preliminary denaturation for 5 min at 94°C followed by 30 cycles of denaturation at 94°C for 1 min, annealing at 54°C for 45 s and primer extension at 72°C for 2 min, and a final extension step for 5 min at 72°C.
3. Analyze 10 µL of each reaction in a 1% agarose gel to control for size and presence of a single band. Amplicons generated by this primer pair vary among the species B1 Ad serotypes (as shown in **Fig. 3**) with hAd3, hAd7, and hAd21 yielding a 670-bp fragment and hAd16 yielding a 772-bp fragment.
4. Purify PCR fragments for sequencing from the remaining 40 µL (two reactions per sample are recommended for best yields) using Montage-PCR columns (Millipore) or QIAquick PCR purification kit as recommended by the manufacturers.
5. Determine the concentration and purity of the PCR fragment spectrophotometrically as described in **Subheading 3.2.1.**
6. Dilute fragment and primers to meet the needs of the sequencing reaction.

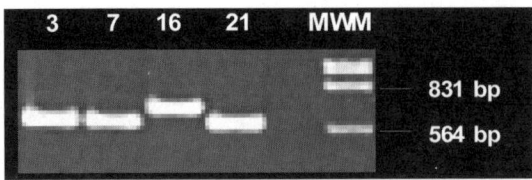

Fig. 3. Polymerase chain reaction amplification of the fiber gene of subspecies B1 human adenoviruses.

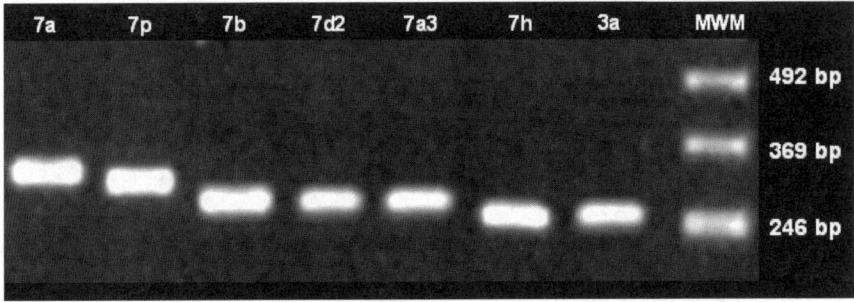

Fig. 4. Polymerase chain reaction amplification of E3 open reading frame 7.7 kDa.

3.4.2. Characterization of Genetic Variability in the E3 Region

Although they display a general organization similar to that of species C Ads, the E3 regions of species B Ads exhibit a distinct array of ORFs. Unique genes, not present in other serotypes and whose functions are still unknown, are positioned between the highly conserved 19K and 10.4K ORFs. All species B hAds that have been examined to date contain ORFs 20.1K and 20.5K (*72*). Subspecies B1 viruses encode an additional ORF in the same location as the species C ADP but with very low homology. This ORF varies in size between Ad3p (9K), Ad7p (7.7K) (*73*), and Ad16p, Ad21p, and Ad50p, and exhibits a marked sequence polymorphism among Ad7 and Ad3 genome types (*74*). The absence of this ORF from its anticipated location in the E3 regions of subspecies B2 Ad11 and Ad35 (*75,76*) has been related to their distinct pathogenic phenotypes. A 16K ORF encoded between 12.5K and 19K has partial homology to Ad5 6.7K (*77*).

3.4.3. PCR Amplification and Sequence Analysis of ORF 7.7K

ORF 7.7K appears to be the most variable coding region among the E3 regions of species B Ads (**Fig. 4**). The biological significance of this polymorphism is unclear.

Table 2
Representative GenBank Entries for Sequence Reference Data

Serotype	Genome type[a]	Strain	Place, year of isolation	Sequence	Accession no.
3	3p	G.B., prototype	United States, 1953	hexon	X76549
				E3	M15952
				fiber	AY380318
7	7p	Gomen, prototype	United States, 1954	genome	AY594255
	7p1	BC3423	China, 1981	hexon (partial)	U75955
	7a	S-1058	United States, 1955	hexon	AF053085
				E3 + fiber	AF104382
				E4 ORF 6/7 + fiber	AY921616
	7a3	55142 vaccine	United States, 1963	hexon	AF065067
	7b	BC30	China, 1958	hexon (partial)	U75951
	7c	37300	Sweden, 1964	hexon (partial)	U75952
	7d	383	Japan, 1992	E3 + fiber	AF104384
				E4 ORF 6/7 + fiber	AY921618
				hexon	AF053086
	7d2	2000026865	Israel, 1993	hexon (partial)	AF321311
	7g	BC25	China, 1985	hexon (partial)	U75956
	7h	87-922	Argentina, 1987	E3 + fiber	Z48954
				hexon (partial)	U75956
	7l	99095	Korea, 1995	hexon	AY769946
				E4 ORF 6/7 + fiber	AY921617
16	16p	Ch. 79, prototype	Saudi Arabia, 1955	hexon	X74662
				E3 + fiber	AB073632
				fiber	U06106
21	21p	AV-1645	Saudi Arabia, 1956	hexon	AY008279
				E3 + fiber	AB073222
				fiber	U06107
50	50p	Wan	The Netherlands, 1988	hexon (partial)	AB099379

[a]Wadell et al., 1980; Li and Wadell, 1986.
The complete genome sequences of Ad3p, Ad16p, Ad21p, and Ad50p have recently become available from GenBank under accession numbers AY599834, AY601636, AY601633, and AY737798, respectively.

We have adopted the primer pair and amplification protocol developed by Hashido et al. *(27)* with minor modifications as detailed below:

1. Perform PCR in 50-µL volumes containing 45 µL of reaction mixture (10 m*M* Tris-HCl, pH 8.3, 1.5 m*M* MgCl$_2$, 50 m*M* KCl, 200 µ*M* of each deoxynucleotide triphosphate, 0.2 µ*M* of each primer, 1 U of Taq DNA polymerase) and 5 µL of DNA extract.
2. Cycling conditions consist of a preliminary denaturation at 94°C for 2 min, followed by 35 cycles of denaturation at 94°C for 30 s, annealing at 50°C for 90 s, and primer extension at 72°C for 2 min, and a final product extension at 72°C for 5 min.
3. Analyze 10 µL of each reaction in a 1% agarose gel to control for size and presence of a single band. Amplicons range in size from 129 to 285 bp depending on the serotype and/or genome type as previously described *(74)*.
4. Purify PCR fragments for sequencing from the remaining 40 µL (two reactions per sample are recommended for best yields) using Montage-PCR columns (Millipore) or QIAquick PCR purification kit as recommended by the manufacturers.
5. Determine the concentration and purity of the PCR fragment spectrophotometrically as described in **Subheading 3.2.1.**
6. Dilute fragment and primers to meet the needs of the sequencing reaction.

4. Notes

1. The Species B fiber primers were developed by Xu and colleagues *(60)* to specifically amplify the species B hAd serotypes as part of a diagnostic multiplex PCR for type-specific identification of Ad3, Ad7, and Ad21.
2. DNA extraction protocols can be interrupted for an undefined period of time at any of the alcohol-precipitation steps and resumed later.
3. Restriction enzymes with a four- or five-nucleotide recognition site are particularly useful for comparison of closely related genome types and for the generation of complex profiles to determine the percentage of comigrating restriction fragments between two genome types. Restriction endonucleases *Cfo*I, *Msp*I, *Sau*3A, *Bfr*I 3I, and *Rsa*I have been successfully used for this purpose *(22,78)*. Viral DNA digests generated with 4- or 5-cutters should be analyzed in 1.5–2% agarose gels.
   A recent paper by Choi and colleagues reports the characterization of Korean Ad7 strains and describes sequence differences mapping to E4 ORF 3 between different genome types *(79)*.
4. In **Table 2**, a list of representative GenBank entries for sequence reference data is provided (to be of value for primer development).

References

1. Simila, S., Ylikorkala, O., and Wasz-Hockert, O. (1971) Type 7 adenovirus pneumonia. *J. Pediatr.* **79,** 605–611.

2. Simila, S., Linna, O., Lanning, P., Heikkinen, E., and Ala-Houhala, M. (1981) Chronic lung damage caused by adenovirus type 7: a ten-year follow-up study. *Chest* **80**, 127–131.

3. Becroft, D. M. (1971) Bronchiolitis obliterans, bronchiectasis, and other sequelae of adenovirus type 21 infection in young children. *J. Clin. Pathol.* **24**, 72–82.

4. Chany, C., Lepine, P., Lelong, M., Le-Tan-Vinh, Satge, P., and Virat, J. (1958) Severe and fatal pneumonia in infants and young children associated with adenovirus infections. *Am. J. Hyg.* **67**, 367–378.

5. Herbert, F. A., Wilkinson, D., Burchak, E., and Morgante, O. (1977) Adenovirus type 3 pneumonia causing lung damage in childhood. *Can. Med. Assoc. J.* **116**, 274–276.

6. Mistchenko, A. S., Robaldo, J. F., Rosman, F. C., Koch, E. R., and Kajon, A. E. (1998) Fatal adenovirus infection associated with new genome type. *J. Med. Virol.* **54**, 233–236.

7. Simila, S., Jouppila, R., Salmi, A., and Pohjonen, R. (1970) Encephaloningitis in children associated with an adenovirus type 7 epidemic. *Acta. Paediatr. Scand.* **59**, 310–316.

8. O'Donnell, B., McCruden, A. E., and Desselberger, U. (1993) Molecular epidemiology of adenovirus conjunctivitis in Glasgow 1981–1991. *Eye* **7 (Pt. 3 Suppl.)**, 8–14.

9. O'Donnell, B., Bell, E., Payne, S. B., Mautner, V., and Desselberger, U. (1986) Genome analysis of species 3 adenoviruses isolated during summer outbreaks of conjunctivitis and pharyngoconjunctival fever in the Glasgow and London areas in 1981. *J. Med. Virol.* **18**, 213–227.

10. Piazza, M. and Paradisi, F. (1968) A familial outbreak of adenovirus type 21 infection (follicular conjunctivitis). *Riv. Ist Sieroter Ital.* **43**, 241–243.

11. Sakata, H., Taketazu, G., Nagaya, K., et al. (1998) Outbreak of severe infection due to adenovirus type 7 in a paediatric ward in Japan. *J. Hosp. Infect.* **39**, 207–211.

12. Knopf, H. L. and Hierholzer, J. C. (1975) Clinical and immunologic responses in patients with viral keratoconjunctivitis. *Am. J. Ophthalmol.* **80**, 661–672.

13. Morgan, P. N., Moses, E. B., Fody, E. P., and Barron, A. L. (1984) Association of adenovirus type 16 with Reye's-syndrome-like illness and pneumonia. *South. Med. J.* **77**, 827–830.

14. De Jong, J. C., Wermenbol, A. G., Verweij-Uijterwaal, M. W., et al. (1999) Adenoviruses from human immunodeficiency virus-infected individuals, including two strains that represent new candidate serotypes Ad50 and Ad51 of species B1 and D, respectively. *J. Clin. Microbiol.* **37**, 3940–3945.

15. Wadell, G., Cooney, M. K., da Costa Linhares, A., et al. (1985) Molecular epidemiology of adenoviruses: global distribution of adenovirus 7 genome types. *J. Clin. Microbiol.* **21**, 403–408.

16. Li, Q. G. and Wadell, G. (1986) Analysis of 15 different genome types of adenovirus type 7 isolated on five continents. *J. Virol.* **60**, 331–335.

17. Golovina, G. I., Zolotaryov, F. N., and Yurlova, T. I. (1991) Sensitive analysis of genetic heterogeneity of adenovirus types 3 and 7 in the Soviet Union. *J. Clin. Microbiol.* **29**, 2313–2321.

18. Azar, R., Varsano, N., Mileguir, F., and Mendelson, E. (1998) Molecular epide-miology of adenovirus type 7 in Israel: identification of two new genome types, Ad7k and Ad7d2. *J. Med. Virol.* **54,** 291–299.

19. Wadell, G., J., de Jong, C., and Wolontis, S. (1981) Molecular epidemiology of adenoviruses: alternating appearance of two different genome types of adenovi-rus 7 during epidemic outbreaks in Europe from 1958 to 1980. *Infect. Immun.* **34,** 368–377.

20. de Silva, L. M., Colditz, P., and Wadell, G. (1989) Adenovirus type 7 infections in children in New South Wales, Australia. *J. Med. Virol.* **29,** 28–32.

21. Li, Q. G., Zheng Q. J., Liu, Y. H., and Wadell, G.(1996) Molecular epidemiology of adenovirus types 3 and 7 isolated from children with pneumonia in Beijing. *J. Med. Virol.* **49,** 170–177.

22. Kajon, A. and Wadell, G. (1994) Genome analysis of South American adenovi-rus strains of serotype 7 collected over a 7-year period. *J. Clin. Microbiol.* **32,** 2321–2323.

23. Gerber, S. I., Erdman, D. D., Pur, S. L., et al. (2001) Outbreak of adenovirus genome type 7d2 infection in a pediatric chronic-care facility and tertiary-care hospital. *Clin. Infect. Dis.* **32,** 694–700.

24. Erdman, D. D., Xu, W., Gerber, S. I., et al. (2002) Molecular epidemiology of adenovirus type 7 in the United States, 1966–2000. *Emerg. Infect. Dis.* **8,** 269–277.

25. Noda, M., Yoshida, T., Sakaguchi, T., Ikeda, Y., Yamahoka, K., and Ogino, T. (2002) Molecular epidemiological analyses of human adenovirus type 7 strains isolated from the 1995 nationwide outbreak in Japan. *J. Clin. Microbiol.* **40,** 140–145.

26. Ikeda, Y., Yamaoka, K., Noda, M., and Ogino, T. (2003) Genome types of aden-ovirus type 7 isolated in Hiroshima City. *J. Med. Virol.* **69,** 215–219.

27. Hashido, M., Mukouyama, A., Sakae, K., et al. (1999) Molecular and serological characterization of adenovirus genome type 7h isolated in Japan. *Epidemiol. Infect.* **122,** 281–286.

28. Tanaka, K., Itoh, N., Saitoh-Inagawa, W., et al. (2000) Genetic characterization of adenovirus strains isolated from patients with acute conjunctivitis in the city of Sao Paulo, Brazil. *J. Med. Virol.* **61,** 143–149.

29. Li, Q. G. and Wadell, G. (1988) Comparison of 17 genome types of adenovirus type 3 identified among strains recovered from six continents. *J. Clin. Microbiol.* **26,** 1009–1015.

30. Adrian, T., Best, B., Hierholzer, J. C., and Wigand, R. (1989) Molecular epidemi-ology and restriction site mapping of adenovirus type 3 genome types. *J. Clin. Microbiol.* **27,** 1329–1334.

31. van der Avoort, H. G., Adrian, T., Wigand, R., Wermenbol, A. G., Zomerdijk, T. P., and de Jong, J. C. (1986) Molecular epidemiology of adenovirus type 21 in the Netherlands and the Federal Republic of Germany from 1960 to 1985. *J. Clin. Microbiol.* **24,** 1084–1088.

32. Kim, Y. J., Hong J. Y., Lee, H. J., et al. (2003) Genome type analysis of adenovirus types 3 and 7 isolated during successive outbreaks of lower respiratory tract infections in children. *J. Clin. Microbiol.* **41**, 4594–4599.

33. McGeoch, D. J. and Davison, A. J. (1999) The molecular evolutionary history of herpesviruses, in *Origin and Evolution of Viruses* (Domingo, E., Webster, R., and Holland, J., eds.), Academic Press, London, pp. 441–465.

34. Upton, C., Slack, S., Hunter, A. L., Ehlers, A., and Roper, R. L. (2003) Poxvirus orthologous clusters: toward defining the minimum essential poxvirus genome. *J. Virol.* **77**, 7590–7600.

35. Kellis, M., Patterson, N., Endrizzi, M., Birren, B., and Lander, E. S. (2003) Sequencing and comparison of yeast species to identify genes and regulatory elements. *Nature* **423**, 241–254.

36. Davison, A. J., Benko, M., and Harrach, B. (2003) Genetic content and evolution of adenoviruses. *J. Gen. Virol.* **84**, 2895–2908.

37. Benko, M., and Harrach, B. (2003) Molecular evolution of adenoviruses. *Curr. Topics Microbiol. Immunol.* **272**, 3–35.

38. Belak, S., Virtanen, A., Zabielski, J., Rusvai, M., Berencsi, G., and Pettersson, U. (1986) Subtypes of bovine adenovirus type 2 exhibit major differences in region E3. *Virology* **153**, 262–271.

39. Dragulev, B. P., Sira, A., Abouhaidar, M. G., and Campbell, J. B. (1991) Sequence analysis of putative E3 and fiber genomic regions of two strains of canine adenovirus type 1. *Virology* **183**, 298–305.

40. Linne, T. (1992) Differences in the E3 regions of the canine adenovirus type 1 and type 2. *Virus Res.* **23**, 119–133.

41. Tamanoi, F. (1986) *On the Mechanism of DNA Replication.* Nijhoff, Boston.

42. Adrian, T., Wadell, G., Hierholzer, J. C., and Wigand, R. (1986) DNA restriction analysis of adenovirus prototypes 1 to 41. *Arch. Virol.* **91**, 277–290.

43. Sambrook, J., Sleigh, M., Engler, J. A., and Broker, T. R. (1980) The evolution of the adenoviral genome. *Ann. NY Acad. Sci.* **354**, 426–452.

44. Williams, J., Grodzicker, T., Sharp, P., and Sambrook, J. (1975) Adenovirus recombination: physical mapping of crossover events. *Cell* **4**, 113–119.

45. Mautner, V. and Boursnell, M. E. (1983) Recombination in adenovirus: DNA sequence analysis of crossover sites in intertypic recombinants. *Virology* **131**, 1–10.

46. Boursnell, M. E. and Mautner, V. (1981) Recombination in adenovirus: crossover sites in intertypic recombinants are located in regions of homology. *Virology* **112**, 198–209.

47. Adrian, T. and Wigand, R. (1986) Adenovirus 3-7, an intermediate strain of subgenus B. *Intervirology* **26**, 202–206.

48. Hierholzer, J. C. and Pumarola, A. (1976) Antigenic characterization of intermediate adenovirus 14-11 strains associated with upper respiratory illness in a military camp. *Infect. Immun.* **13**, 354–359.

49. Norrby, E. and Skaaret, P. (1968) Comparison between soluble components of adenovirus types 3 and 16 and of the intermediate strain 3-16 (the San Carlos agent). *Virology* **36**, 201–211.

50. Crandell, R. A., Dowdle, W. R., Holcomb, T. M., and Dahl, E.V. (1968) A fatal illness associated with two viruses: an intermediate adenovirus type (21-16) and influenza A2. *J. Pediatr.* **72**, 467–473.
51. Kajon, A. E. and Wadell, G. (1996) Sequence analysis of the E3 region and fiber gene of human adenovirus genome type 7h. *Virology* **215**, 190–196.
52. Meuth, M. (1989) Illegitimate recombination in mammalian cells, in *Mobile DNA* (Berg, D. E. and Howe, M. M., eds.), Am. Soc. Microbiol., Washington, DC, pp. 833–853.
53. Efstratiadis, A., Posakony, J. W., Maniatis, T., et al. (1980) The structure and evolution of the human beta-globin gene family. *Cell* **21**, 653–668.
54. Kunkel, T. A. (1990) Misalignment-mediated DNA synthesis errors. *Biochemistry* **29**, 8003–8011.
55. Kunkel, T. A. and Soni, A. (1988) Mutagenesis by transient misalignment. *J. Biol. Chem.* **263**, 14,784–14,789.
56. Crawford-Miksza, L. K. and Schnurr, D. P. (1996) Adenovirus serotype evolution is driven by illegitimate recombination in the hypervariable regions of the hexon protein. *Virology* **224**, 357–367.
57. Crawford-Miksza, L. K., Nang, R. N., and Schnurr, D. P. (1999) Strain variation in adenovirus serotypes 4 and 7a causing acute respiratory disease. *J. Clin. Microbiol.* **37**, 1107–1112.
58. Li, Q. and Wadell, G. (1999) Genetic variability of hexon loops 1 and 2 between seven genome types of adenovirus serotype 7. *Arch. Virol.* **144**, 1739–1749.
59. Kajon, A. E., Xu, W., and Erdman, D. D. (2005) Sequence polymorphism in the E3 7.7K ORF of subspecies B1 human adenoviruses. *Virus Res.* **107**, 11–19.
60. Xu, W. and Erdman, D. D. (2001) Type-specific identification of human adenovirus 3, 7, and 21 by a multiplex PCR assay. *J. Med. Virol.* **64**, 537–542.
61. Wadell, G. (1984) Molecular epidemiology of human adenoviruses. *Curr. Topics Microbiol. Immunol.* **110**, 191–220.
62. Wadell, G., Hammarskjold, M. L., Winberg, G., Varsanyi, T. M., and Sundell, G. (1980) Genetic variability of adenoviruses. *Ann. NY Acad. Sci.* **354**, 16–42.
63. Shinagawa, M., Matsuda, A., Ishiyama, T., Goto, H., and Sato, G. (1983) A rapid and simple method for preparation of adenovirus DNA from infected cells. *Microbiol. Immunol.* **27**, 817–822.
64. Deryckere, F. and Burgert, H. G. (1997) Rapid method for preparation of adenovirus DNA. *Biotechniques* **22**, 868–870.
65. Wadell, G. and Varsanyi, T. M. (1978) Demonstration of three different subtypes of adenovirus type 7 by DNA restriction site mapping. *Infect. Immun.* **21**, 238–246.
66. Adrian, T., Best, B., and Wigand, R. (1985) A proposal for naming adenovirus genome types, exemplified by adenovirus type 6. *J. Gen. Virol.* **66**, 2685–2691.
67. Crawford-Miksza, L. and Schnurr, D. P. (1996) Analysis of 15 adenovirus hexon proteins reveals the location and structure of seven hypervariable regions containing serotype-specific residues. *J. Virol.* **70**, 1836–1844.
68. Blasiole, D. A., Metzgar, D., Daum, L. T., et al. (2004) Molecular analysis of adenovirus isolates from vaccinated and unvaccinated young adults. *J. Clin. Microbiol.* **42**, 1686–1693.

69. Sarantis, H., Johnson, G., Brown, M., Petric, M., and Tellier, R. (2004) Comprehensive detection and serotyping of human adenoviruses by PCR and sequencing. *J. Clin. Microbiol.* **42,** 3963–3969.
70. Mei, Y. F., Lindman, K., and Wadell, G. (1998) Two closely related adenovirus genome types with kidney or respiratory tract tropism differ in their binding to epithelial cells of various origins. *Virology* **240,** 254–266.
71. Mei, Y. F. and Wadell, G. (1995) Highly heterogeneous fiber genes in the two closely related adenovirus genome types Ad35p and Ad34a. *Virology* **206,** 686–689.
72. Hawkins, L. K. and Wold, W. S. M. (1995) A 20,500-Dalton protein is coded by region E3 of subgroup B but not subgroup C human adenoviruses. *Virology* **208,** 226–233.
73. Hong, J. S., Mullis, K. G., and Engler, J. A. (1988) Characterization of the early region 3 and fiber genes of Ad7. *Virology* **167,** 545–553.
74. Rekosh, D. M., Russell, W. C., Bellet, A. J., and Robinson, A. J. (1977) Identification of a protein linked to the ends of adenovirus DNA. *Cell* **11,** 283–295.
75. Mei, Y. F. and Wadell, G. (1992) The nucleotide sequence of adenovirus type 11 early 3 region: comparison of genome type Ad11p and Ad11a. *Virology* **191,** 125–133.
76. Flomenberg, P. R., Chen, M., and Horwitz, M. S. (1988) Sequence and genetic organization of adenovirus type 35 early region 3. *J. Virol.* **62,** 4431–4437.
77. Hawkins, L. K., Wilson-Rawls, J., and Wold, W. S. M. (1995) Region E3 of subgroup B human adenoviruses encodes a 16-kilodalton membrane protein that may be a distant analog of the E3-6.7K protein of subgroup C adenoviruses. *J. Virol.* **69,** 4292–4298.
78. Kajon, A. E. and Wadell, G. (1992) Characterization of adenovirus genome type 7h: analysis of its relationship to other members of serotype 7. *Intervirology* **33,** 86–90.
79. Choi, E. H., Kim, H. S., Eun, B. W., et al. (2005) Adenovirus type 7 peptide diversity during outbreak, Korea, 1995–2000. *Emerg. Infect. Dis.* **11,** 649–654.
80. Bailey, A. S. and Richmond, S. J. (1986) Genetic heterogeneity of recent isolates of adenovirus types 3, 4, and 7. *J. Clin. Microbiol.* **24,** 30–35.
81. Niel, C., Moraes, M. T., Mistchenko, A. S., Leite, J. P., and Gomes, S. A. (1991) Restriction site mapping of four genome types of adenovirus types 3 and 7 isolated in South America. *J. Med. Virol.* **33,** 123–127.

Index